MW01506356

# Chemistry in the Earth System

To obtain permission(s) or for inquiries, submit a request to:
Discovery Education, Inc.
4350 Congress Street, Suite 700
Charlotte, NC 28209
800-323-9084
Education_Info@DiscoveryEd.com

ISBN 13: 978-1-68220-650-8

Printed in the United States of America.

3 4 5 6 7 8 9 10 CWD 23 22 21 20 19 B

**Acknowledgments**
Acknowledgment is given to photographers, artists, and agents for permission to feature their copyrighted material.

Cover and inside cover art: Kevin Key / Shutterstock.com

# Table of Contents

## UNIT 5 | Chemistry and the Biosphere

## Dear Student,

You are about to experience science like you never have before! In this class, you'll be using California Science Techbook™—a comprehensive science program developed by the educators and designers at Discovery Education. Science Techbook is full of Explorations, videos, Hands-On Activities, digital tools, reading passages, animations, and more. These resources will help you learn scientific concepts and procedures, and apply them to the world around you. California Science Techbook allows you to work at your own pace and investigate questions you may have related to science. You'll even be able to monitor your progress in real time using the Student Learning Dashboard.

The Student Edition accompanies the digital Science Techbook. You have access to Science Techbook's core text—the key ideas and details about each scientific concept—even when you do not have access to a device or the Internet. You can use this resource to explore important ideas, make connections to the digital content, and develop your own scientific understanding.

This Student Edition is organized by concept and includes the following:

- OVERVIEW: What's it all about? Lesson Questions, Lesson Objectives, and key vocabulary will help you prepare for each science concept.
- ENGAGE: What do you already know about the topic? Follow a link to uncover your prior knowledge about each concept.
- EXPLORE: What are the main ideas in the concept? The Explore pages include core text and images to help you answer each of the concept's Lesson Questions. Use evidence to write a scientific explanation and answer questions to check for understanding.
- STEM IN ACTION: How is science used in the real world and in STEM careers? Read more in this section to find out how the knowledge you're building applies to real-world situations—both today and in the future.

Throughout this Student Edition, you'll find QR codes that take you to the corresponding online section of Science Techbook for that concept. To use the QR codes, you will need a QR reader. Readers are available for almost any device. The reader will scan the code and direct you to the correct page or resource in Science Techbook.

Enjoy this deep dive into the exciting world of science!

**Sincerely,**

**The Discovery Education Science Team**

## Dear Parent/Guardian,

This year, your student will be using California Science Techbook™, a comprehensive science program developed by the educators and designers at Discovery Education. Science Techbook is an innovative program that offers engaging, real-world problems to help your student master key scientific concepts and procedures. In class, students experience dynamic content, Explorations, videos, digital tools, and game-like activities that inspire and motivate scientific learning and curiosity.

This Student Edition allows students to explore the core Techbook content when the Internet is not available. Students are encouraged to use this resource to read about key ideas, seek connections, think about scientific questions, and develop their own scientific understanding.

This Student Edition is organized by concept and includes the following:

- OVERVIEW: Students preview a concept's Lesson Questions, Lesson Objectives, and key vocabulary to help them make connections to the science content.
- ENGAGE: Students answer questions to activate their prior knowledge of a concept's essential ideas, and begin making connections to the Explain Question.
- EXPLORE: Students deepen their understanding of the concept by exploring the core text related to each Lesson Question. Online, students have access to additional resources, Hands-On Activities, and interactives. They will also complete scientific explanations and answer questions to check for understanding.
- STEM IN ACTION: Students connect the skills and knowledge they are building in each concept with real-world applications. Online, they can explore related videos and resources, and complete additional activities.

Within this resource, you'll find QR codes that take you and your student to a corresponding section of Science Techbook. Once in Techbook, students will have access to the Core Interactive Text of each concept, as well as thousands of resources and activities that build deep conceptual scientific understanding. Additionally, tools and features such as the Interactive Glossary and text-to-speech functionality allow Science Techbook to target learning for students of a variety of abilities.

To use the QR codes, you'll need a QR reader. Readers are available for phones, tablets, laptops, desktops, and virtually any device in between.

We encourage you to support your student in using California Science Techbook. Together, may you and your student enjoy a fantastic year of science!

**Sincerely,**

**The Discovery Education Science Team**

CONCEPT 1.1

# Heat



## LESSON OVERVIEW

### Lesson Questions

- What is heat?
- What is the difference between thermal energy and temperature?
- How can problems involving conversions between joules calories, and kilocalories be solved?
- How does thermal-energy transfer (heat) cause substances to change state?
- How can two objects of different masses have the same temperature but different amounts of thermal energy?
- What are some examples that show how thermal energy is transferred by means of conduction, convection, and radiation?
- How can applying the law of conservation of energy help solve calorimetry problems?

### Key Vocabulary

Which terms do you already know?

- ☐ atom
- ☐ average kinetic energy
- ☐ boiling point
- ☐ calorie
- ☐ calorimetry
- ☐ Celsius
- ☐ change of state
- ☐ chemical
- ☐ closed system
- ☐ conduction
- ☐ conductor
- ☐ conservation of energy
- ☐ convection
- ☐ density
- ☐ distance
- ☐ electromagnetic spectrum
- ☐ electromagnetic wave
- ☐ electron

dlc.com/ca9009s

## Lesson Objectives

By the end of the lesson, you should be able to:

- Model that heat is the transfer of thermal energy from an object of higher temperature to an object of lower temperature until an equilibrium temperature is reached.
- Explain the difference between thermal energy and temperature.
- Provide evidence of how two objects of different masses can have the same temperature but different amounts of thermal energy.
- Describe the energy transfer that occurs when substances change state.
- Use examples to explain how thermal energy is transferred via conduction, convection, and radiation.
- Solve problems involving conversions between joules, calories, and kilocalories.
- Solve calorimetry problems by applying the law of conservation of energy.

## Key Vocabulary continued

- [ ] Fahrenheit
- [ ] force
- [ ] freezing point
- [ ] freezing point depression
- [ ] frequency
- [ ] gamma ray
- [ ] gas
- [ ] heat
- [ ] heat energy
- [ ] insulator
- [ ] inversely proportional
- [ ] joule
- [ ] kelvin
- [ ] latent heat of fusion
- [ ] latent heat of vaporization
- [ ] liquid
- [ ] mass
- [ ] melting point
- [ ] microwave
- [ ] molecule
- [ ] phase change
- [ ] plasma
- [ ] pressure
- [ ] radiation
- [ ] rotation
- [ ] solid
- [ ] specific heat
- [ ] temperature
- [ ] thermal energy
- [ ] thermal equilibrium
- [ ] transport
- [ ] visible light
- [ ] wavelength
- [ ] work
- [ ] X-ray

# Investigating Heat

Have you ever been to a sandy beach on a hot, sunny summer day? Was the sand so hot that it burned your bare feet as you ran out to the cooling water?

dlc.com/ca9010s

**EXPLAIN QUESTION**

**What is heat, how is it related to temperature and thermal energy, and how does the transfer of thermal energy affect matter?**

**SPLISH-SPLASH**

Would you expect the people walking on the sand or in the water to have warmer feet?

## What Is Heat?

**Heat** is the transfer of **thermal energy** from a high-**temperature** object to a low-temperature object that are in *thermal contact.* Matter does not contain heat. Heat is energy in transit. Heat transfer will continue until the two objects are at the same temperature. When the objects reach the same temperature, they are said to be in **thermal equilibrium**.

In spontaneous processes, thermal energy moves (as heat) in only one direction, from higher-temperature areas to lower-temperature areas. Sometimes people speak of hot and cold, but in truth, these words are relative terms used to describe only the temperature of a substance, not the total amount of thermal energy of a substance. There is no physical process in which thermal energy is transferred from a lower-temperature object to a higher-temperature object unless outside influences, such as **work**, **force** it to do so.

People sometimes express the movement of thermal energy by heat incorrectly. When they are talking about cold weather, they may say that their house needs better insulation "to keep the cold out." What they mean is that they want to keep the thermal energy in. Thermal energy always flows from regions of higher temperature to regions of lower temperature. If you hold a frozen snowball in your hand, heat flows from your hand to the snowball. The snowball melts because it gains thermal energy; your hand gets cold because it loses thermal enery.

**A SUNNY DAY IN ANTARCTICA**

A sunny day in Antarctica looks and feels very different from a sunny day at a beach in Florida. How would objects in these locations differ due to heat transfer?

### Heat and Conservation of Energy

The law of **conservation of energy** applies to isolated systems. An isolated system is a distinct region that is closed off from its surroundings from which no matter or energy can be added or removed. Objects within the system may gain or lose thermal energy, but the system's total energy is constant. Any thermal energy lost by one object must be gained by another object.

The total energy of an isolated system is conserved. One way to express this principle is to state that energy can neither be created nor destroyed, but it can be transformed to a different form of energy or transported somewhere else. During the manufacturing process, power plants and factories generate a lot of extra thermal energy. This "waste heat" can be recycled in one of two ways. It can be transferred to another type of useful energy such as electrical energy. A common method for converting thermal energy into electrical energy is to use the energy to make the steam that drives a steam turbine. A steam turbine is a machine that generates electricity. Extra thermal energy can also be transported away from the factory or power plant and used somewhere else. The thermal energy is used to heat up water. The hot water is then pumped into nearby buildings where it is used as a source of heat.



## What Is the Difference between Thermal Energy and Temperature?

### Thermal Energy versus Temperature

*Thermal energy* and *temperature* are terms that have specific definitions in physics. The **thermal energy** of an object is the sum of all the kinetic energy of its molecules. Its **temperature** is a measure of their **average kinetic energy**. Different objects with the same temperature can have very different thermal energy. For example, seawater in the ocean and seawater in a bucket can have the same temperature. However, the ocean will have more thermal energy, because it has more molecules.

Molecules of water in that bucket have kinetic energy. This would be low if that water were frozen. This would be higher if it were a **liquid**, and still higher if it was a vapor. For the same **mass** of a substance, molecules of vapor have more thermal energy and higher temperature than those of a liquid. Molecules of a liquid have more thermal energy and higher temperature than those of a **solid**.

## Units for Temperature

Temperature is typically measured using one of three common scales:

**Fahrenheit** scale: This scale is in everyday use in the United States. It uses the **freezing point** (32°F) and the **boiling point** (212°F) of water as reference points. It divides the range between them into 180 equal degrees. Although almost all of the rest of the world has abandoned it, the **Fahrenheit** scale does have the advantage of having divisions that are smaller compared with those in the other two major scales. This allows for greater resolution of temperature. The scale is specified by using either the notation °F or the phrase *degrees Fahrenheit*.

**Celsius** scale: This scale is in everyday use throughout most of the world. It is used in science and increasingly in industry in the United States. It also uses the freezing and boiling points of water as reference points. It sets them at convenient temperatures of 0 and 100°C, respectively, dividing the range into 100 equal parts. Temperatures below the freezing point of water are negative. The scale is specified by using either the notation °C or the phrase *degrees* **Celsius**.

**Kelvin** scale: This scale is in universal scientific use. Its divisions are the same size as the degrees in the Celsius scale. The lowest possible temperature is set to absolute zero. At absolute zero, there is no energy from molecular motion available for transfer to other objects. Absolute zero on the **Kelvin** scale (0 K) is equivalent to −273.15°C. It is very difficult to produce the conditions required to observe matter at this temperature, even if it is possible. A laboratory in Finland has been able to supercool a piece of rhodium to 0.0000000001 K, or 100 pK above absolute zero. In contrast with the other two scales, the word *degrees* and the symbol ° are not used with the Kelvin scale. Examples are 2,300 K and 273.15 kelvin.

**LORD KELVIN**

Lord Kelvin was a British scientist who established the absolute, or Kelvin, scale of temperature.

# How Can Problems Involving Conversions between Joules, Calories, and Kilocalories Be Solved?

## Units for Thermal Energy

**Thermal energy** and **temperature** are related, but they have different units. Thermal energy uses the energy units joules (J) or calories (cal). It can also be expressed using kilojoules (kJ) and kilocalories (kcal).

A **joule** (J) is an amount of energy equal to the **work** done by a **force** of 1 newton (N) acting through a **distance** of 1 meter (m).

One **calorie** (cal) is the energy required to raise the temperature of 1 gram of water by 1°C.

The prefix *kilo* means 1,000, so one kilojoule (kJ) is equal to 1,000 J, and one kilocalorie (kcal) is equal to 1,000 cal.

Another word for a kcal is Calorie, denoted by a capital C. This unit is used on food labels to indicate the amount of energy available in the food.

Units of joules can be converted to calories, and vice versa, by using the following conversion factors:

- 1 cal = 0.001 kcal = 4.186 J
- 1 kcal = 1,000 cal = 4186 J
- 1 J = 0.239 cal = 0.000239 kcal

## Units for Thermal Energy Sample Problem

Convert 516 calories (cal) into joules (J) and kilojoules (kJ).

**Solution:**

$$(516 \text{ cal})\ (4.186 \text{ J/cal}) = 2.160 \text{ J}$$

To convert to kilojoules, divide by 1,000:

$$(2{,}160 \text{ J})\ (1 \text{ kJ/1,000 J}) = 2.16 \text{ kJ}$$

# How Does Thermal-Energy Transfer (Heat) Cause Substances to Change State?

## Effect of Thermal Energy

When **thermal energy** is added to or removed from a substance, the effect is either to change the substance's **temperature** or to change its state of matter. A **change of state** is also called a **phase change**. It occurs when a substance changes among its **solid**, **liquid**, **gas**, and **plasma** states.

If thermal energy is transferred to a substance, under most circumstances, its atoms and molecules gain kinetic energy. This increase in **average kinetic energy** increases the temperature of the substance. When the substance's melting or **boiling point** is reached, however, adding more thermal energy does not immediately result in a further temperature increase. It causes the substance to change state instead.

In a substance, the electrical potential energy between atoms and molecules is stored in the form of intermolecular forces. During a change of state, the additional thermal energy allows particles in the substance to overcome the intermolecular forces holding them together. The average kinetic energy of the particles does not change. The temperature of the substance remains constant. Only when the phase change is complete does additional thermal energy cause the temperature increase to resume.

The principle is also true in reverse. As thermal energy is removed from a substance, the average kinetic energy of its particles decreases. Therefore, the temperature also decreases until the substance reaches a phase-change point. Then the temperature remains constant while more thermal energy is removed. Only when the phase change is complete will further removal of thermal energy cause the temperature to resume its decrease.

For rising temperatures, the temperatures at which phase changes take place are called the substance's **melting point** and boiling point. For falling temperatures, these same temperatures are known as the substance's **freezing point** and condensation point.

Phase change temperatures are not the same for all substances. For example, while the solid/liquid and liquid/vapor phase-change temperatures for water are 0°C and 100°C, respectively, those for silver are approximately 962°C and 2,212°C.

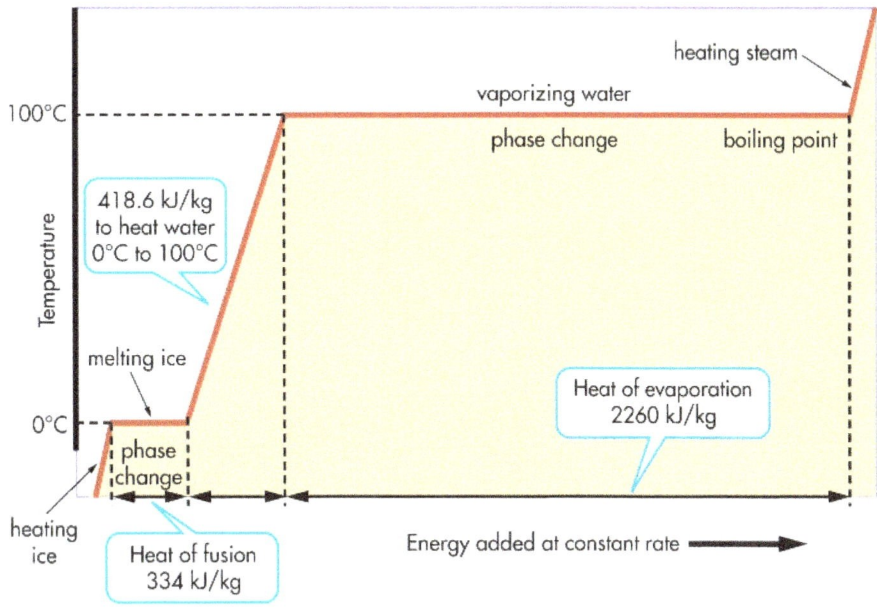


**HEATING CURVE FOR WATER**

As energy is added to water, the temperature does not increase at a constant rate. What happens to the relationship between the energy added and the temperature during a phase change?

Not only must a substance be at a certain temperature in order to change phase, it must also gain or lose a certain amount of thermal energy. The **heat** needed to change a certain amount of a substance from the solid to the liquid phase without changing its temperature is its **latent heat of fusion**. The heat required to change a certain amount of a substance from the liquid to the vapor (**gas**) phase without changing its temperature is the **latent heat of vaporization**.

Water, for example, requires about 334 kJ/kg of thermal energy to transition between solid and liquid form. It needs 2,260 kJ/kg of thermal energy to transition between liquid and gaseous states.

The amount of thermal energy a substance gains in a phase change is the same amount of energy it loses in the reverse phase change. For example, the same amount of thermal energy added to a solid to melt it completely must be removed from the liquid to freeze it to the solid phase.

Whether a substance changes state at a given temperature is affected by various factors. Under special conditions, thermal energy can be added or removed from a substance such that it does not undergo a phase change as it normally would. The temperature continues to increase or decrease past the phase-change points. The substance is then said to be superheated or supercooled, and it is unstable. The phase change is likely to occur suddenly when conditions shift slightly.

**BOILING WATER**
You can boil water for tea on a gas stovetop. What happens to the amount of thermal energy in the water as it starts to boil?



# How Can Two Objects of Different Masses Have the Same Temperature but Different Amounts of Thermal Energy?

## Temperature Is Average Kinetic Energy

Recall the earlier discussion about the difference between **temperature** and **thermal energy**. The question of how two objects of different masses can have the same temperature but different amounts of thermal energy is a question of comparing an average to a total.

Assume a 500 L section of ocean and a 1 L bucket of seawater. If the particles in both samples move with the same range of speeds and have the same masses, then the range of kinetic energies will also be the same. The average of these kinetic energies must also be the same. Therefore, both the 500 L and 1 L samples of seawater have the same temperature.

On the other hand, the two masses of water will not have the same total thermal energy. This is because there are many more particles in the 500 L sample than in the 1 L sample. The total kinetic energy, or thermal energy, of the ocean water is about 500 times greater than that of the water in the bucket.

## Specific Heat

The amount of total energy in the section of ocean and the amount of total energy in the bucket of seawater at the same temperature were different because of the different amounts of **mass**. Even if two objects made of different substances had the same temperature and mass, they still might have different amounts of total energy.

The abilities of materials to absorb and release thermal energy vary. The amount of energy required to change the temperature of 1 g of a given substance by 1°C (or 1 K) is its **specific heat**. Specific heats are typically given in units of J/(g × K) or kJ/(kg × K).

For equal masses, different materials require different amounts of energy to achieve the same temperature change. Conversely, given the same amount of thermal energy, equal masses may achieve different temperature changes if the masses are not made of the same substance.

Consider two playground slides. One is made from steel and the other from plastic. Steel has a specific **heat** of about 0.5 kJ/(kg × K), while plastic has a specific heat of about 2.0 kJ/(kg × K). The steel slide's significantly lower specific heat causes it to achieve a higher temperature than does the plastic slide.

If a substance has a low specific heat, it takes relatively little energy to change its temperature. If it has a high specific heat, it takes a lot of energy to change its temperature. A substance with high specific heat resists changes in its temperature.

Among common substances, water is very unusual in that it has a high specific heat of 4.186 kJ/(kg × K). In other words, it would take 4,186 J to raise the temperature of 1 kg of water by 1°C. Changing the temperature of water involves a lot of energy.

**HOT SAND, COOL WATER**

Why is the temperature of the sand so much higher than the temperature of the water on a sunny day at the beach?

# What Are Some Examples That Show How Thermal Energy Is Transferred by Means of Conduction, Convection, and Radiation?

## Three Methods of Heat Transfer

**Heat** transfer occurs in three main ways: **conduction**, **convection**, and **radiation**.

In conduction, **thermal energy** is transferred between different areas of the same substance. Thermal energy applied in one part of the substance causes atoms and molecules in the substance to move more rapidly. This causes collisions with neighboring particles, causing them to move faster as well. Further collisions transfer kinetic energy to particles farther away, and so forth. Thus, thermal energy spreads throughout the substance.

An example of conduction is the heating of one end of an iron bar. Conduction causes thermal energy to travel from the area of higher **temperature** to the area of lower temperature at the other end of the bar. Heat transfers energy until **thermal equilibrium** is reached.

Conduction works best in solids because the atoms and molecules are close together. Metal solids make particularly good conductors of thermal energy for the same reason that they are good conductors of electricity. Their atoms have free electrons that can move easily through the material and transmit the energy to other particles through collisions. Materials with tightly held electrons, such as wood and paper, make poor conductors. These materials are called insulators. A poor **conductor** is a good **insulator**, and vice versa.

**INSULATOR**

Insulation materials are widely used in building construction. What property of insulation materials makes them useful in buildings?

Conduction is not possible in a vacuum. There are no particles in a vacuum to transfer energy. Air, as a **gas**, has particles, but they are spread farther apart than in liquids or solids. Materials that have air spaces in them, such as fiberglass insulation, plastic foam, and the feathers in down comforters, effectively slow the transfer of thermal energy.

Convection is the transfer of thermal energy by way of moving molecules in fluids. Fluids include liquids and gases.

Convection works because of differences in fluid **density**. Less dense fluids tend to rise, while more dense fluids sink. When thermal energy is applied to one part of a **mass** of fluid, nearby particles become more energetic and spread apart. The fluid in that area becomes less dense. Less dense fluid rises away from the heat source, while more dense fluid sinks to take its place. Eventually, the heated fluid cools, becomes denser again, and sinks. The resulting cycle of moving fluid transports thermal energy.

The upward flow of hot air above a room heater is an example of convection. Hot air moves away from the heater, and cool air moves to replace it. The air in the room circulates. Similarly, convection circulates air in the atmosphere, affecting the weather. The sun transfers energy to Earth's surface. Air near the surface warms, becomes less dense, and rises. Cooler air rushes in underneath, causing wind.

Convection is also important in the world's ocean currents. Cold, dense water near Earth's poles sinks and moves along the bottom of the sea floor. Water warmed near the equator and pushed by wind and Earth's **rotation** moves along the surface toward the poles. Warm water passes near landmasses, warming them and significantly affecting their climate.

Radiation is the transfer of energy by electromagnetic waves. Energy from the sun arrives on Earth by radiation. Unlike conduction and convection, it can travel in a vacuum. Thermal interaction by radiation does not require actual physical contact.

Radiant energy includes radiation from across the **electromagnetic spectrum**, from the short wavelengths and high frequencies of gamma rays and X-rays, through **visible light**, to the long wavelengths and low frequencies of **microwave** and radio energy. All objects are at a temperature above absolute zero and therefore they emit some form of electromagnetic radiation. The radiation **wavelength** an object emits most (its peak wavelength) is **inversely proportional** to its absolute temperature in kelvins.

**Electromagnetic Spectrum**

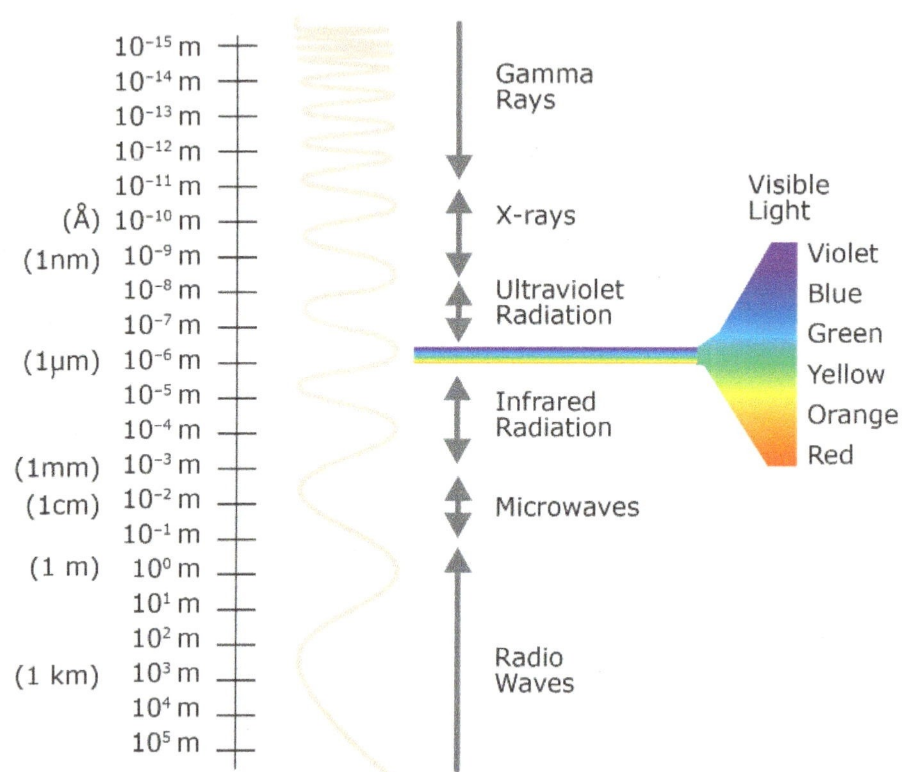


**ELECTROMAGNETIC SPECTRUM**

Electromagnetic radiation ranges from gamma rays to radio waves. Radiation with short wavelengths has the highest energy. What kind of electromagnetic radiation has the lowest energy?

Some objects, such as the sun and incandescent light bulbs, emit radiation that includes the visible portion of the electromagnetic spectrum. Moving from the blue-violet end of the visible light spectrum toward the red end, the wavelengths of electromagnetic radiation get longer and the frequencies get lower. At frequencies just below those for red light, in the infrared range, our eyes no longer perceive the radiation as light. However, we can feel it as heat. A hot fire emits some of its radiant energy as the visible flames and glowing embers and some as invisible infrared energy that is felt as warmth.

# How Can Applying the Law of Conservation of Energy Help Solve Calorimetry Problems?

## Calorimetry

The total energy in an isolated system is conserved. This makes it possible to solve problems in which an amount of **thermal energy** is transferred from one substance to another with known thermal properties. **Calorimetry** is the study of thermal energy changes that occur in a nearly isolated system.

The **heat energy** transferred by an object or substance going through a **temperature** change is expressed by this equation: $Q = mc\Delta T$, where $Q$ is the amount of **heat**, $m$ is the **mass** of the substance, $c$ is its **specific heat**, and $\Delta T$ is the temperature change.

This equation, combined with the understanding that energy is conserved, can be used to solve calorimetry problems.



## Calorimetry Sample Problem

Consider 2.50 kg of hot copper at 98.0°C, which is dropped into a tank containing 10.0 kg of water at 15.0°C. As thermal energy transfers from copper to water, the metal's temperature decreases while the water's temperature increases. Because energy is conserved, the thermal energy removed by heat from the copper will equal the heat transfer to the water. Both copper and water will arrive at the same final temperature—the temperature of **thermal equilibrium**: $T_{eq}$. Find $T_{eq}$.

**Solution:**

Using the formula above, and recalling that the specific heat of water is 4,186 J/(kg × °C), the amount of heat received by water is

$$\begin{aligned} Q_{water} &= (10.0\text{ kg})\,(4{,}186\text{ J/(kg} \times {}^\circ\text{C)})\,(T_{eq} - 15.0^\circ\text{C}) \\ &= (41{,}860\text{ J/}^\circ\text{C})\,T_{eq} - 627{,}900\text{ J} \end{aligned}$$

Meanwhile, copper, which has a specific heat of 386 J/(kg × °C), has cooled from 98.0°C to $T_{eq}$:

$$\begin{aligned} Q_{copper} &= (2.50\text{ kg})\,(386\text{ J/(kg} \times {}^\circ\text{C)})\,(98.0^\circ\text{C} - T_{eq}) \\ &= 94{,}570\text{ J} - (965\text{ J/}^\circ\text{C})T_{eq} \end{aligned}$$

Because of the **conservation of energy**, $Q_{water} = Q_{copper}$.

Thus, $(41{,}860\text{ J/}^\circ\text{C})T_{eq} - 627{,}900\text{ J} = 94{,}570\text{ J} - (965\text{ J/}^\circ\text{C})T_{eq}$.

Solving algebraically for $T_{eq}$ and rounding for significant digits:
$T_{eq} = 16.9°C$.

Intuitively, this makes sense, because $T_{eq}$ should be a temperature that is between the initial temperatures of copper and water. It is closer to the initial temperature of water, which has the higher specific heat and greater mass.

### Consider the Explain Question

**What is heat, how is it related to temperature and thermal energy, and how does the transfer of thermal energy affect matter?**

dlc.com/ca9011s

Go online to complete the scientific explanation.

### Check Your Understanding

**What is the difference between thermal energy and temperature?**

dlc.com/ca9012s

## Applying Heat

Have you ever noticed a strong breeze near a large body of water on a hot day? If you go back at night, you will notice that the direction of the wind has changed. **Convection** and the **specific heat** of water both play a role in the changing direction of winds at the beach from day to night.

Water has a high specific **heat**, and, therefore, it resists changes in **temperature**. In addition, the ocean represents a huge thermal **mass**. It warms slowly when receiving **radiation** from the sun and cools slowly when the sun sets below the horizon at night.

Land, by contrast, gains and loses heat more easily. During the day, the sun warms the land, which transfers energy to the air above it. The warmed air rises, and cooler air from over the ocean moves in under it. At ground level, the convection cycle of air blows from the ocean toward the land. This is called a sea breeze.

At night, the process reverses. The land cools off more quickly than the water. At night, air above the water is warmer than air above the land, so it rises. Air from over the land moves out to replace it. The direction of the wind at ground level is now from land to water. This is called a land breeze.

**Sea and Land Breezes**

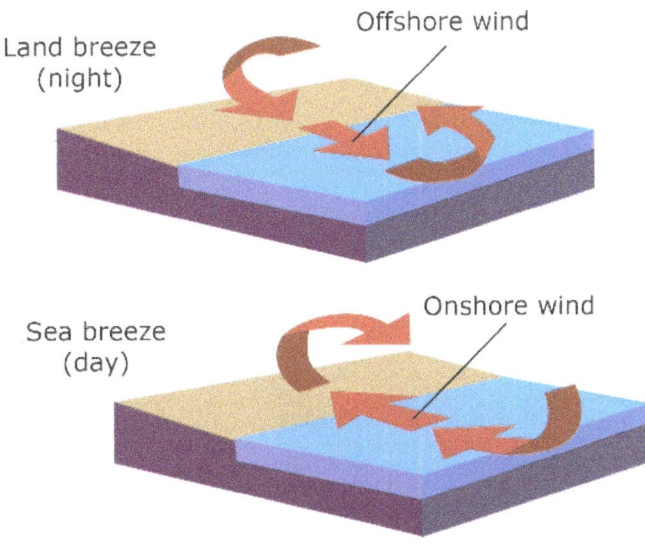


**LAND AND SEA BREEZES**

Differential heating leads to density differences in the air, and daily breezes occur. Does the temperature of the water have to change for offshore and onshore winds to occur?

## STEM and Heat

When civil engineers develop designs for products and projects, they often must consider the thermal properties of the materials used in their manufacturing and construction. They often **work** with engineers who develop new materials or new applications for old materials. Materials engineers might specialize in include metals, ceramics, plastics, or other innovative materials.

When objects are heated, the transferred **thermal energy** causes the molecules of the object's material to move faster and to move farther apart. As temperatures rise, most materials expand. As temperatures fall, most contract. Depending on the material and the shape of the object, the change can be significant.

Water represents a special case, especially if the temperature is near one of its phase-change points. For example, at around 4°C, a given volume of water is at its most compact. As the temperature drops just below the **freezing point** (0°C), the water (ice) expands to take up more space than at any other temperature below the **boiling point**. In areas that experience freezing temperatures, water commonly damages concrete and masonry structures as it infiltrates, changes to ice, and expands.

Civil engineers are particularly concerned with designing roads and bridges that accommodate the thermal expansion and contraction of concrete pavements. The design must include features that allow the pavement to remain level, aligned, and crack-free as it changes size.

To allow the pavement to slide easily over the ground as it changes size, workers can install polyethylene sheeting on a carefully leveled asphalt base before the concrete is placed on top.



**GROUND PREPARATION**

Plastic sheeting allows the concrete to slide easily over the ground as it expands and contracts. How could expansion and contraction of the ground cause problems in the over-lying concrete?

The decks of bridges expand and contract more than other pavements because of the open space underneath the bridge. Why is thermal expansion and contraction not as much of a problem for pavement that lies over the ground?

In addition, road designs typically include gaps between pavement slabs to allow the concrete to shrink or grow without cracking or buckling. In some places, such as bridges, the pavement is expected to move so much that the gap must be 75–100 millimeters wide. In this case, an interlocking "armored" metal joint protects the gap. For smaller expected movements, a doweled expansion joint is sufficient.

**EXPANSION JOINT**

What features of particular metals do you think are important to civil engineers when they design expansion joints?



When the concrete slabs are poured, steel dowel rods implanted along their edges connect them to neighboring slabs across a gap. The rods keep the slabs aligned vertically and horizontally as they change size. They also help to smoothly distribute the load from one slab to the next as heavy trucks roll over them. On one or both ends of the rods inside the concrete, there is a cap with a space in the end so that dowels can move in and out of their holes slightly.

The designed size of the gap depends partly on the extremes of the temperatures the road will be experiencing and partly on the temperatures on the day the road is constructed. Normal shrinkage of the concrete as it cures is also accounted for. Finally, the gap is sealed at the top with polyurethane and silicon to keep water out. As the road ages and the gap seals deteriorate, maintenance crews can remove and replace them. This is much easier than replacing concrete that has been damaged by thermal expansion and contraction.

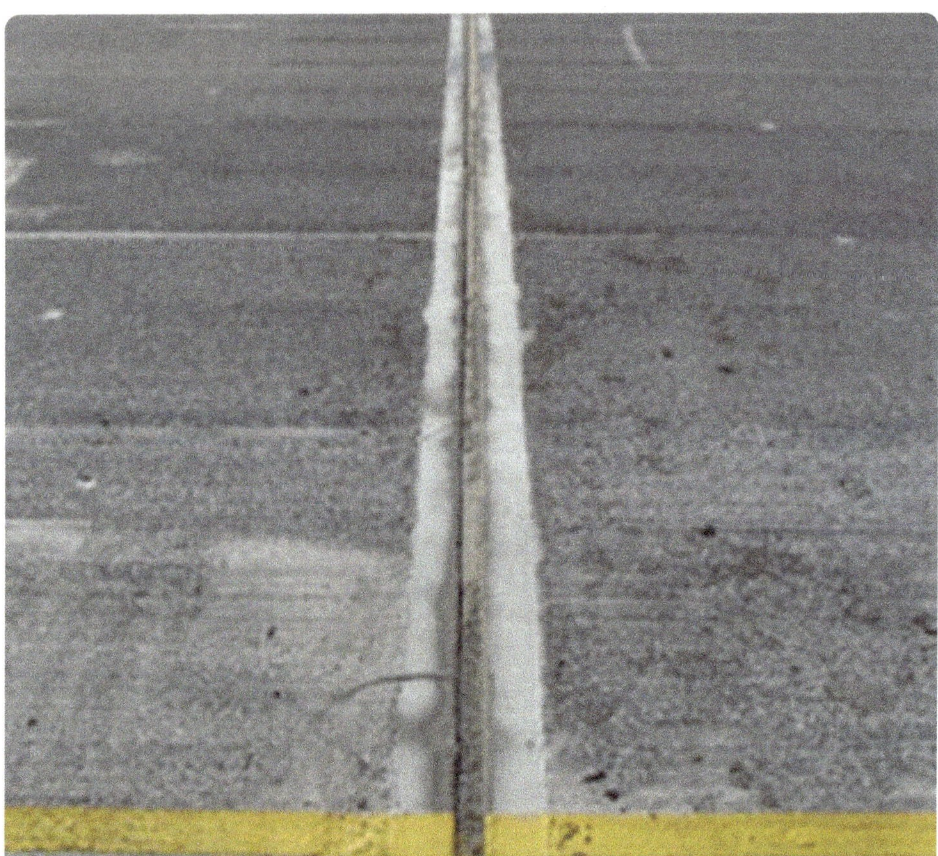

**SEALED EXPANSION JOINT**

The finished joint allows the concrete slabs to expand and contract while supporting traffic and excluding potentially troublesome water. What might happen if the expansion joint were to fail?

# Calorimetry

**Use thermodynamics to solve the problems below.**

**1.** A 0.380-kg sample of aluminum (with a specific heat of 910.0 J/(kg·K)) is heated to 378 K and then placed in 2.40 kg of water that is at 293 K. If the system is left to reach its equilibrium state, what will the equilibrium temperature be? Assume that no thermal energy is lost to the external environment. Use 4.186 J/(g·K) as the specific heat of water.

**2.** A 0.860-kg sample of tin (with a specific heat of 210.0 J/(kg·K)) is heated to 525 K and then placed in 1.20 kg of water that is at 303 K. If the system is left to reach its equilibrium state, what will the equilibrium temperature be? Assume that no thermal energy is lost to the external environment. Use 4.186 J/(g·K) as the specific heat of water.

**3.** In the previous problem involving a heated metal that is placed in water, name at least three factors that would have increased the final equilibrium temperature of the system.

CONCEPT 1.2

# Thermochemistry



## LESSON OVERVIEW

dlc.com/ca9113s

### Lesson Questions

- How do you distinguish between exothermic and endothermic reactions and processes?
- What is the relationship between a system and its surroundings?
- What is the relationship between physical changes and changes in energy?
- What is the relationship between chemical changes and changes in energy?

### Key Vocabulary

Which terms do you already know?

- ☐ activation energy
- ☐ average kinetic energy
- ☐ calorimetry
- ☐ chemical potential energy
- ☐ combustion reaction
- ☐ conduction
- ☐ convection
- ☐ critical point
- ☐ deposition (phase change)
- ☐ electromagnetic radiation
- ☐ endothermic
- ☐ energy
- ☐ enthalpy
- ☐ exothermic
- ☐ first law of thermodynamics
- ☐ heat energy
- ☐ Hess's Law
- ☐ joule

## Lesson Objectives

By the end of the lesson, you should be able to:

- Distinguish between exothermic and endothermic reactions and processes
- Explain the relationship between a system and its surroundings
- Explain the relationship between physical changes and changes in energy
- Explain the relationship between chemical changes and changes in energy
- Determine the change in enthalpy of a chemical reaction

## Key Vocabulary continued

- [ ] kinetic energy
- [ ] Le Chatelier's Principle
- [ ] metric
- [ ] non-spontaneous
- [ ] phase diagram
- [ ] second law of thermodynamics
- [ ] specific heat
- [ ] spontaneous
- [ ] standard heat of formation
- [ ] sublimation
- [ ] surroundings
- [ ] system
- [ ] temperature
- [ ] triple point

## Thinking about Thermochemistry

dlc.com/ca9114s

**After a long, strenuous workout, you probably feel exhausted and sore. After a nice long cool down and stretch, you are uncomfortable because your muscles ache. Can a heating pad or hot pack work to relieve some soreness? Or should ice be used to help you feel better?**

**RUNNERS**

After a long run, how can hot and cold therapies help your body recover?

**EXPLAIN QUESTION**

**How are endothermic and exothermic processes identified, and how does the energy of the system change with each?**

# How Do You Distinguish between Exothermic and Endothermic Reactions and Processes?

## Releasing Energy

As wood burns, **energy** is released. Combustion of the wood releases energy in the form of light and thermal energy. Likewise, the synthesis reaction of sodium metal with chlorine gas also releases energy. Chemical reactions that release energy are called **exothermic** reactions. The **combustion reaction**, such as that of methane with oxygen or propane with oxygen, is another exothermic reaction. During each reaction, energy is released as light and thermal energy.

Where does this energy come from? The energy released during a reaction results from a decrease in the **chemical potential energy** as reactants change into products. Chemical bonds between atoms store chemical potential energy. In an exothermic reaction, the products of the reaction have less chemical potential energy than the reactants. The difference in potential energy is released as light or thermal energy.

**FIREWORKS EXPLODING**
What is happening to the energy of reactants and products of the chemical reactions that occur during a fireworks show?



## Absorbing Energy

Some chemical reactions and physical processes release energy. Others absorb energy. Energy is absorbed during an **endothermic** reaction or process. Toasting marshmallows and popping corn are endothermic processes. The substances absorb energy as they undergo changes. Boiling water and melting steel are also endothermic. The water or steel must absorb energy to undergo the change. Instant cold packs work on the basis of an endothermic reaction when a solid is dissolved in water. Energy is absorbed by reactants during the reaction, so the pack becomes cold to the touch.

Another example of an endothermic process is photosynthesis. Energy is added to the **system** from sunlight and helps the plant convert carbon dioxide and water into glucose and oxygen. The energy absorbed remains in the glucose molecules. The decomposition of sodium chloride into sodium and chlorine is the reverse reaction of the exothermic processes described previously. Energy must be absorbed to break the chemical bond between sodium and chlorine. The potential energy stored in the bond is greater than the chemical potential energy of the products, sodium and chlorine. The reactant must absorb the difference in energy to undergo the reaction.

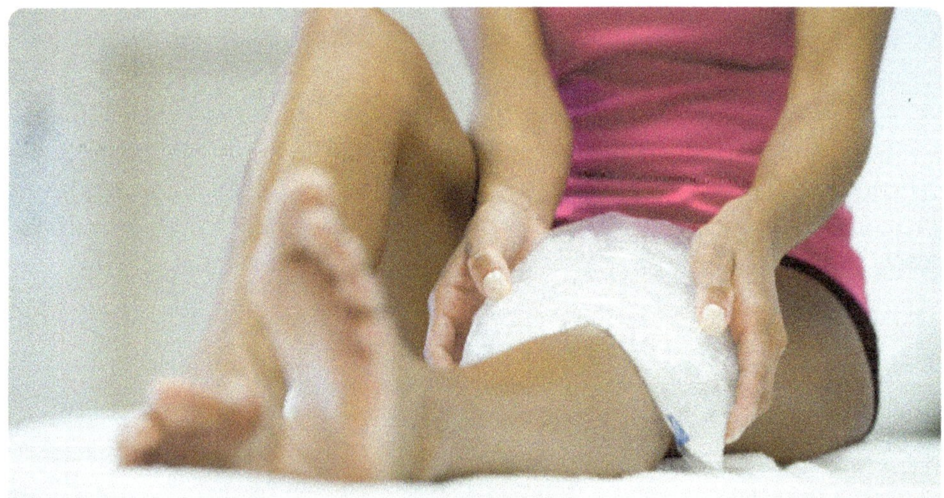

**ICE PACK**

This girl is using an ice pack to relieve inflammation in her knee. How does an ice pack work?



## What Is the Relationship between a System and Its Surroundings?

### Conservation of Energy

When an **exothermic** reaction releases **energy**, people might describe the reaction as "producing" energy. Likewise, an **endothermic** reaction does not destroy the energy it absorbs. Energy can be transformed from one form into another, but it cannot be created or destroyed during chemical and physical changes.

For exothermic reactions, the chemical energy stored in the bonds of the reactants is transformed into thermal energy that makes the surrounding matter warmer. Sometimes chemical energy changes into light energy, which happens when you make a light stick glow. The opposite happens during endothermic reactions. Thermal energy, or even light, is transformed into chemical energy that is stored in the bonds of the products. Energy is often transformed to a useful form of energy, but usually some energy also becomes a less useful form.

Energy is transformed from one useful form into another useful form in a car engine when the energy released by burning gasoline is transformed into **kinetic energy**. However, not all of the thermal energy released by the reaction makes the car move. Some of the thermal energy also heats the parts of the engine, which is why the hood of a car that has just been driven feels warm. In this case, the transformation to thermal energy that heats the car parts is not a useful transformation. People may say that some of the released energy is lost during combustion. This does not mean that it was destroyed. The "lost" energy changed to a form of energy that was not used to make the car run.

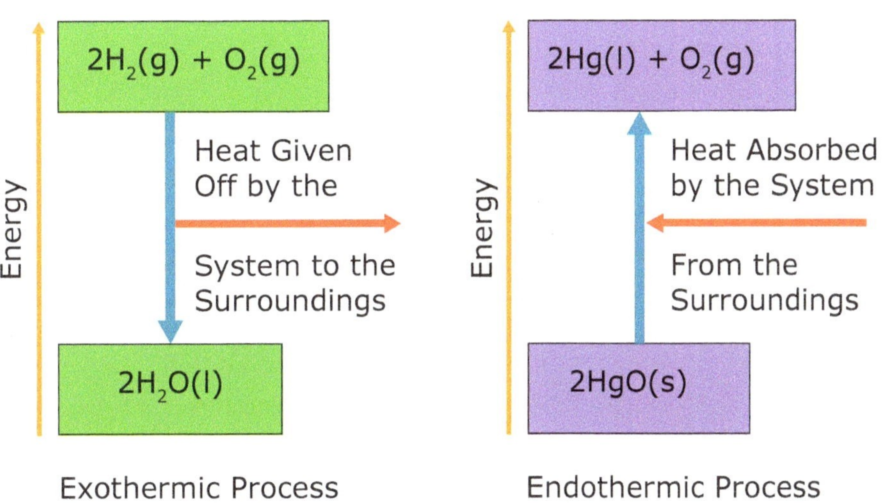


**ENERGY IN CHEMICAL REACTIONS**

Not all of the energy in a chemical reaction is useful when it transforms. Consider this diagram: Are the energy transformations in a car endothermic or exothermic?



In addition to being transformed from one form into another, energy can be transported from one place to another and transferred between systems. This movement of energy is what makes your hand warm when you touch the hood of a car when exothermic **combustion reaction** is happening in the engine.

The statement that energy can be transferred or transformed but is not created or destroyed during chemical and physical changes is the **first law of thermodynamics**. This law is also called the law of conservation of energy.

## System versus Surroundings

To get a more complete view of the energy changes associated with a chemical or physical change, consider two regions. The **system** is the part studied. Everything outside of the system is the **surroundings**. A barrier separates the system's interconnected parts from the surroundings. When considering systems, one must consider if the system represented is an open, closed, or isolated system.

- In an open system, both mass and energy are exchanged with its surrounding. Matter can be added easily like adding spices to a cooking pot. Energy can be exchanged through heat and work, like turning up the heat on the cooking pot.
- In a closed system, energy is exchanged with its surrounding, but there is no transfer of mass. If one places a lid on the cooking pot, matter cannot be added, but energy in the form of heat can still be added to the system.
- In an isolated system, there is no transfer or mass or energy across the boundaries of the systems. Real-world isolated systems are rare. One example of an isolated system is an insulated closed container. If hot liquid is poured into the container and the lid is placed on the container, there is no additional energy or mass applied to the system.

Cooling an injury with an ice pack can help reduce swelling. Consider the ice pack as the system—everything outside of the ice pack is the surroundings. The ice warms up and melts within the system because it absorbs energy from the surroundings. The injured leg and the air around the ice pack release energy to the system and are cooled. The total energy of the universe remains unchanged, which is consistent with the first law of thermodynamics.

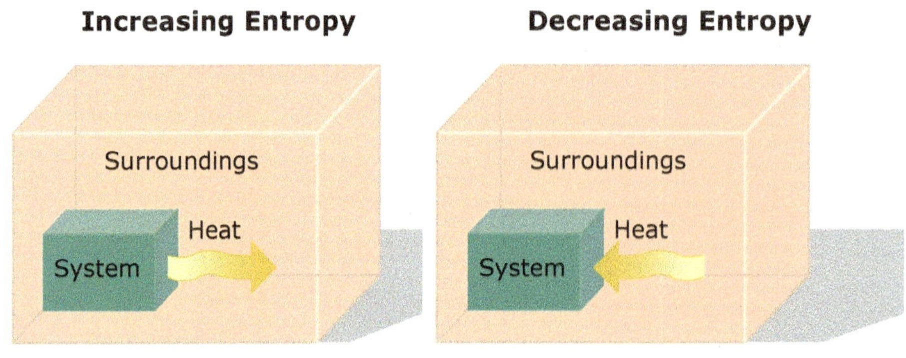


**CHANGING ENTROPY**
The flow of heat into or out of a system changes the system's entropy. How is the entropy of a system related to the entropy of its surroundings?

## From Hot to Cold

Heat transfer occurs from the surroundings into a system during an endothermic change. Heat transfer occurs from the system to the surroundings during an exothermic change. In each case, thermal energy naturally moves from a warmer area to a cooler area during the change. This is just as the **second law of thermodynamics** would indicate. The second law of thermodynamics states that in any natural process, the total entropy of a system and its surrounding environment must increase.

Energy transfer always favors a more stable state, meaning the energy distribution becomes more uniform. In other words, the warmer parts become cooler and the cooler parts become warmer. For example, when hot water is poured into a cup, the water transfers heat to the cup until the cup and the water are the same **temperature**. The cup becomes warmer, and the water becomes slightly cooler. In a closed system, the cup and the water would reach a stable state and no more transfer would occur between them. However, a cup sitting on the table is not in a closed system; it is in an open system. An open system allows the exchange of matter and energy with its surroundings. The system including the cup and water will continue to transfer heat to the surrounding air until the cup and the water inside it reach the same temperature as the air.



Temperature is a measure of the **average kinetic energy** of the particles in matter. The higher the temperature, the faster the particles move. It makes sense to think that energy moves from warm areas to cool ones. In other words, energy moves from an area where particles are moving quickly to an area where they are slower. But how does this heat transfer happen?

Thermal energy will move from a high-temperature object to a low-temperature one. It will end when the objects are at the same temperature (second law). The total energy released by the warmer object will be the same as the total energy absorbed by the cooler one (first law).

Heat transfer happens in three ways: through **convection**, radiation, or **conduction**.

- In convection, currents in air, water, or other fluids carry thermal energy from one place to another. These currents form as warmer fluids rise and cooler fluids sink. For example, a warm ocean current allows heat transfer between one part of the ocean and another.

- In radiation, energy travels as **electromagnetic radiation**. The radiation is in the form of visible or invisible light. For example, a red-hot stove burner transfers heat by radiating infrared and red light energy.
- In conduction, heat transfer occurs between a higher-temperature object and a lower-temperature object that are in contact. The thermal energy is transferred directly through the collision of the moving particles. For example, heat transfer takes place when you put your cold feet on a hot-water bottle.

## What Is the Relationship between Physical Changes and Changes in Energy?

### Energy Released in Freezing and Condensing

Freezing temperatures can destroy an orange crop. How can a farmer protect the unharvested fruit from freezing? A farmer can protect the fruit by spraying water on and around the trees in the orchard. As water cools and then freezes, it releases **energy**. Freezing is an **exothermic** process. Thermal energy released by the water is transferred to the air and the fruit. The fruit is warmed as the water freezes. The ice on the fruit also insulates the fruit against air temperatures below 0°C, the freezing point of water.

The change of a gas to a liquid, or condensing, is also an exothermic physical change. A gas must release energy to change into a liquid. So when a mirror fogs up, it absorbs the energy released by the condensing water vapor.

Another exothermic process is deposition. Deposition is a physical change that results in a gas changing into a solid, without first changing into a liquid. Deposition can be observed on windows during cold weather when water vapor changes into ice crystal.

### Energy Absorbed in Melting and Vaporizing

A change that is exothermic in one direction is **endothermic** in the reverse direction. Freezing is exothermic, so melting must be endothermic. An ice cube melts when it is taken out of a freezer. Thermal energy flows into the ice cube from the warmer air around it, so it melts. Condensing is exothermic, so vaporizing is endothermic. Before it vaporizes, water usually boils. For example, when water is boiled when cooking pasta, energy must be added from a heat source, such as the burner underneath the pot.

## Calorimetry Calculations

To understand how substances change **temperature** with gain or loss of energy chemists measure the amount of energy that is absorbed or released by a substance. They use **calorimetry** to determine **specific heat** and other quantities related to this process. The chemist conducts the energy transfer in a calorimeter. For example, the chemist puts a sample of hot metal in cool water inside a calorimeter. The metal and water reach the same temperature. Then, the chemist finds the change in energy of the water from the temperature change. The decrease in energy of the metal must equal the increase in energy of the water for energy to be conserved. The chemist can use these data to calculate the specific heat of the metal. The equation that relates the heat ($q$) that a substance absorbs or releases to its specific heat ($c$), mass ($m$), and temperature change ($\Delta T$) is $q = cm\Delta T$.

Using this equation, chemists can determine the change in energy as measured in joules (J).

A calorimetry reaction that is exposed to its **surroundings** occurs at constant pressure. The thermal energy absorbed or released by the surroundings is called the **enthalpy**. Its value is the same as $q$, but its sign is opposite:

$$\Delta H = -cm\Delta T$$

If a substance releases heat, then $q$ is positive, $\Delta H$ is negative, and the reaction is exothermic. If a substance absorbs heat, then $q$ is negative, $\Delta H$ is positive, and the reaction is endothermic.

When a hot metal gives off heat when placed in cool water, then $\Delta H$ is negative, because **heat energy** is lost by the metal. It is similar to purchasing an item from the store. If a person pays for an item in cash, then a specific amount of money is lost from the wallet. That same amount of money is added to the amount of money in the cash register. Cash flow for the shopper is negative, but it is positive for the store. Heat, $q$, is the same way. Heat released by a **system** is $-q$, something has to gain that heat and will be the same number but a positive value. In the example above, the heat lost by the metal, $-q$, must be equal in value to the heat gained by the water, but positive.

## Calorimetry Calculations: Sample Problem

A calorimeter holds 145 g water at 22.08C. A sample of hot aluminum is added to the water, and the final temperature of the water and aluminum is 28.08C. How much heat was released by the aluminum to make this change? The specific heat of water is 4.19 J/g°C.

**Solution:**

The given information includes

$$m = 145 \text{ g}$$
$$\Delta T = 6.0°\text{C}$$
$$c = 4.19 \text{ J/g°C}$$

Substitute into the equation and solve for $q$.

$$q = -Cm\Delta T$$
$$q = -(4.19 \text{ J/g°C})(145 \text{ g})(6.0°\text{C})$$
$$q = -3{,}637$$

Rounding to the correct number of significant digits, the heat released is $-3.6 \times 10^3$ joules.

## Changing Temperature, Phase Change, and Specific Heat

Temperature is related only to the kinetic energy of particles, but particles also have potential energy. Thermal energy is the sum of the kinetic energy and potential energy of the particles within a substance. When a substance is heated, the thermal energy of that substance is increased. Usually this thermal energy changes to kinetic energy of the particles. The increase in the particles' kinetic energy increases the temperature of the substance.

The addition or removal of thermal energy in a substance can result in a phase change, like when a solid turns into a liquid. During a phase change, the thermal energy changes to potential energy of the particles. The increase in the particles' potential energy changes the phase of the substance.

When a hot object is in contact with a cold object, thermal energy flows from the hot object to the cold object. If an object absorbs thermal energy, its temperature increases. So the cold object becomes warmer. If an object releases thermal energy, its temperature decreases. So the hot object becomes colder. Eventually, the two objects will reach the same temperature, and the flow of thermal energy will stop.

Think about baking a cheese pizza. The hot air in the oven transfers energy to the pizza and bakes it, raising its temperature. When the pizza is taken out, all of its ingredients are at the same temperature. After a few minutes, the crust is cool enough to touch, but the cheese is still hot. Why is the cheese hotter than the crust when they both came out of the oven at the same time?

**HOT PIZZA**

Differences in specific heats let the pizza crust cool faster than the cheese. Why do these substances have different specific heats?



Specific heat is the amount of energy needed to change the temperature of 1 g of a material by 1°C. When a substance with a high specific heat changes temperature, it absorbs or releases a large amount of energy per gram.

Likewise, a substance with a low specific heat absorbs or releases less energy per gram for the same change in temperature of a substance with higher specific heat. So the pizza effect is explained because the cheese has a higher specific heat than the crust. The cheese remains hotter than the crust because more energy per gram of cheese must be released compared to the same change in temperature for an equal mass of crust.

Another example of specific heat is observed at the beach. During the hot summer months, the sand along the water is hotter than the adjacent water. This is because the specific heat of the sand is less than the specific heat of the water. Less energy from the sun is required to increase the temperature of the sand than is required to increase the temperature of the water.

## Phase Diagrams

Matter can exist in a variety of phases depending upon the conditions of temperature and pressure. For example, water changes from ice to liquid and then to steam when its temperature changes. Substances can also be made to change state by changing pressure. For example, when compressed enough, the gases in air will eventually turn into a liquid. The actual phase a substance exists in can therefore be plotted using two axes—one for temperature (*x*) and one for pressure (*y*). The diagrams produced using these axes are called phase diagrams.

The diagram below shows a typical simple **phase diagram** for a pure substance.

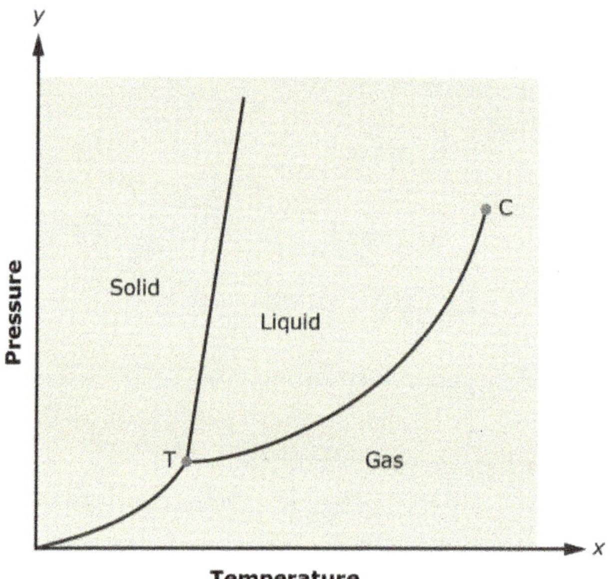


**PHASE DIAGRAM**

In a typical phase diagram, temperature is plotted on the *x*-axis and pressure is plotted on the *y*-axis. How do the lines represent where phases exist in equilibrium?



The diagram is split into three areas that represent the three phases of solid, liquid, and gas. The lines plotted on the graph, therefore, represent the conditions of temperature and pressure where more than one phase can exist. They are lines of equilibrium between two phase changes. There are many different sets of conditions that lead to a phase change of a substance. Whenever the equilibrium lines are crossed, at different temperatures and pressures, the substance undergoes a phase change. A substance can turn from a solid to a liquid (melting), from a liquid to vapor (evaporation), and from a solid to a gas (**sublimation**). At these same temperatures and pressures, this substance can also turn from a liquid to a solid (freezing), a vapor to a liquid (condensation), and a gas to a solid (deposition). At just one set of conditions the substance can exist in all three forms. For this reason, the particular temperature and pressure is called a **triple point** and is marked T.

In the previous phase diagram, the **critical point** is marked with the letter C. The temperature and pressure coordinates for the critical point are known as the critical temperature and pressure, respectively. Below the critical point, it is possible to condense a gas into a liquid by increasing the pressure. Above the critical point, it is impossible to condense the gas by increasing the pressure. The more pressure applied, the more compressed the gas gets, but it does not condense. This is because the gas particles are too energetic for intermolecular attraction to enable a liquid to be formed.

## Sample Problem: Interpreting Phase Diagrams

Let's look at the phase diagram for carbon dioxide and use it to answer some questions.

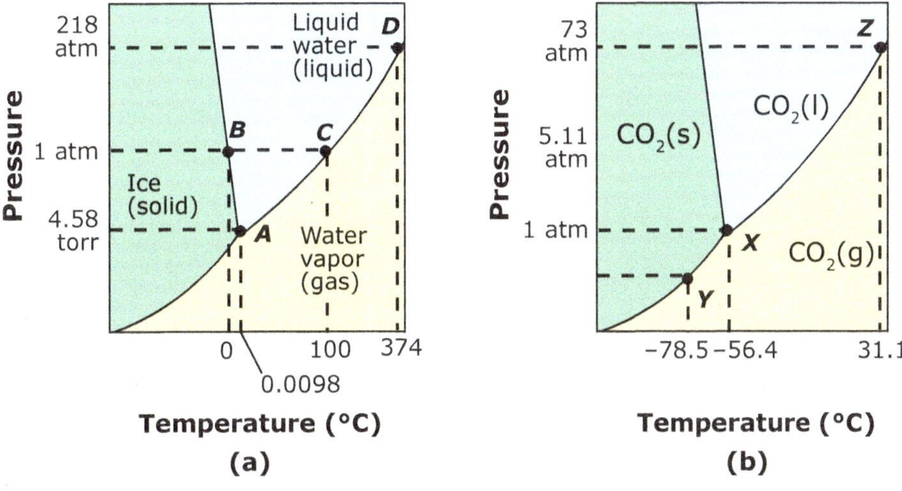


**PHASE DIAGRAMS**
Use this phase diagram to answer the following sample questions.

- At about what temperature and pressure do the three phases of carbon dioxide exist?

  *At about 57° Celsius, 5 atmospheres (the triple point).*

- What is the approximate temperature and pressure of the critical point?

  *About 31° Celsius, 73 atmospheres.*

- What is/are the phase(s) of carbon dioxide at 0° Celsius and a pressure of 10 atmospheres (atm)?

  *Carbon dioxide is a gas.*

- At about what temperature would carbon dioxide at 100 atm pressure exist as both a solid and a liquid?

  *About −58° Celsius.*

# What Is the Relationship between Chemical Changes and Changes in Energy?

## Activation Energy

Burning a strip of magnesium metal is an **exothermic** reaction. As the reactants change into products, there is a decrease in **enthalpy** as **energy** is released. This change in enthalpy is the heat of reaction. Enthalpy (H) is the amount of energy exchanged between a **system** and its **surroundings**. This energy is usually thermal energy. The energy of the product, magnesium oxide, is lower than the combined energy of the reactants.

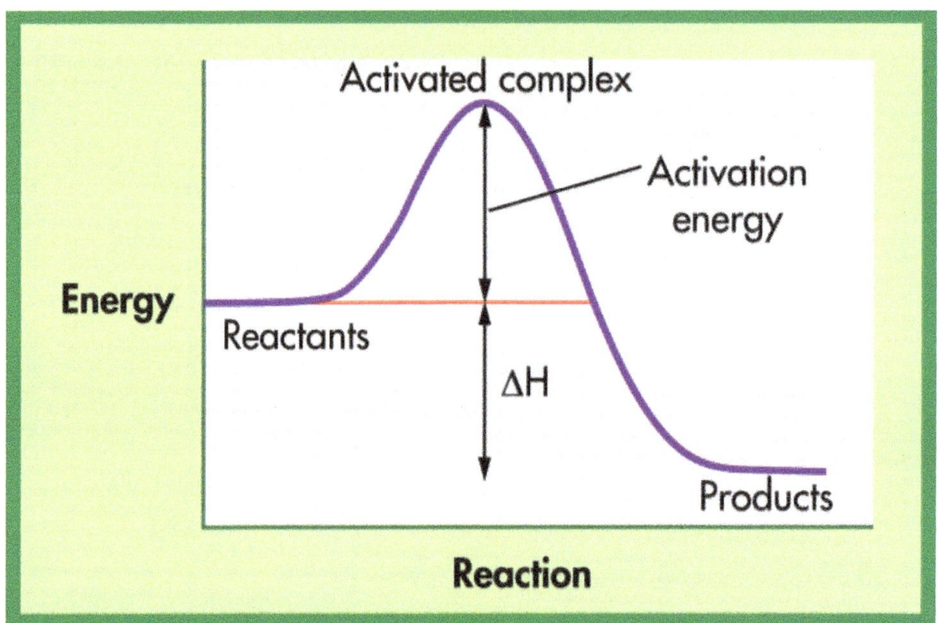


**EXOTHERMIC REACTION**
To start the exothermic chemical reactions of a sparkler, a flame must first supply the activation energy. Why is the energy of the products lower than the energy of the reactants?

Although the reaction is exothermic overall, some energy must be supplied to the magnesium to begin the reaction. **Activation energy** is the energy needed to begin a reaction. In the case of magnesium, an open flame will supply the energy.

What about the reverse reaction? In the reverse direction, the activation energy is much higher than in the forward direction because the products are at a lower energy than the reactants. In the reverse direction, the reaction's change in enthalpy has the same value, but the opposite sign as that of the forward reaction. If the forward reaction is exothermic, the reverse reaction is **endothermic**, but the amount of the enthalpy change is the same. Only the sign is reversed.

## Hess's Law

In calorimetry we could do experiments to calculate the heat associated with chemical and physical changes, but not all reactions happen that quickly. There are two other methods to calculate heat changes with chemical reactions. One is Hess's Law, which will be examined here. The method that will be looked at in the next section is Heat of Reactions. Imagine two hikers climbing to the top of a mountain. One hiker goes straight up the slope of the mountain. The other reaches the top by a path that meanders up and down the sides of the mountain. In the end, the net change in position of each hiker is the same.

Chemical reactions present a similar situation. Hess's Law states that the enthalpy change of a reaction is the same no matter how many intermediate steps occur. Like the hikers, the reaction will have the same net enthalpy change whether it occurs as a single reaction or as the result of several reactions.

When combining equations, keep in mind the following points:

- If you reverse an equation, change the sign of the enthalpy change from negative to positive, or positive to negative.
- If you multiply the coefficients in an equation, multiply the enthalpy change.

Example: Determine the enthalpy for the reaction of sulfur with oxygen to form sulfur trioxide:

$$2S(s) + 3O_2(g) \rightarrow 2SO_3(g) \quad \Delta H = ?$$

The following reactions are known and can be used.

1. $S(s) + O_2(g) \rightarrow SO_2(g) \quad \Delta H = -297\ kJ$
2. $2SO_2(g) \rightarrow 2SO_2(g) + O_2(g) \quad \Delta H = 198\ kJ$

Multiply the first reaction (a) by 2: $2S(s) + 2O_2(g) \rightarrow 2(g)$
$\Delta H = 2\ (-297\ kJ)$

Reverse the second (b) reaction: $2SO_2(g) + O_2(g) \rightarrow 3SO_3(g)$
$\Delta H = -198\ kJ$

Finally, combine the two reactions. Make sure that the combined reaction makes the wanted reaction. Then add the enthalpies:

$$2S(s) + 3O_2\ (g) \rightarrow 2SO_3(g) \quad \Delta H = -792\ kJ$$

In this example, notice that when the second equation was reversed, the enthalpy changed from 198 kJ to −198 kJ.

## Heat of Reactions

It's practically impossible to gather data on the enthalpy change of every reaction. There are too many reactions, and the change in enthalpy of a reaction is subject to a number of factors, including **temperature**. To help standardize enthalpy change values, chemists have assembled data for many reactions. In these standardized reactions, the reactants and products are at specified thermodynamic conditions of 25°C and 1 atm. (These conditions differ from the standard temperature and pressure used in the study of gases.) The symbol for the enthalpy change of a reaction at standard conditions is $\Delta H^\circ_{rxn}$. The symbol ° indicates standard conditions.

To make it easier, chemists use tables of the heats of reactions. The heat of reaction is the change in the enthalpy of a chemical reaction that occurs at a constant pressure in which 1 mol of a substance is formed from its elements. These values are called standard heats of formation, $\Delta H^\circ f$. The **standard heat of formation** of an element is defined as 0 kJ/mol.

Using standard heats of formation, the heat of reaction is calculated by subtracting the total heat of formation of the reactants from the total heat of formation of the products. The formula for calculating the heat of a reaction can be written $\Delta H^\circ_{rxn} = \Delta H^\circ_{products} - \Delta H^\circ_{reactants}$.

Because the standard heats of formation are expressed in kilojoules per mole, the heat of formation for a specific compound must be multiplied by the coefficient of the substance in the balanced equation.

The use of standard heats of formation combines the equations for the formation of products in the reaction. Each of these equations is multiplied by the coefficient in the desired equation, and the equations for the reactants are reversed. Hence, their total enthalpy change is subtracted. The effect is the same as breaking down the reactants to elements and then combining the elements to form the products.

**STANDARD HEAT TABLE**

Table comparing different standard heats of formation. What standard conditions are assumed?

| Substance | $\Delta H^\circ f$ (kJ/mol) |
|---|---|
| $C_2H_5OH(l)$ | −277.0 |
| $O_2(g)$ | 0 |
| $CO_2(g)$ | −393.5 |
| $H_2O(g)$ | −241.8 |

## Standard Heats of Formation: Sample Problem

Calculate the standard heat of reaction for the complete combustion of ethanol, $C_2H_5OH(l)$.

$$C_2H_5OH(l) + 3O_2(g) \rightarrow 2CO_2(g) + 3H_2O(g)$$

**Solution:**

The standard heat of reaction is calculated using Hess's Law:

$$\Delta H^\circ_{rxn} = \Delta H^\circ_{products} - \Delta H^\circ_{reactants}$$

For this reaction, the equation becomes

$$\Delta H^\circ_{rxn} = [2 \text{ mol } \Delta H^\circ f(CO_2(g)) + 3 \text{ mol } \Delta H^\circ f(H_2O(g))] - [1 \text{ mol } \Delta H^\circ f(C_2H_5OH(l)) + 3 \text{ mol } \Delta H^\circ f(O_2(g))]$$

Substitute the values from the table into the equation, and then solve.

$$\Delta H^\circ_{rxn} = [2 \text{ mol } (-393.5 \text{ kJ/mol}) + 3 \text{ mol } (-241.8 \text{ kJ/mol})] - [1 \text{ mol } (-277.0 \text{ kJ/mol}) + 3 \text{ mol } (0 \text{ kJ/mol})]$$
$$= -1235.4 \text{ kJ}$$



### Consider the Explain Question

**How are endothermic and exothermic processes identified, and how does the energy of the system change with each?**

dlc.com/ca9115s

Go online to complete the scientific explanation.

### Check Your Understanding

**Can you describe the energy changes of a system and its surroundings associated with exothermic changes and with endothermic changes?**

dlc.com/ca9116s

## Applying Thermochemistry

Today we take refrigeration for granted, but in the past fortunes were made shipping large blocks of ice around the world as rich people sought ways to preserve food from spoiling. When you think of a refrigerator or air conditioner, do you think of cold? Devices such as these also use heat. A refrigerator transfers heat away from its inner compartment, which is the area where food is stored. This heat is then transferred to the region outside. Hence, this is why the back of a refrigerator is warm.

A device inside the refrigerator, called an evaporator, is used to capture the heat from the inside compartment. The evaporator contains a refrigerant fluid, such as ammonia or R-134a (tetrafluoroethane), that readily liquefies. Warm air from inside the refrigerator flows over cold low-pressure coils of pipe in the refrigerator that are filled with this refrigerant. As a result, the refrigerant absorbs some of the thermal **energy** from the inside air and in the process evaporates from a liquid to a gas. The pipe takes the refrigerant to a compressor, which compresses the gas inside the pipe. This compressed gas, now at very high pressure and very hot, is pumped into radiator coils which are on the outside back of the refrigerator. These radiator coils are often called condensers. As the gas moves through the condenser coils (radiator), it cools and becomes a room **temperature** liquid. The heat from the hot gas is released into the air behind the refrigerator as the refrigerant condenses into a liquid. Now the cooler room temperature liquid flows through an expansion valve. The liquid refrigerant immediately boils, vaporizes, and becomes an extremely cold gas, which flows through the coils ready to repeat the process again.

Air conditioners work in a similar way. Hot air from a room flows into an evaporator, and a compressor circulates refrigerant from the evaporator to a condenser (which is just like the radiator in a refrigerator). Behind the evaporator is a fan, which draws in hot air from the room, and another fan pushes heat from the condenser to the outside. As with a refrigerator, the back of an air conditioner is hot because it is moving heat from the area to be cooled.

**REFRIGERATOR**
A refrigerator works due to the principles of thermochemistry. How does a refrigerator keep food cold?



## STEM and Thermochemistry

Efforts to develop clean, renewable energy resources have often focused on the sun. Photovoltaic cells, such as those that power solar-powered calculators, convert light energy directly into electrical energy. Typical solar panels only generate electricity during daylight hours. During the night, electricity comes from the utility grid. Solar engineers have been hard at work creating systems to store heat absorbed during the day to provide a steady stream of electricity day or night.

Using a different approach to solve the problem, scientists have used nanotechnology to develop super-thin solar films. These films have embedded nanoantennas, which absorb infrared energy released by the earth for several hours after the sun sets. Because they absorb energy from both sunlight and the earth itself, these tiny cells are more efficient than conventional solar cells.

Power plants have been built around the world which store energy and generate electrical energy even when the sun is not shining. One method designed by engineers uses large arrays of mirrors to shine sunlight on a tower. In the tower, a liquid of molten salts absorbs the energy and is heated to over 1,000°C. The hot, molten liquid is stored in well-insulated tanks. When energy is needed, thermal energy from the salt is transferred to water, which becomes superheated steam. The steam drives turbines that generate electrical energy. The cooled liquid is stored in another tank for return to the tower where it is reheated. Molten salt thermal energy storage is the most economical method to store solar energy.

Recently, companies have completed experimental grid-style projects, which use batteries to store energy from solar panels. This new battery technology was initially developed for electric cars. The batteries allow power systems to dispense electricity in the evening after it was stored by solar panels during the day.

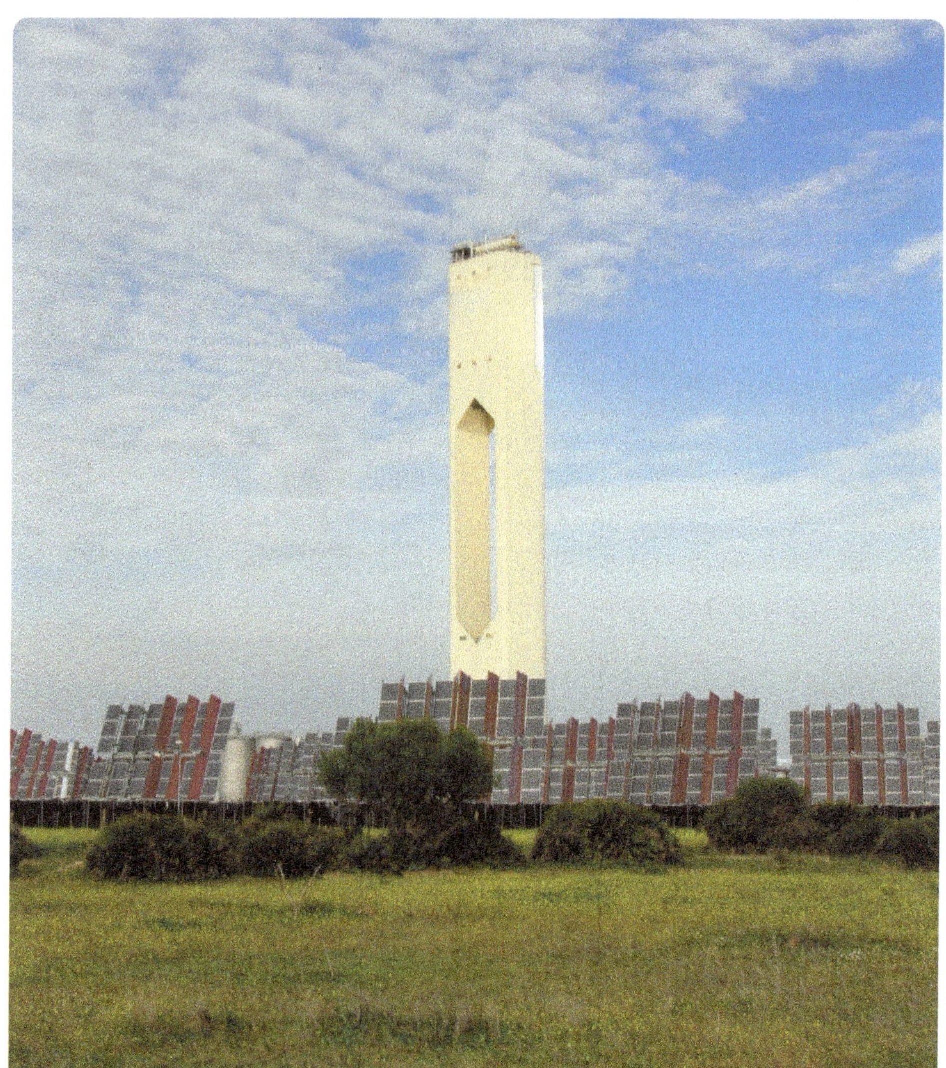

**SOLAR POWER TOWER**

This concentrating solar power tower in Spain uses molten salts to capture the sun's energy. How could this process advance the solar power industry?

# Standard Heats of Formation Calculation

**Use Hess's Law to solve the problems below.**

**1.** Use the standard heats of formation provided to calculate the standard heat of reaction for the reaction between water and sulfur trioxide shown. Is this reaction endothermic or exothermic?

$$H_2O(g) + SO_3(g) \rightarrow H_2SO_4(g)$$

| Substance | $\Delta H^\circ_f$ (kJ/mol) |
|---|---|
| $H_2O(g)$ | −241.8 |
| $SO_3(g)$ | −395.8 |
| $H_2SO_4(g)$ | −735.1 |

**2.** Use the standard heats of formation provided to calculate the standard heat of reaction for the reaction between water and carbon dioxide shown. Is this reaction endothermic or exothermic?

$$2H_2O(g) + 4CO_2(g) \rightarrow 2C_2H_2(g) + 5O_2(g)$$

| Substance | $\Delta H^\circ_f$ (kJ/mol) |
|---|---|
| $H_2O(g)$ | −241.8 |
| $CO_2(g)$ | −393.5 |
| $C_2H_2(g)$ | 226.7 |
| $O_2(g)$ | 0 |

**3.** Use the standard heats of formation provided to calculate the standard heat of reaction for the reaction between hydrogen sulfide and oxygen shown. Is this reaction endothermic or exothermic?

$$2H_2S(g) + 3O_2(g) \rightarrow 2H_2O(l) + 2SO_2(g)$$

| Substance | $\Delta H^\circ_f$ (kJ/mol) |
|---|---|
| $H_2S(g)$ | −20.6 |
| $O_2(g)$ | 0 |
| $H_2O(l)$ | −285.8 |
| $SO_2(g)$ | −296.8 |

CONCEPT 1.3

# The Cycling of Matter and Energy

dlc.com/ca9014s

## LESSON OVERVIEW

### Lesson Question

- How do energy and matter interact as they cycle through Earth's systems?

### Lesson Objective

By the end of the lesson, you should be able to:

- Understand the interactions of matter and energy among various Earth systems.

### Key Vocabulary

Which terms do you already know?

- [ ] biofuel
- [ ] carbon cycle
- [ ] carbon reservoir
- [ ] condensation
- [ ] conduction
- [ ] convection
- [ ] cycle
- [ ] energy
- [ ] evaporation
- [ ] geochemical cycle
- [ ] matter
- [ ] nitrogen cycle
- [ ] phosphorous cycle
- [ ] photosynthesis
- [ ] radiation
- [ ] radioactive decay
- [ ] residence time
- [ ] rock cycle
- [ ] solar energy
- [ ] thermal energy
- [ ] water cycle

## Tracing the Cycling of Matter and Energy

**A walk down the beach offers a myriad of seashells dotting the sand, while the fall brings a cascade of colorful leaves down to the ground. Trees grow old and fall in the forest, lying on the forest floor. These are all examples of objects that seem no longer useful to our environment. Or are they? What do you think will happen to all of these materials?**

dlc.com/ca9015s

**EXPLAIN QUESTION**

**How do the different cycles of energy and matter on Earth interact with each other?**

**SEASHELLS ON THE BEACH**

Seashells wash up on the shore when they are no longer part of a living animal. What do you think happens to all the empty shells?

# How Do Energy and Matter Interact as They Cycle through Earth's Systems?

## Thermal and Solar Energy

**Matter** is any substance that takes up space and has mass. Most matter on Earth exists in one of three states: solid, liquid, or gas. (A fourth phase of matter, plasma, consists of highly energized, charged particles.) The addition or removal of **energy** is needed to transform matter from one state to another. Important compounds and elements that **cycle** through Earth's different spheres include water, carbon, nitrogen, and phosphorus. While they cycle, they sometimes undergo a change of state or a change of chemical identity.

**Thermal energy** is internal energy responsible for the temperature of an object. Nuclear energy is generated during nuclear reactions. Fission describes energy released when an unstable atom splits into smaller particles and ionizing **radiation** (including alpha, beta, gamma, and neutron radiation). This **radioactive decay** transforms an unstable atom to a more stable configuration. Fission occurs in Earth's interior. The heat it generates is called geothermal energy.

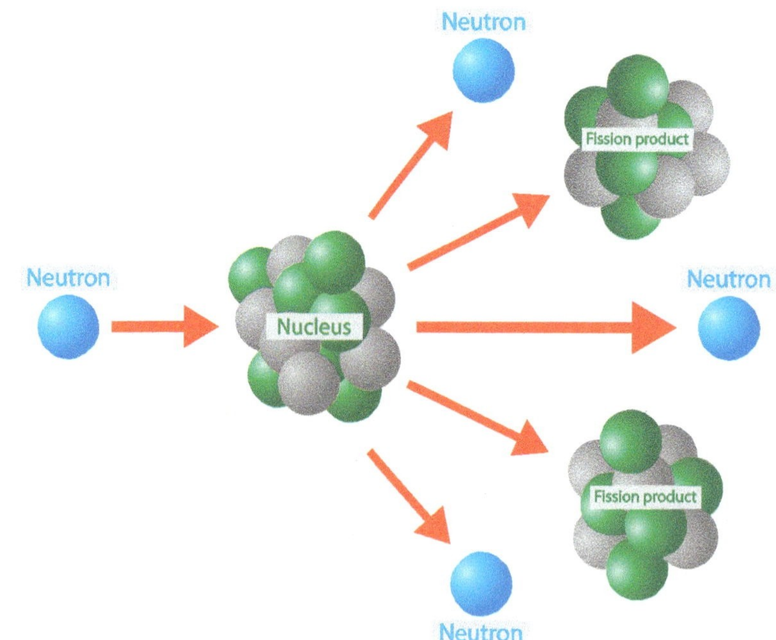


**NUCLEAR FISSION**
During fission, the nucleus of an unstable atom releases energy when it splits into smaller atoms. How have humans harnessed the energy created by nuclear fission?

Fusion is the process in which atoms combine to form larger atoms. This process occurs in the sun. The sun is primarily hydrogen. Inside the sun, four hydrogen atoms fuse to produce one helium atom, which contains two protons and two neutrons. The helium atom's mass is smaller than the four hydrogen atoms combined. The loss in mass is converted into **solar energy**, the light and heat that power the sun. We can harness this energy to power our devices—and our bodies.

## Thermal Energy Movement

Thermal energy travels in three ways:

- Radiation describes energy that travels in the form of electromagnetic waves. Solar radiation enters Earth's atmosphere in the form of short-wavelength (high energy) electromagnetic waves on the daylight side of the planet. On the dark, nighttime side, much of this energy is radiated back into space in the form of long-wavelength (low energy) radiation. This process removes large amounts of heat from the atmosphere. If Earth did not expel much of this solar energy back into space, the planet would overheat.
- **Conduction** is the transfer of heat from molecule to molecule or atom to atom through direct contact. One particle collides with another, transferring some of its energy as heat to the other particle. Conduction is the transfer of energy you experience when you touch a hot object. When hot magma comes in contact with other rocks, its thermal energy is transferred from the point of contact by conduction.
- **Convection** is the transfer of heat through primarily vertical movement in fluids. Convection occurs most commonly in gas and liquid substances because heat moves with the fluid, establishing convection cells. (Convection cells also have horizontal components as rising particles spread out, cool, and sink back down to produce a cycling motion.) This is illustrated by placing a pot of water on the stove. As the water heats up, it becomes less dense and rises in the pot. As the water cools, it sinks back to the bottom of the pot, completing the cycle. Convection also can happen in fluid-like solids such as the material in Earth's upper mantle, the asthenosphere.

## Uneven Distribution of Heat

Although its incoming and outgoing solar radiation is balanced, Earth does not heat evenly, because the planet tilts at 23.5°. At any given time, one region of the planet is exposed more directly to the sun's rays. When Earth's southern hemisphere tilts toward the sun, it is exposed to solar radiation for a longer period. The sunlight is more concentrated on the southern hemisphere and dispersed over a wider area in the northern hemisphere. This corresponds to summer in the southern hemisphere and winter in the northern hemisphere. When the northern hemisphere tilts toward the sun, it is exposed to solar radiation for a longer period and receives more direct sunlight; at the same time, sunlight in the southern hemisphere is dispersed over a wider area. This corresponds to summer in the northern hemisphere and winter in the southern hemisphere. These changes are magnified in the northern hemisphere because more land is situated in this hemisphere. Land absorbs and loses heat more quickly than water does.

Earth's uneven warming affects air near the planet's surface. Radiation heats the surface of Earth. Heat imparts energy to air through conduction, or direct contact with the warmed surface. Warm air rises, leaving behind a region of lower air pressure. Cold air sinks, adding cold air to the lower levels and increasing the air pressure. This cycle of rising and sinking air forms convection cells, moving warm air away from the equator and cold air away from the poles. Global bands of air pressure form at the equator (low pressure), 30° (high pressure), 60° (low pressure), and 90° (high pressure) north and south latitude.

Winds form between the pressure gradient set up by the global bands of air pressure. (A pressure gradient is the extent to which pressure changes over a given distance.) Air moves from the high- to the low-pressure bands. The winds distribute heat and moisture around the planet. Winds also move water across the ocean's surface, producing surface currents that distribute heat, moisture, and nutrients around the planet. More importantly, surface currents carry warm, saline water to higher latitudes. This water freezes at the poles, increasing the density of the water beneath. This produces density currents that sink into the deep ocean. Thermohaline circulation describes the flow of density currents throughout the global ocean. This circulation mixes ocean waters and helps regulate Earth's climate. In summary, convection, conduction, and radiation transfer heat through Earth's systems, affecting weather patterns, climate, ocean circulation, and tectonic plate movement.

## The Water Cycle

Earth contains many systems that interact with each other and that cycle matter and energy. Scientists have analyzed each of these systems in detail, determining the boundaries of each system, the initial conditions of each system in Earth's history, and the inputs and outputs of each system. Diagrams and models help describe the functions of a system.

The uneven distribution of heat on Earth drives several crucial cycles of matter and energy. The **water cycle** describes how fresh water moves between the land, ocean, and the atmosphere, passing through the three states of matter at or close to the planet's surface. Each transition, or phase change, requires the addition or removal of energy. With the addition of energy, ice melts to form liquid water, and liquid water evaporates to form water vapor. With the removal of energy, water vapor condenses to form liquid water, and liquid water freezes to form ice. When a great deal of energy is added, ice sublimes directly to water vapor. When a great deal of energy is removed, water vapor deposits directly as ice. As water evaporates from the planet's surface, the gaseous water vapor in the atmosphere produces humidity. When water vapor condenses, it forms clouds. When the concentration of water droplets or ice crystals in clouds is high enough, the water falls from the atmosphere back to the planet's surface as precipitation—primarily rain, snow, or hail. The precipitation recharges oceans, rivers, and lakes. It also percolates through the rocks and soil, recharging aquifers.



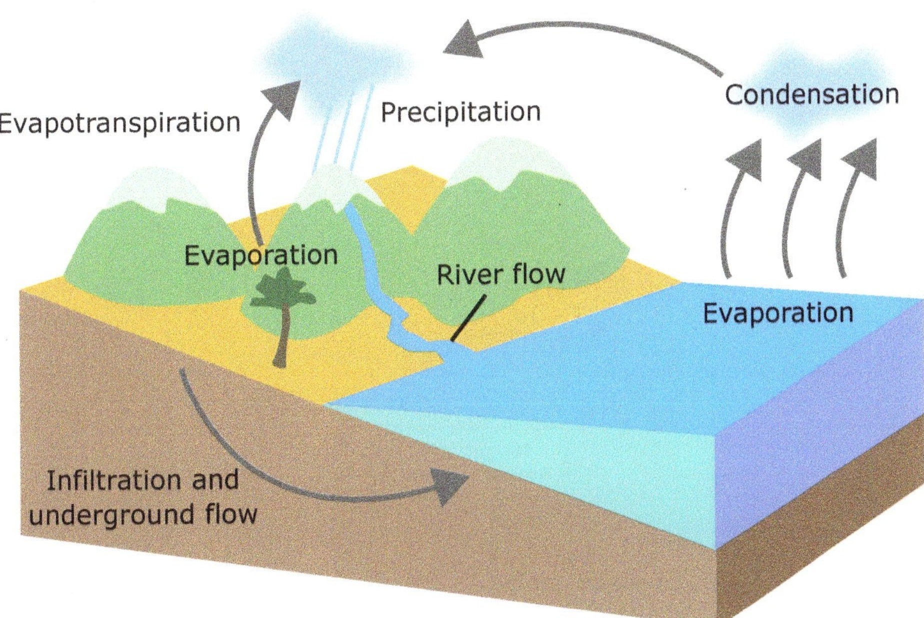


**THE WATER CYCLE**
Water cycles between the hydrosphere, atmosphere, and geosphere. At what points in the water cycle does water change states?

Greenhouse gases in the atmosphere, including carbon dioxide, trap heat from the sun that would otherwise escape back into space, warming Earth. This warming of the atmosphere by the heat absorbed by greenhouse gases is known as the "Greenhouse Effect." The increased temperature has an impact on the water cycle. **Evaporation** rates increase, which in turn increases moisture in the troposphere, or lower atmosphere, leading to more precipitation. Because temperatures are higher, this precipitation is more likely to occur as rain than snow. Changes in global climate patterns caused by global warming also impact precipitation patterns. For example, models predict that, as climate change continues, precipitation will increase slightly in most of the United States, but that summers—particularly in the south—will become hotter and have longer periods without rain. The same model predicts minor increases in precipitation in the northwest and northeast of the country.

## The Rock Cycle

The **rock cycle** describes how rocks change over time. Like all cycles, the rock cycle does not have a true beginning or an end. Igneous rocks are formed as magma cools beneath Earth's surface and lava cools at Earth's surface. Rocks on Earth's surface are broken down by chemical and physical weathering, producing sediments. These sediments are deposited and compressed to form sedimentary rocks. When sedimentary or igneous rocks are subjected to changes in temperature or pressure conditions, metamorphic rocks form. If the temperature increases sufficiently, the rocks will melt and form magma that gradually cools to form more igneous rock.

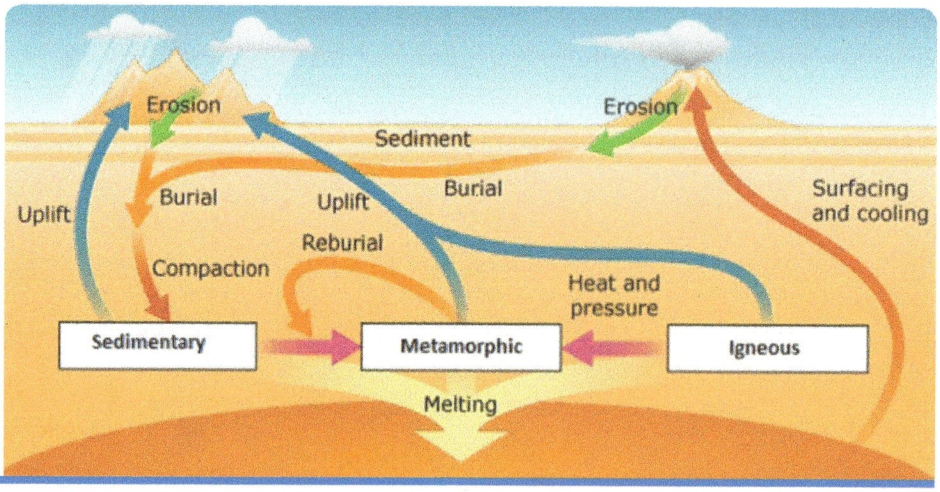

**THE ROCK CYCLE**

The rock cycle describes the formation and transformation of rocks. How are igneous, sedimentary, and metamorphic rocks formed and transformed?

# How Do Energy and Matter Interact as They Cycle through Earth's Systems?

## Geochemical Cycles

Geochemical cycles describe the flow of elements and compounds between living organisms and the physical environment. In addition to the **water cycle**, geochemical cycles that are important to the study of Earth science include the **carbon cycle**, the phosphorus **cycle**, and the **nitrogen cycle**.

## The Carbon Cycle

The carbon cycle graphically represents the movement of carbon within and between Earth systems. Carbon on Earth flows between the primary reservoirs on the planet, including the atmosphere, the terrestrial biosphere, the ocean, and sediments. As carbon travels between reservoirs, it moves between sources and sinks. A source is the reservoir that supplies carbon to a region. A sink is the reservoir that removes the carbon.



Carbon can be found on the planet as carbon dioxide in the atmosphere or oceans, or incorporated as compounds in the bodies of organisms. It can be found in many molecules as it moves between different reservoirs on the planet. In the atmosphere, carbon most commonly is found as carbon dioxide ($CO_2$) and carbon monoxide (CO). These gases can dissolve in water, producing several carbon molecules and ions in the ocean, including carbonic acid ($H_2CO_3$), bicarbonate ion ($HCO_3^-$), and carbonate ion ($CO_3^{2-}$). On land, inorganic carbon is stored in rocks and minerals such as calcium carbonate ($CaCO_3$) and diamond (C). Carbon is also a fundamental component of living organisms. Plants and other organisms, by capturing $CO_2$ and releasing $O_2$, are able to gradually change the carbon composition of the atmosphere. For example, deforestation removes old growth trees that have captured and stored a great deal of carbon. Not only do these trees no longer capture and store $CO_2$, but the stored $CO_2$ within them may be released if the wood from the trees is used as fuel. Changes in the atmosphere due to human activity also have increased carbon dioxide concentrations and thus affect climate.

In the carbon cycle, carbon compounds, primarily carbon dioxide, are incorporated into living organisms through **photosynthesis** and passed along food chains. The carbon is returned to the atmosphere by respiration, the decay of dead organisms, and more recently, the burning of fossil fuels.

## The Phosphorus Cycle

The **phosphorous cycle** describes how phosphate salts transfer between water, organisms, rocks, and minerals. (There is no gaseous form of phosphorus at Earth's surface.) Phosphate salts are removed from rocks and sediments through chemical and physical weathering on land. Phosphorous in the form of phosphate ($PO_4^{3-}$) is carried to the oceans by rivers. Organisms take up the phosphorous and incorporate it into their organic tissue. When an organism dies or excretes waste, its phosphorus is released back into the environment. The phosphorous from organisms cycles through Earth's systems more rapidly than the phosphorous released from rocks and minerals on land.

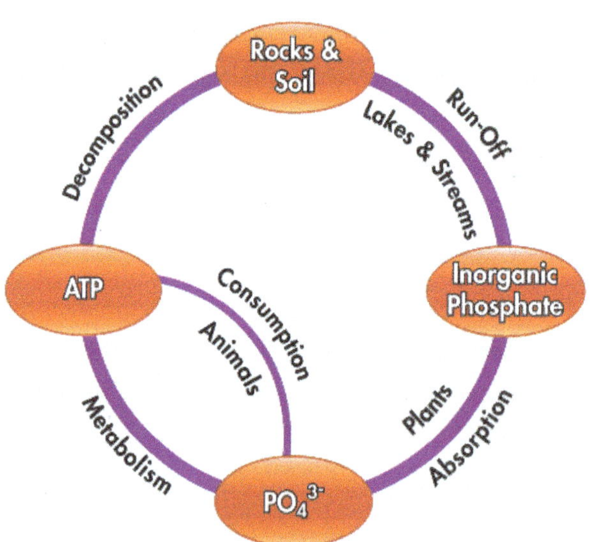


**THE PHOSPHORUS CYCLE**

The phosphorous cycle represents the movement of the element phosphorus through various reservoirs on the planet. How does phosphorous move through the levels of an ecosystem's food chain?



## The Nitrogen Cycle

Nitrogen can be found in solid, liquid, and gaseous forms on the planet's surface. The nitrogen cycle illustrates how nitrogen travels between different reservoirs on the planet, including the atmosphere, rocks, ocean, and biosphere. Not all of the nitrogen on Earth is available to organisms for life processes. The most common forms of nitrogen used by organisms include nitrate ions ($NO_3^-$), ammonium ion ($NH_4^+$), and urea ($CO(NH_2)_2$). Nitrogen is essential for many biological processes, including the production of organic compounds such as amino acids, proteins, and the nucleic acids that make up DNA and RNA. The excess nitrogen that is not used in biological functions ultimately finds its way back into the land and ocean in the form of ammonia ($NH_3$).

## Photosynthesis and the Carbon Cycle

Photosynthesis takes place in the chloroplasts of photosynthetic autotrophs such as plants, algae, and photosynthetic bacteria. Autotrophs are organisms that make their own food from simple substances—in this case, carbon dioxide and water. Sunlight provides the **energy** for the photosynthesis reaction.

$$6CO_2 + 6H_2O \text{ (+light energy)} \rightarrow C_6H_{12}O_6 + 6O_2$$

During the reaction, colored pigments—such as the green pigment chlorophyll—absorb certain wavelengths of sunlight and use them to provide energy to combine carbon dioxide and water to form carbohydrates—solid compounds rich in carbon. These energy-rich compounds include the substances glucose and starch.

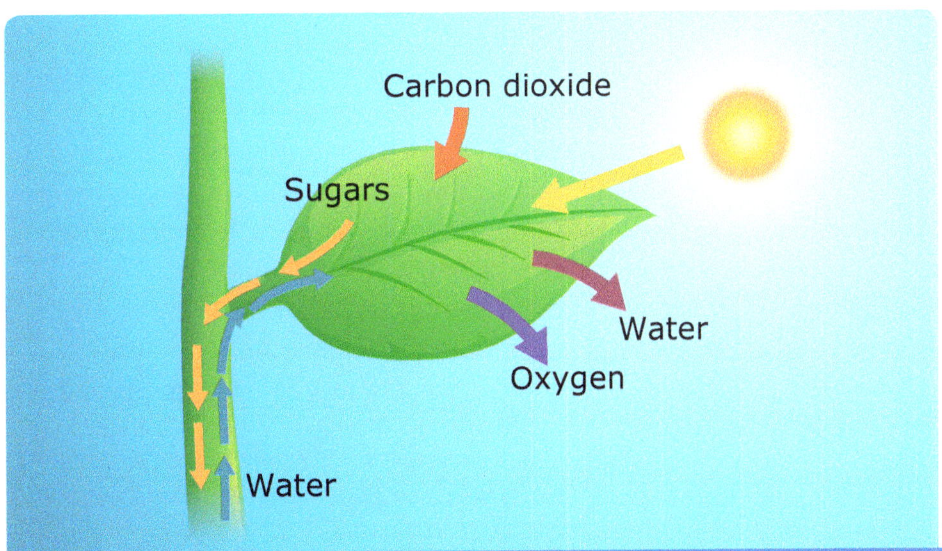


**PHOTOSYNTHESIS**
Plants produce the energy they need for life through the process of photosynthesis. Why is the sun necessary for photosynthesis to occur?

Photosynthesis forms the base of the food chains upon which nearly all other organisms depend on for survival. The energy captured by photosynthetic autotrophs moves up a food chain, from autotrophs to heterotrophs; for example, when plants are eaten by herbivores. Herbivores may in turn be eaten by carnivores. At each step in the food chain, some of the food consumed is converted back into carbon dioxide and water as it is used in the process of respiration. Organisms obtain energy from the process of respiration. In terms of chemistry, respiration is the reverse of photosynthesis.

$$C_6H_{12}O_6 + 6O_2 \rightarrow 6CO_2 + 6H_2O \text{ (+energy)}$$

Eventually all organisms die and are entirely, or partly, decomposed. At this point, the carbon they contain may be released by decomposition into the atmosphere or water, however, some carbon may remain and enter soil or sediment as organic **matter**. Some of the carbon eventually may get locked up in rocks. For example, the carbon in the shells of some marine organisms may become locked in sediments when the shells sink to the ocean floor. Once on the floor, the sediments may become compressed to form limestone. In this way, carbon dioxide from photosynthesis becomes locked away in rocks, sometimes for hundreds of millions of years.

## Residence Time

**Residence time** describes the amount of time a particle remains in a region or reservoir. Residence time can be calculated by comparing the initial concentration of the particles in a system to the rate at which it is supplied or removed from the reservoir. Residence time can be described as days, months, or years, depending on how quickly the material cycles through the sphere. In the equation below, residence time of a particle is represented by the Greek letter tau ($\tau$). The capital letter $V$ represents the initial concentration of the particles in the reservoir. The lowercase letter $v$ represents the rate of change of the particles. For example, particles may leave a reservoir at the rate of 2 kilograms per minute (2 kg/min).

$$\tau = \frac{V}{v}$$

Residence time often is described using a box model such as the diagram shown. This simplified model contains one or multiple boxes. The boxes are connected by pipes that show the flow of different parameters, such as temperature, into and out of the box. The box model is useful because it provides a simplified way to look at a complex problem by focusing on only one or two parameters.

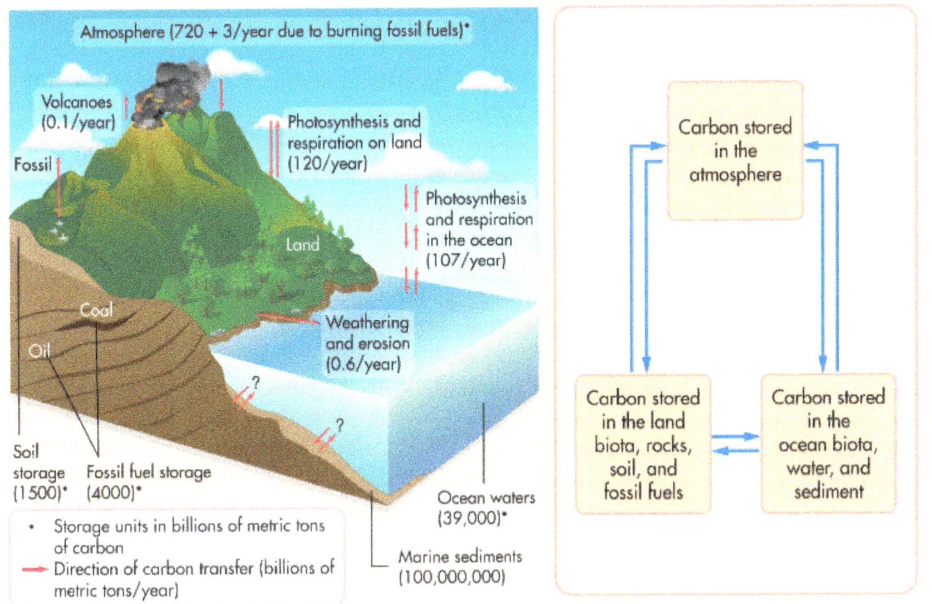


**RESIDENCE TIME**

The residence time of carbon in a reservoir varies. How does this model depict the movement of carbon between the primary reservoirs on the planet?

## Feedbacks between the Biosphere and Earth's Other Systems

The biosphere includes all places on Earth where life exists. The biosphere is a segment of the complex of Earth's systems that have been changed over the history of our planet. Through the interrelations of the biosphere and other spheres of Earth, life adapted through evolution. Some organisms dominated Earth's early biosphere and other organisms are more widespread today. Interactions within and around Earth have shaped its history and will predict its future.

Because the atmosphere, the hydrosphere, and the geosphere all change, the biosphere is also undergoing a constant evolutionary process. As a result, all living organisms take part in a simultaneous coevolution with Earth's systems. For example, photosynthetic life altered the atmosphere by producing oxygen, which in turn increased weathering rates and allowed for the evolution of animal life; microbial life on land increased the formation of soil, which allowed for the evolution of land plants; the evolution of corals produced reefs that altered patterns of coastal erosion and deposition and provided habitats for the evolution of new life forms.

The sustainability of Earth directly depends on the interactions between the physical, chemical, and biological components of the environment, including their effects on all types of organisms.

### Consider the Explain Question

**How do the different cycles of energy and matter on Earth interact with each other?**

dlc.com/ca9016s

Go online to complete the scientific explanation.

### Check Your Understanding

**What is the role of thermal energy in natural processes?**

dlc.com/ca9017s

## Applying the Cycling of Matter and Energy

Scientists speculate that the world's oceans currently absorb 30–50 percent of the carbon dioxide produced by burning fossil fuels. Can you think of a process that would bring atmospheric $CO_2$ into the hydrosphere? Certainly the reaction between $CO_2$ and water in the atmosphere creates carbonic acid, which the rains carry to the oceans. But scientists think there are several ways carbon can be absorbed by the oceans, and they are looking at similar processes that store carbon on land.

Tiny, aquatic plants called phytoplankton are some of the oldest types of organisms on the planet. For the past 3.5 billion years, phytoplankton have made use of **photosynthesis**. This process has profoundly changed the composition of the atmosphere by producing oxygen, which made life for all other organisms on the planet possible. Scientists speculate that 90 percent of the oxygen in Earth's atmosphere was produced by phytoplankton during photosynthesis.

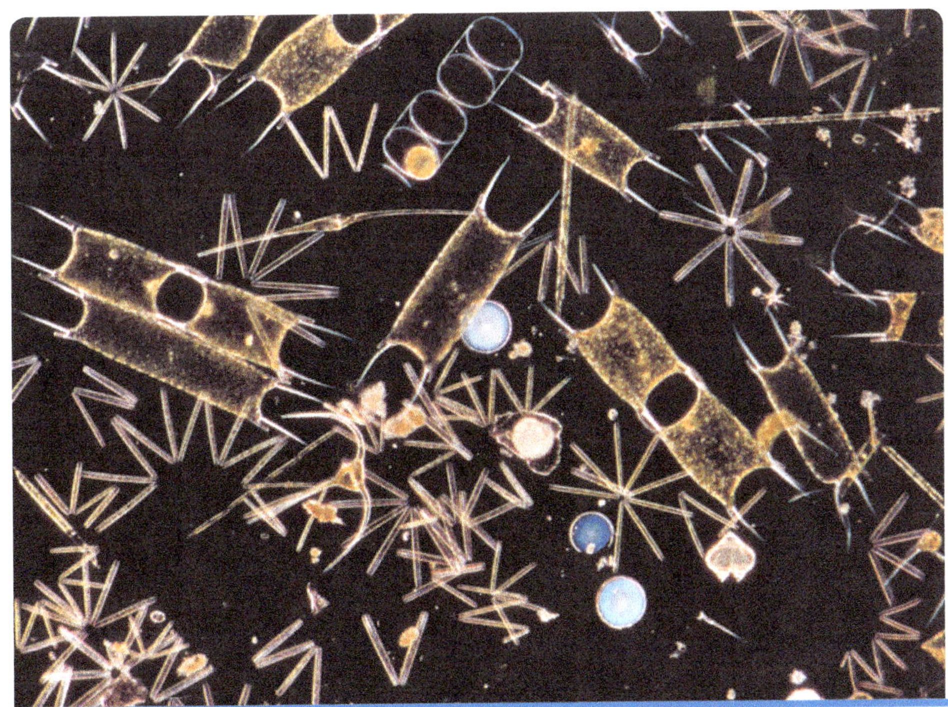

**MARINE PLANKTON**
Under the microscope, tiny phytoplankton can be seen. What are the benefits of phytoplankton to the oceans?

Phytoplankton also control the concentration of carbon dioxide in the atmosphere. Carbon dioxide, a greenhouse gas, is linked to climate change, such as global warming. Scientists are hopeful that phytoplankton may play an important role in reversing or stalling the increase of this gas in the atmosphere. As carbon dioxide is taken up during photosynthesis, it is removed from the atmosphere. Some phytoplankton, like coccolithophores, also participate in reducing the concentration of atmospheric carbon dioxide by building shells made of calcium carbonate ($CaCO_3$).

Most carbon in the ocean is recycled in the first 200 meters of the water column. However, a small percentage escapes to deeper regions of the ocean. Scientists think that transferring carbon to the deep ocean keeps it from interacting with the atmosphere. Any process that increases or speeds up this transfer will decrease the concentration of atmospheric carbon dioxide and slow the warming of the planet.

## STEM and the Cycling of Matter and Energy

**Solar energy** is being studied as a source of electrical **energy**, but not in the way you might think. We already have solar cell technology in place, converting light energy directly to electricity and reducing our carbon production as we decrease our need to burn fossil fuels. But scientists are also looking directly at the process of photosynthesis as an alternative source of electricity. Working in interdisciplinary groups, microbiologists first considered the amount of energy that a plant produces in the form of glucose during the process of photosynthesis. They studied a plant's complete **cycle**, including the organic material it excreted through its roots. They discovered microorganisms were breaking down that material in the soil to absorb energy, creating an interesting product in the reaction—electrons. Engineers developed equipment to harness those electrons. They found if they placed an electrode in the area of the microorganisms to attract the electrons, the electrons could be stored and used for power. Studies have shown that harvesting the electrons in this process does not affect plant growth. A Dutch company already has a working model that powers their headquarters with this plant power.

In a similar process, but under the sea, carbon compounds fall to the ocean floor, and bacteria break them down through a series of aerobic and anaerobic reactions. These reactions release electrons. As the electrons build up in the sediment, an electric potential is produced between the electron-rich sediments and the electron-poor overlying water. Scientists have discovered a way to harness this potential source of electrons to generate energy to power their instruments.

Electrical engineers developed a device consisting of two graphite disks. They buried one disk in sediment on the ocean floor and placed the other above the sediment in the water column. Bacteria growing in the sediment release electrons as they break down carbon compounds. The electrons migrate upward to the overlying water, producing an electric current. The current can generate up to 0.02 milliwatts of power, which is enough to operate simple sensors in oceanographic equipment. Oceanographers use data from this equipment to evaluate how small changes in one variable affect the greater system. Unfortunately, most of these instruments run on batteries. The need for batteries limits where scientists can place instruments in the ocean. They must be able to locate the equipment in order to replace the batteries. This process can cost from $10,000 to $17,000 for every piece of equipment left in the ocean. Improvements in the efficiency of this power supply may provide an indefinite, environmentally friendly power source for moored equipment in the ocean.

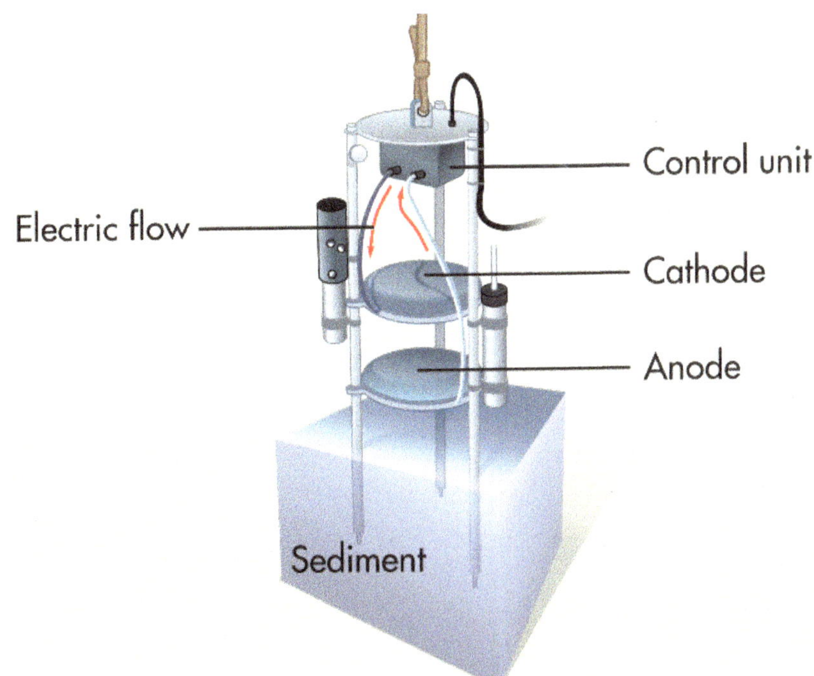


**FUEL CELL**

Electrons released by microbes are the source of current for this underwater fuel cell. What uses could you think of for an underwater fuel cell?

Understanding how microbes participate in the **carbon cycle** to release electrons is one example of how Earth's natural cycles can benefit the human race.

CONCEPT 1.4

# Earth's Interior

dlc.com/ca9019s

## LESSON OVERVIEW

### Lesson Question

- How do physical conditions and mineral composition change with depth beneath Earth's surface?

### Lesson Objective

By the end of the lesson, you should be able to:

- Identify and compare Earth's compositional and structural layers.

### Key Vocabulary

Which terms do you already know?

- [ ] asthenosphere
- [ ] continental crust
- [ ] convection
- [ ] core
- [ ] heat energy
- [ ] inner core
- [ ] lithosphere
- [ ] magma
- [ ] mantle
- [ ] mid-ocean ridge
- [ ] oceanic crust
- [ ] outer core
- [ ] thermal energy

## Seeing inside Earth's Interior

**What comes to mind when you consider Yellowstone National Park? Maybe you think of Yellowstone's famous geysers, such as Old Faithful, or perhaps you wonder when the Yellowstone supervolcano will erupt. Where does the geothermal heat in these hotspots come from?**

dlc.com/ca9020s

**CASTLE GEYSER**

Geysers, such as Castle Geyser in Yellowstone National Park, are driven by thermal energy from Earth's interior. Why does Earth's internal heat reach the surface in some places, but not in others?

### EXPLAIN QUESTION

**How are the composition and structure of Earth's layers different as you go deeper into Earth?**

# How Do Physical Conditions and Mineral Composition Change with Depth beneath Earth's Surface?

## Earth's Crust

The outer, rocky layer of Earth that we live on and observe directly is called the crust. The crust is made up of the continents, or **continental crust**, and the ocean floor, or **oceanic crust**. Together with the top part of the underlying layer, the **mantle**, these two types of crust make up the brittle part of Earth called the **lithosphere**. The brittle nature of the lithosphere causes it to crack under stress. The lithosphere is broken into large pieces called tectonic plates.

Overall, the crust is rich in familiar elements such as oxygen, silicon, aluminum, iron, calcium, magnesium, and sodium. Many of these elements are important for living things. Aluminosilicate minerals are common building blocks of crustal rocks.

The two kinds of crust differ somewhat in composition and density. Both continental and oceanic crust are made mostly of silicate minerals. However, continental crust is characterized by a higher concentration of the alkali elements sodium and potassium, while oceanic crust is characterized by higher concentrations of the heavier elements iron and magnesium. The higher concentration of iron and magnesium makes oceanic crust more dense than continental crust. Continental crust ranges from 30 kilometers to 100 kilometers thick; it is thickest under mountain ranges. Oceanic crust is only 5 kilometers to 10 kilometers thick. We can think of the continents as lighter, thicker rocks floating on a denser base similar to oceanic crust.

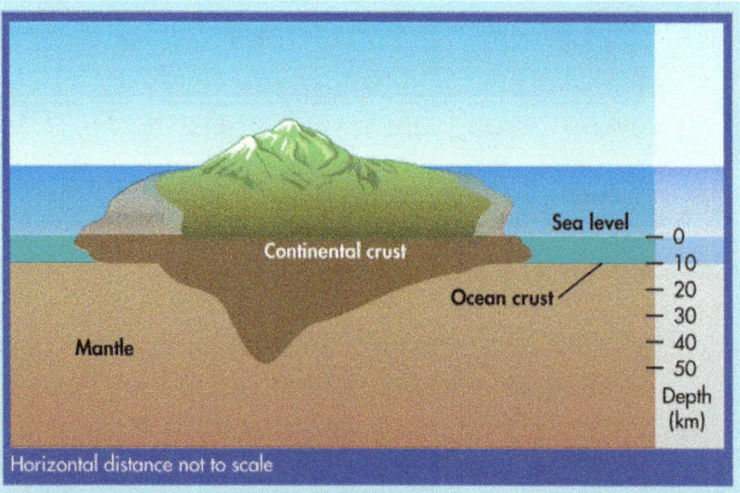


**EARTH'S LITHOSPHERE**
Earth's lithosphere consists of both types of crust and the top part of the mantle. Continental crust is much thicker than oceanic crust. Which type is denser?

Earth formed about 4.6 billion years ago from collisions of rock and dust in the disc of matter circling the young sun. The crust formed and solidified over the next few billion years as Earth slowly cooled. It continues to form as lava rising from the mantle cools into hard rock. Oceanic crust is continually generated along mid-ocean ridges and is recycled as it sinks back into the mantle in deep trenches.

## Earth's Mantle

The mantle is about 2,900 kilometers thick. This thick layer makes up two-thirds of Earth's mass. Remember that the lithosphere includes the crust and the top part of the mantle. Much of the lava that forms Earth's oceanic crust comes from a part of the mantle called the **asthenosphere**, which is located below the lithosphere. Rock in the asthenosphere is able to flow because of high heat and pressure generated in this layer. The rock of the asthenosphere is soft like putty and flows slowly.



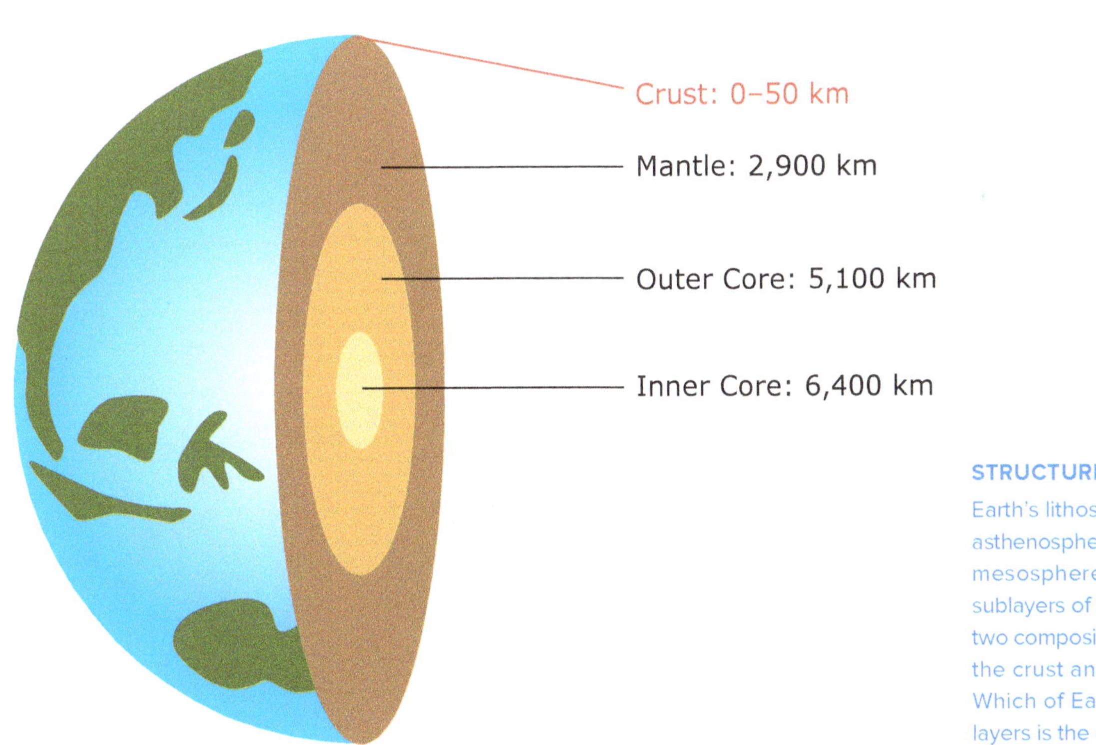


**STRUCTURE OF EARTH**
Earth's lithosphere, asthenosphere, and mesosphere are sublayers of Earth's top two compositional layers, the crust and mantle. Which of Earth's top layers is the deepest?

The flow of rock in the asthenosphere is due to **convection**, a type of heat transfer. Convection occurs in liquids and gases, and also in solid, but soft and plastic, rock. It is the same process that causes the circular, rolling motion of boiling water.

How does mantle convection occur? **Thermal energy** from deep inside Earth heats rock at the bottom of the asthenosphere. This heat comes from two sources. **Heat energy** left over from Earth's formation is continually released from the deep interior. Much of it is stored in radioactive elements, which release energy as their isotopes undergo nuclear disintegration. Heat is also produced by pressure in the interior. As early Earth became larger, its gravitational field became stronger, compressing materials into rock and producing heat. The release of this internal heat takes place even today as Earth continues to cool. When this internal heat rises toward the surface, it eventually encounters the asthenosphere and converts the solid rock into mobile, plastic-like material which flows.

**EARTH'S LAYERS**

Each of Earth's layers has distinctive properties based on mineral composition, temperature, and pressure. How do these layers vary with depth?

Heat flow from Earth's interior produces thermal convection cells. Circulation of heat takes place in the upper mantle because warm, buoyant rock rises and cooler, denser rock sinks. Imagine a pot of a thick liquid such as oatmeal slowly heating on the stove, with bubbles sometimes rising. Rock in the mantle is close to its melting point and actually behaves like hot corn pudding. Rock expands when heated. Expansion lowers the density of the hot rock. As a result, the hot rock flows upward. As rock moves upward away from the heat source, it begins to cool. As a result, the rock contracts, becomes denser, and sinks. It is then drawn downward by gravity. As the rock sinks to the bottom of the asthenosphere, it heats up and begins the cycle again.

Scientists use seismic tomography to map the mechanisms of thermal convection. They produce three-dimensional images of the forces at work below Earth's surface. Like medical CT scans that use a rotating X-ray beam to produce an image from a variety of angles, scientists synthesize information from earthquakes in Earth's crust. Seismic waves generated at the focal point of an earthquake travel through Earth. They spread out in every direction. Detectors in different locations on Earth measure the speed and intensity of these seismic waves, which are affected by the composition of the materials they pass through. The phase, density, pressure, temperature, and rigidity of these materials affect the velocity of the waves and provide a record of where they have passed. Warm material slows down the waves by absorbing seismic energy. Cold material speeds them up. By synthesizing data from earthquakes around the world, the structure of Earth's mantle can be determined.



Mantle convection drives the formation of oceanic crust and the movements of tectonic plates. Where new crust is formed at mid-ocean ridges, scientists map the properties of the material brought up from far below the crust. Measurements of the changes in the rocks' magnetic fields also provide evidence of a long history of constant flow from below. It leaves a record showing that Earth's magnetic field has been continuously changing direction. Volcanism is also caused by mantle convection, when molten rock erupts through the Earth's surface, transferring heat from the mantle to the crust.

Beneath the asthenosphere lies the mesosphere, a layer in which intense pressure due to depth prevents the flow of rock and heat in convection cells. Instead, heat is transferred in the mesosphere when huge bodies of **magma**, called plumes, rise to the top of the crust. The composition of the asthenosphere and mesosphere is similar, containing minerals rich in iron and magnesium.

## Earth's Core

The innermost part of Earth's interior is called the **core**. The core is mostly metal—mainly iron and nickel—though it possibly contains small amounts of sulfur. It is 6,856 kilometers in diameter and makes up approximately one-third of Earth's mass. (The crust makes up less than 1 percent of Earth's mass!)

As scientists study recordings of earthquake waves, or seismograms, they observe a shadow zone through which some waves—called S waves—do not pass. Scientists know that S waves cannot travel through liquid. Therefore, they infer the presence of an outer, liquid portion of the core, called the **outer core**.

The outer core is liquid because its temperature exceeds the melting points of iron and nickel. However, the centermost part of the core—the **inner core**—is solid. Pressure from the cumulative weight of the overlying layers of Earth is intense and prevents the inner core from melting.

### Consider the Explain Question

**How are the composition and structure of Earth's layers different as you go deeper into Earth?**

Go online to complete the scientific explanation.

dlc.com/ca9021s

### Check Your Understanding

**Can you describe the chemical and physical properties of each of Earth's compositional layers?**

dlc.com/ca9022s

# STEM in Action

## Applying Earth's Interior

Volcanologists risk their lives in order to better understand the dynamics of these explosive phenomena. Although some data can be gathered remotely, eruptions have destroyed sensors that were placed in advance, interrupting the data flow. Also, some kinds of data—such as lava, ash, and gas samples—can only be collected on the scene. Special equipment is required to protect the scientists from extreme heat and poisonous gases. Even without an impending eruption, collecting volcano data can mean traveling to remote areas with rugged road conditions, or no roads at all, where outside help can take days to arrive in case of a medical emergency. Remoteness has other challenges; far from civilization, volcanologists may be robbed of their equipment, get lost in the wilderness, or have vehicle breakdowns that they must repair themselves.

**VOLCANOLOGIST RECORDING DATA**

A volcanologist records data. What kinds of data collected will help scientists to predict a volcanic eruption?

Most of the work volcanologists do is much quieter than their field work. Once the data and samples are collected, they spend years analyzing it. Analysis requires many hours in the lab and at the computer, whatever the goal: improving predictions of the timing and type of volcanic eruptions; better understanding of the composition and source of the material ejected; or finding answers to other questions. Many volcanologists hold positions with universities. Their responsibilities also include instructing college students and advising graduate students who are preparing for careers in the field. Like all scientists, volcanologists regularly share results of their research by writing and publishing papers in scientific journals, and by participating in conferences.

Volcanologists need a strong background in science and mathematics, an advanced degree, and an ability and willingness to endure difficult and sometimes dangerous conditions in their field work. Although volcanologists risk their own lives, they also help to save the lives of others by expanding the understanding of volcanoes and the threat they pose to people living and traveling nearby.

## STEM and Earth's Interior

Scientists could not do the work of understanding Earth's interior without the cooperation of a wide range of technological and computer science specialists. The availability of the Internet has made it practical to set up large-scale data collection programs like the Transportable Array. This continent-wide seismic grid uses hundreds of identical seismometers that, once deployed by field technicians, require only occasional visits to keep them running. The seismometers were developed by mechanical and electrical engineers. These instruments run on electricity from solar panels designed by engineers in that field. The seismometers send around-the-clock data directly to the scientists' computers. Software engineers write the code that graphs the data in real time, so the scientists can quickly analyze it. Computer hardware specialists develop the servers that archive the data for future use, where other scientists can analyze it.

Seismometers provide important information about the seismological status of areas. Geophysicists continually analyze both past and real-time data to spot trends as well as unexpected events. When a significant seismic event occurs, the scientists characterize the progression of the ground motion in order to make predictions about the nature of the motion. These predictions can save lives by warning the public about the possibility of dangerous earthquakes.

**DIAMOND FORMATION** Rough diamonds have a crystal shape that is cut and polished into the familiar gems. What temperatures and pressures are required to form a diamond?

Diamonds are made of just one element: carbon. But carbon also forms a very different substance, graphite, and it is a primary ingredient in all forms of coal. What conditions in Earth's interior turn carbon into diamond instead of one of these other Earth materials?

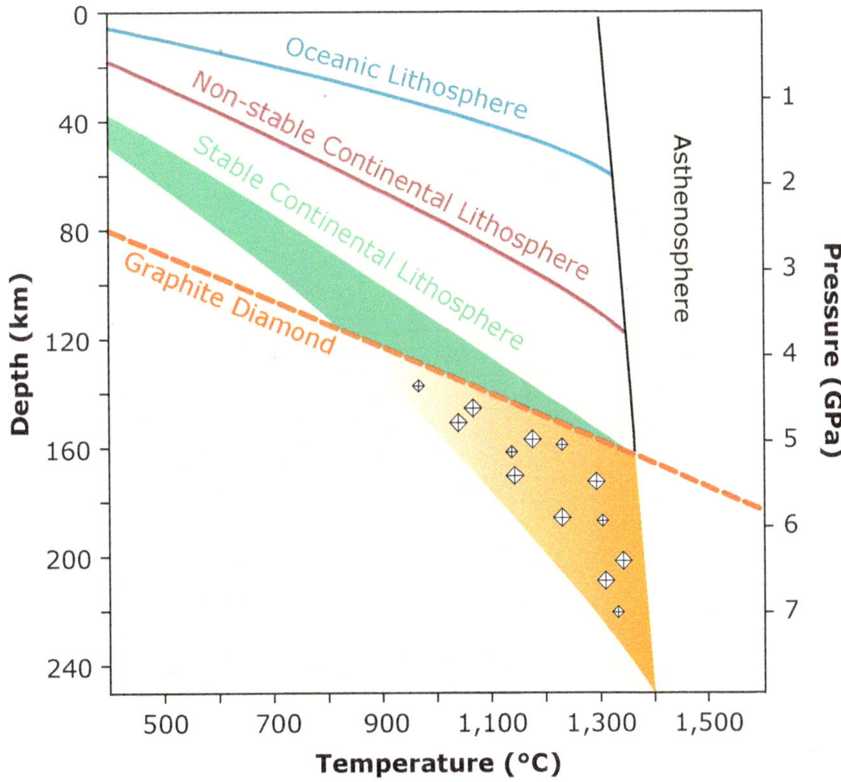


**CONDITIONS FOR DIAMOND FORMATION** The graph shows the temperature, pressure, and depth conditions necessary for diamond formation. Diamonds can form in conditions below the dashed line. How would you describe the conditions that allow diamonds to form?

CONCEPT 1.5

# Tectonic Plate Interactions



## LESSON OVERVIEW

dlc.com/ca9024s

### Lesson Questions

- What are the three types of plate boundaries?
- What features in the landscape are associated with tectonic plates?
- How has plate tectonics affected evolution, ocean currents, and global climate?
- What is the supercontinent cycle?

### Lesson Objectives

By the end of the lesson, you should be able to:

- Identify the three main types of plate boundaries.
- Identify and describe structures that result from plate motion.
- Explain how changes in the arrangements of tectonic plates over time impacted evolution, ocean currents, and global climate.
- Explain the supercontinent cycle.

### Key Vocabulary

Which terms do you already know?

- ☐ asthenosphere
- ☐ compression
- ☐ convergent boundary
- ☐ divergent boundary
- ☐ fault-block mountain
- ☐ folded mountain
- ☐ Gondwana
- ☐ mid-ocean ridge
- ☐ Pangaea
- ☐ plate motion
- ☐ rift valley
- ☐ rift zone
- ☐ seafloor spreading
- ☐ subduction
- ☐ subduction zone
- ☐ supercontinent
- ☐ supercontinent cycle
- ☐ transform boundary
- ☐ volcanic arc

## Looking at Tectonic Plate Interactions

Have you ever felt an earthquake? One of the largest earthquakes in recorded history happened on March 11, 2011. The earthquake, which was a magnitude 9.0 and the fourth-largest earthquake since 1900, struck Japan's main island, Honshu.

dlc.com/ca9025s

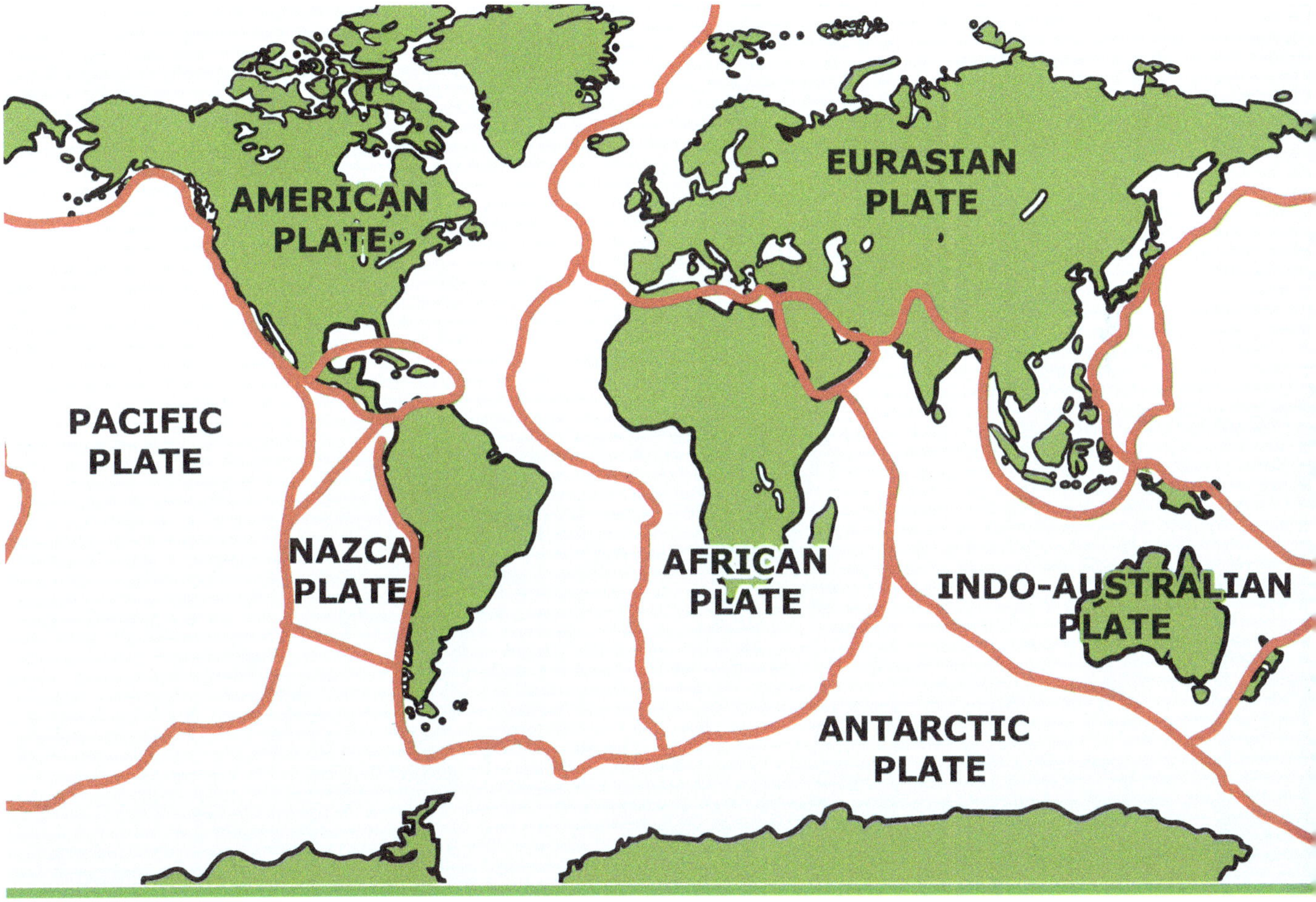

AMERICAN PLATE
EURASIAN PLATE
PACIFIC PLATE
NAZCA PLATE
AFRICAN PLATE
INDO-AUSTRALIAN PLATE
ANTARCTIC PLATE



**SOME TECTONIC PLATE BOUNDARIES**

Tectonic plates are not static. Where are some places in the world where the movement of tectonic plates could cause serious problems?

**EXPLAIN QUESTION**

What risks and benefits do humans endure as the result of plate tectonics?

# What Are the Three Types of Plate Boundaries?

## Convergent Boundary

A **convergent boundary** forms when two plates move together. This type of boundary forms between two oceanic plates, two continental plates, or an oceanic and a continental plate.

Features along a convergent boundary result from compressional forces. When an oceanic plate converges with a continental plate, the oceanic plate sinks below the continental plate. Oceanic plates are made from mainly basaltic rocks. They are less dense than continental plates that are made mainly of felsic or granitic rocks. Oceanic plates form at mid-ocean ridges and may be destroyed where they meet other plates. This convergence produces a feature called a **subduction zone**, which forms when an oceanic plate meets and goes beneath a less dense continental plate.



**OCEANIC-CONTINENTAL CONVERGENCE**

A subduction zone forms when a tectonic plate with oceanic crust converges with a plate with continental crust or when two plates with oceanic crust converge. Why does the oceanic plate move beneath the continental plate and not the other way around?

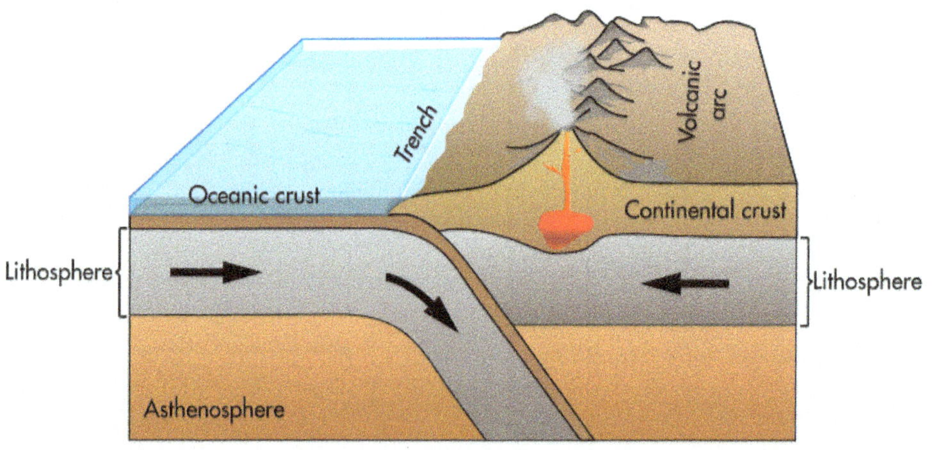


A convergence zone also forms when two oceanic plates collide. One plate passes beneath the other. The convergence of two continental plates produces a mountain-building event. During the event, rocks and sediment are thrust together, tilting and deforming rock layers and causing wide-scale metamorphism. Geologists study the location and depth of earthquakes along **subduction** zones to understand how converging plates behave and deform.

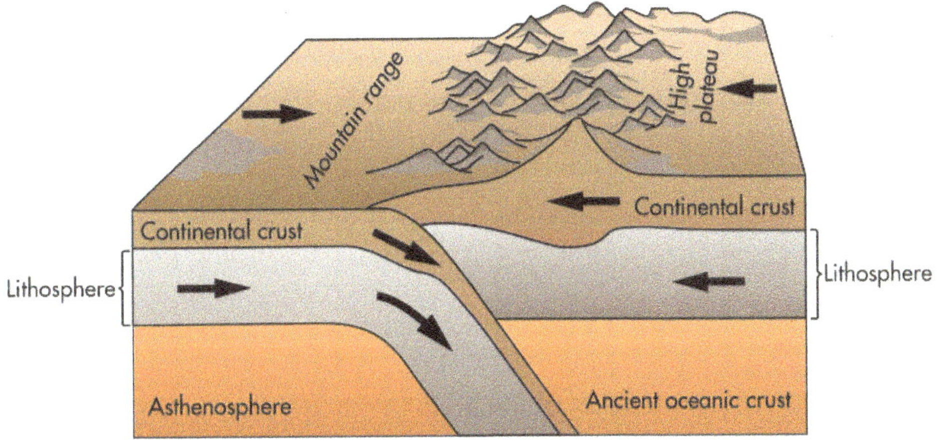


**CONTINENTAL-CONTINENTAL CONVERGENCE** Mountains form when continental crust converges with continental crust. What events occur as they push together?

## A Case Study in Convergence: The Himalayas

The Indian subcontinent has been on the move for 71 million years. Beginning 10 million years ago, the Indian Plate collided with the Eurasian Plate producing the Himalayas and the Plateau of Tibet. The event affected an area larger than 2,400 sq km. Mount Everest, part of the Himalayas mountain chain, towers 8,848 m above sea level.

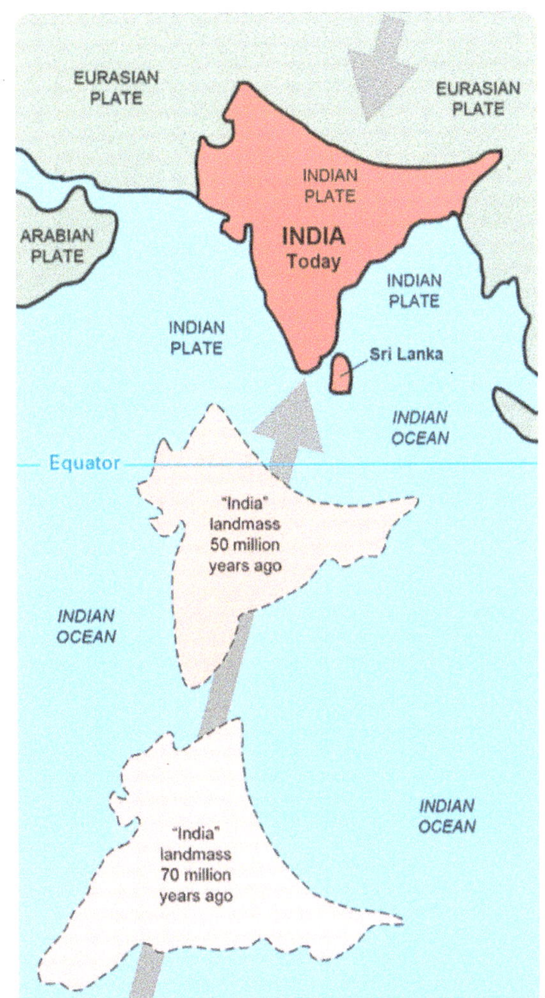


**EURASIA** The Himalayan Mountains and the Tibetan Plateau formed when the Indian Plate collided with the Eurasian Plate. What type of plate boundary is this?

The region is still experiencing uplift as the Indian Plate continues to collide with the Eurasian Plate. The western region of the Himalayas mountain chain, called Nanga Parbat, is currently experiencing the highest rate of uplift—approximately one centimeter per year. Scientists think that over the next 5 to 10 million years, the continued collision will force India another 180 km farther inland.

## Divergent Boundary

A **divergent boundary** forms when two oceanic plates or two continental plates move apart. Divergence does not occur between an oceanic plate and a continental plate.

Features along divergent boundaries result from tension or extensional forces. A spreading center marks the divergent boundary between two oceanic plates.

A triple junction in the shape of a "Y" marks divergence of three continental plates. Each arm of the Y becomes a long rift, or crack, in Earth's crust. Two such rifts separate Africa and Arabia, forming the Gulf of Aden and the Red Sea. The stem of the Y forms the Great African **Rift Valley**, which is ripping the interior of East Africa apart. This valley is characterized by volcanic activity and deep lakes. As the rift opens wider, it will deepen further until the surrounding ocean floods in, creating a new gulf, or sea.

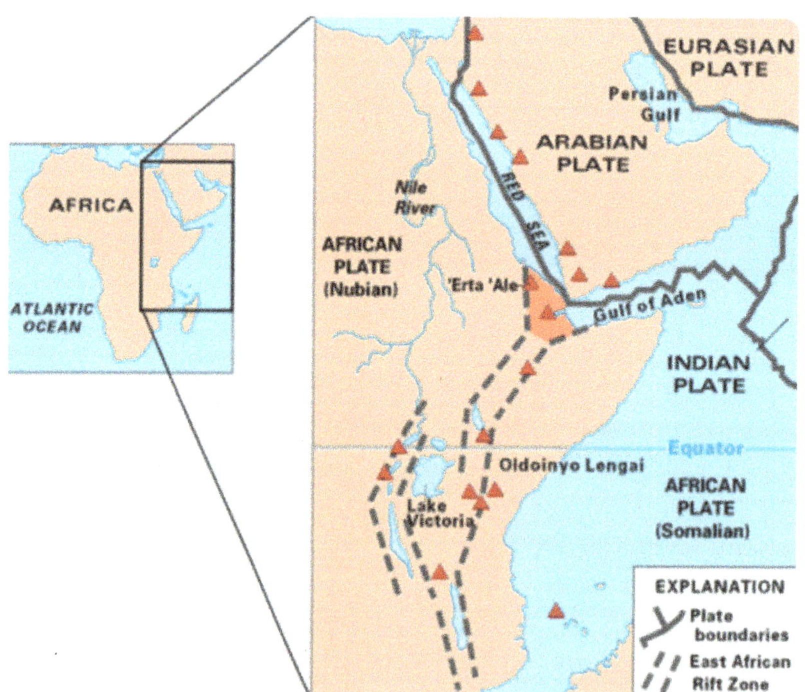


**AFRICA**
A triple junction forms divergent boundaries between three continental plates. What new geologic features form at this type of boundary as plates move apart?

## A Case Study in Divergence: The Mid-Atlantic Ridge

A **mid-ocean ridge** system consists of a series of volcanoes and fissures that form along a spreading center at a divergent boundary. During **seafloor spreading**, the plates move apart. The heat from the magma makes the surrounding lithosphere less dense. As a result, the lithosphere rides higher on the **asthenosphere**, accentuating the height of the mid-ocean ridge system. Magma wells up from Earth's interior and rises to the surface through volcanoes and fissures, forming new ocean crust. The crust slowly moves away from the fissures in two opposing directions, eventually creating new ocean floor.

Beginning in the northern Atlantic basin, the Mid-Atlantic Ridge bisects the length of the Atlantic basin. As the Mid-Atlantic Ridge curves around the southern margins of Africa and South America, it enters the Indian Ocean from the east and the west.



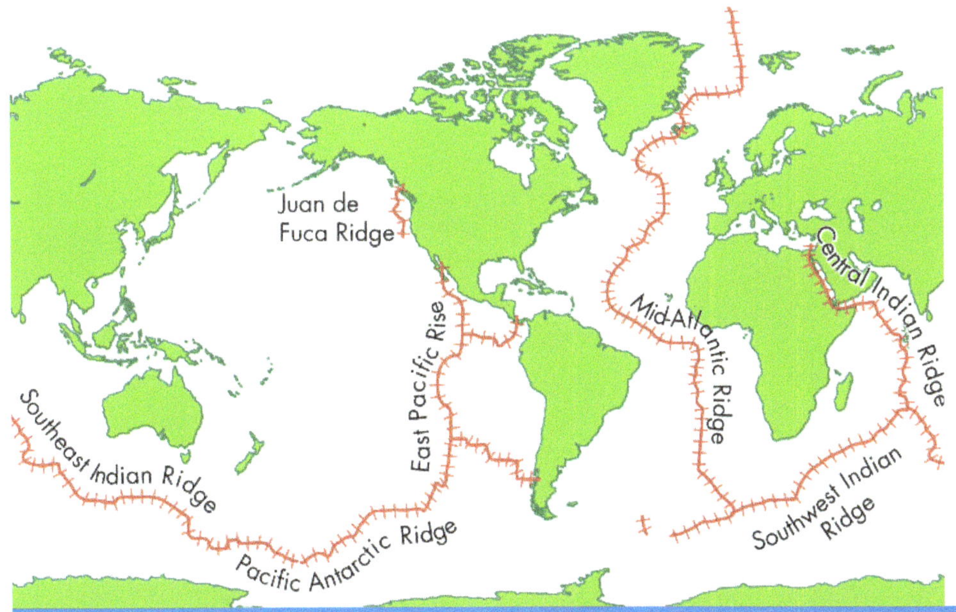


**MID-OCEAN RIDGES**
The Mid-Atlantic Ridge is one of several mid-ocean ridges. What type of boundary is a mid-ocean ridge?

## Transform Boundary

A **transform boundary** forms when two plates slide past one another parallel to the plate boundary. Transform boundaries can occur under the ocean or on land. At a transform boundary, crust is neither made nor destroyed. Instead, the plates push against each other, producing enormous stress that builds up potential energy. When the stress reaches a critical threshold, potential energy suddenly transforms into kinetic energy, and the plates slide parallel to each other.

### Transform Boundary Case Study: The San Andreas Fault

Along the coast of California, the Pacific Plate and the North American Plate are in direct contact. The two plates moved laterally past one another, parallel to a transform fault. This transform boundary in southern California is called the San Andreas Fault. The fault is 1,300 km long and spans the southwest section of California in the United States. The city of Los Angeles is located on the Pacific Plate, Sacramento sits on the North American Plate, and San Francisco sits near the San Andreas Fault, where the two plates meet. The Pacific Plate is slowly migrating north along the fault. At some point in the distant future, the cities of Los Angeles and San Francisco will likely be neighbors.

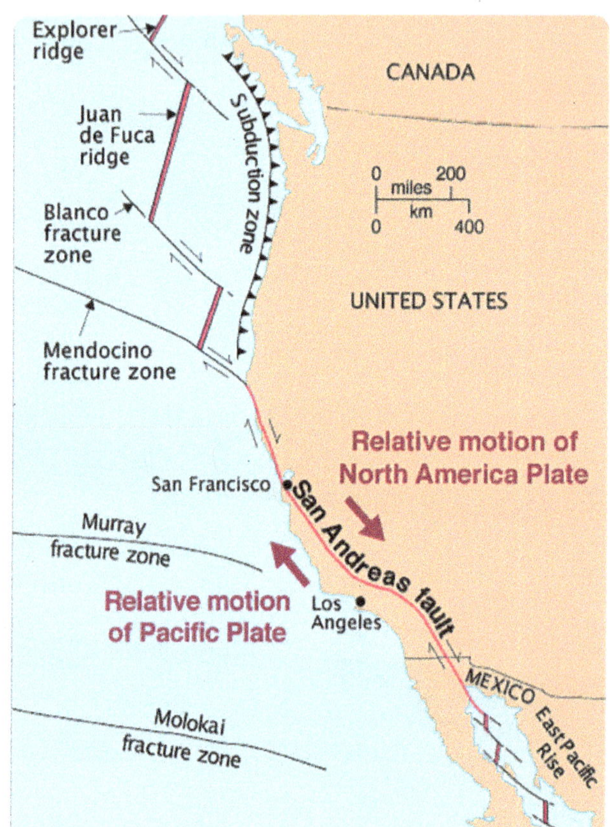


**SAN ANDREAS FAULT**

The San Andreas Fault is a transform boundary where the Pacific Plate moves laterally against the North American Plate. What are some results of plate movement at transform boundaries?



## What Features in the Landscape Are Associated with Tectonic Plates?

### Subduction Zone

When an oceanic plate converges with a continental plate, the denser oceanic crust sinks below the continental crust along a unique region called a **subduction zone**. When oceanic plates converge, the older, denser plate subducts below the plate that is less dense.

As the denser plate is subducted, it is recycled back into the mantle. As the oceanic lithosphere slips into the mantle, it rubs against the overriding continental lithosphere. The friction resulting from **subduction** breaks the subducting plate into smaller pieces. The friction also produces a series of earthquakes that follow the subducting plate into the mantle. Some earthquakes have been recorded at a depth of 700 km. Some of the material from the subducting plate also melts and rises up through the overlying continental crust to form volcanoes.

## Trench

A trench is a deep valley that outlines a subduction zone. Most trenches on Earth encircle the Pacific Ocean. This region is commonly called the Ring of Fire because it boasts 40,000 km of oceanic trenches, volcanic arcs, and volcanic belts. This region is responsible for 75 percent of the world's active and dormant volcanoes, as well as 90 percent of the world's earthquakes.

The profile of a trench is an asymmetric "V" controlled by the downward-moving plate and sediment eroding from land. As the downward-moving plate approaches the subduction zone, it bows upward, producing the outer trench swell. This slight deformation allows the plate to descend into the mantle. The descending plate forms the outer trench wall. It can take millions of years for oceanic lithosphere to move from the region of the trench swell to the mantle. The overriding plate produces the steep inner trench wall, which is constantly being deformed as the two plates interact.

## Volcanic Arcs

A **volcanic arc** is a series of volcanoes that form behind an ocean-ocean subduction zone. The volcanic islands in an arc are parallel to the subduction zone. Because the subducting plate contains ocean sediments and water that end up in the rising magma, volcanism tends to be more explosive than the volcanism at spreading centers. The Aleutian Islands that extend away from Alaska are a volcanic arc system. The countries of Japan, Tonga, the Philippines, and the Solomon Islands formed along volcanic arcs.

## Rift Valley

A **rift valley** forms between tectonic plates that are moving apart. They are found at spreading centers under the ocean and on continents. Impressive river valleys or deep lakes can be found in some continental rift valleys. The East African rift valley is an example of a rift valley where two continental plates are moving apart.

Some rift valleys are considered failed and are no longer active. Examples of failed rift valleys converted to river valleys include the Red Sea in eastern Africa, the Rio Grande rift valley in the southwestern United States and Mexico, and the Niger River valley in western Africa. Rift valley lakes are also common; examples include Lake Baikal in Siberia and Lake Superior in North America. The region surrounding a failed rift is often dotted with volcanoes where molten magma has reached Earth's surface. The lava erupted from these volcanoes is a combination of the molten material from Earth's interior and melted, silica-rich continental crust. Volcanoes that form in this manner are explosive.

## Folded Mountains

Convergent boundaries impose compressional forces on the rocks along the colliding plates. The rocks crumple and deform, and the pressure imposed on rocks during deformation can result in metamorphism. These forces reshape the landscape, creating features ranging from small crinkles to massive folded mountains. Examples of folded mountains include the Andes and Himalaya mountains.

Typical structures formed in the folds of rock layers include anticlines and synclines. An anticline is a folded structure with the oldest rocks in the center of the fold. A syncline is a folded structure with the youngest rocks in the center of the fold.

## Fault-Block Mountains

Fault-block mountains are formed by extensional forces that identify locations where rifting is developing. The rifting may ultimately result in the formation of rift valleys and the inception of tectonic plate divergence. They often are visible along the flanks of active and failed rift valleys. They can also form deep in the interior of plates when hot magma in the mantle rises and causes the crust to buckle and initiate rifting.

The extensional forces may originate from the movement of plates along convergent or divergent boundaries. When divergent plates move, they can cause faulting when the plate interiors resist movement because of their weight. When converging plates move, their interiors can resist movement because of drag forces deep in the interior. In both cases, the forces break the rock into segments or blocks. If the formation of fault blocks releases sufficient pressure deep in the crust and upper mantle, igneous activity and rifting will commence.

Fault blocks move vertically in relation to one another. Fault-block mountains form from the blocks of crust that rise above the surrounding rock. Blocks that fall below the surrounding rock produce grabens. Fault-block mountains and grabens are conspicuous features of the Basin and Range region of the western United States.



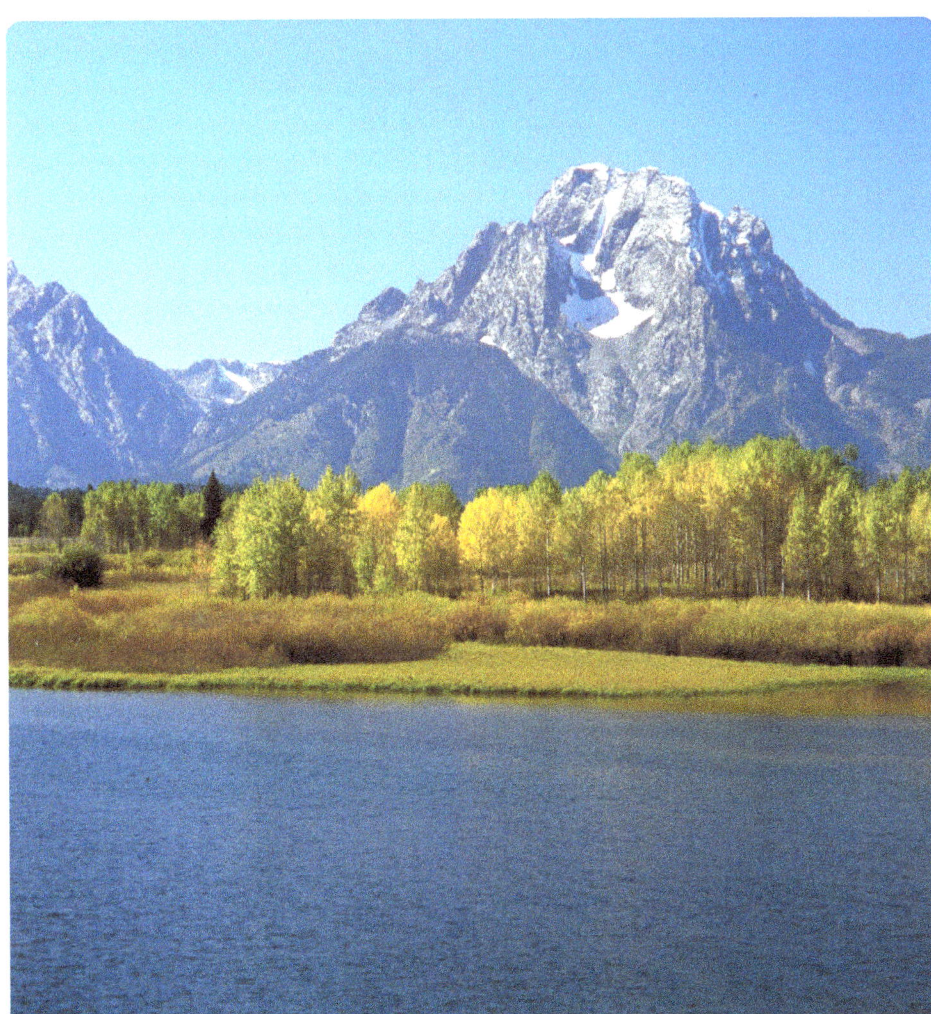

**THE TETONS**

The Tetons in Wyoming are an example of a fault-block mountain range. What defining features of fault-block mountains can you see in this image?

## Accreted Terranes

In geology, accretion describes the process of adding rocks onto crust. During subduction, a volcanic arc, seamount, or landmass on an oceanic plate may be scraped off and accrete, or stick, to a continental plate. The land added to the continent is called an accreted terrane. The accreted terrane is younger than the continental crust it attaches to. Examples of accreted terranes in the geologic record include the large sections of North America west of the Rocky Mountains.

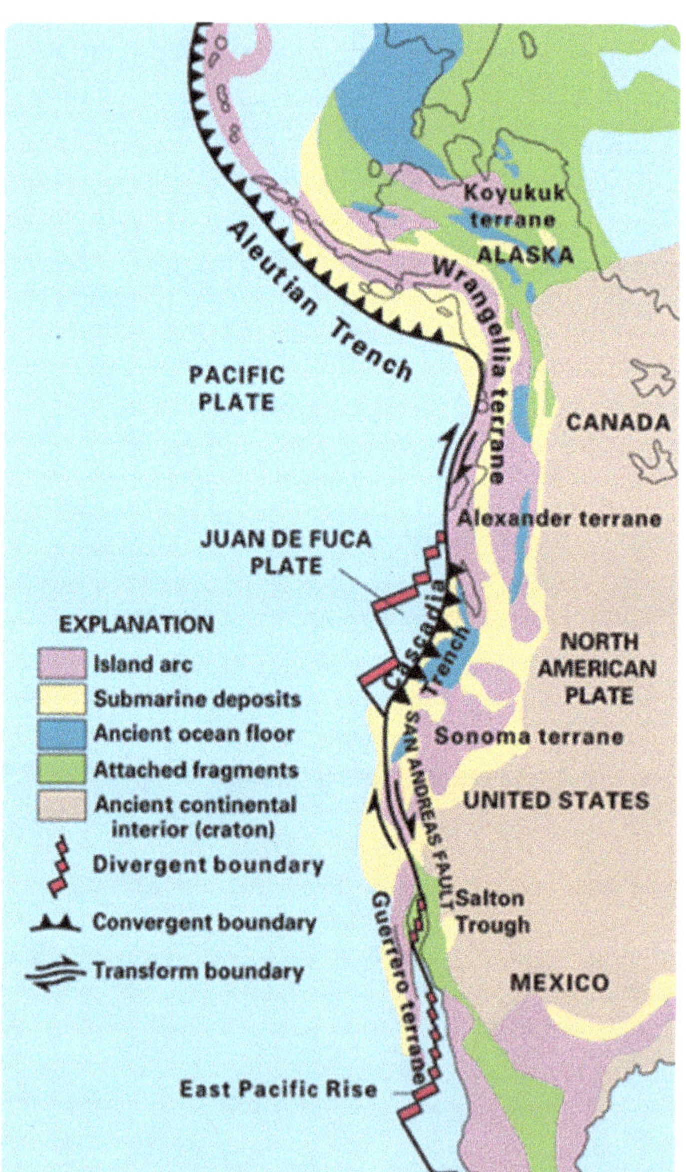


**ACCRETED TERRANE**

Accreted terranes form along coasts adjacent to subduction boundaries. Where does the material from an accreted terrane come from?

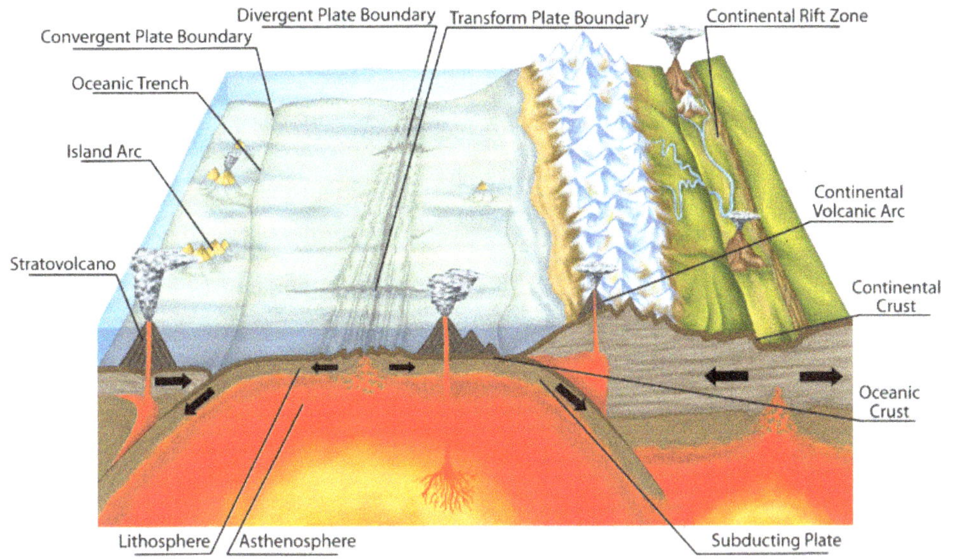


**INTERACTIONS OF TECTONIC PLATES**

Tectonic plate interactions have formed many of Earth's major geological features. What landscape features are produced at each type of plate boundary?



# How Has Plate Tectonics Affected Evolution, Ocean Currents, and Global Climate?

## Global Impacts on Earth's Subsystems

The processes associated with plate tectonics have influenced not only the lithosphere but also most of the other spheres of Earth's systems. The gradual redistribution of the continents over Earth's surface causes long-term global changes in the hydrosphere, the atmosphere, the cryosphere, and the biosphere. Influences on the biosphere become apparent as scientists piece together the details of evolution. Influences on the atmosphere affect weather, climate, and glacial activity. The interaction between landmasses and the oceans are reflected in the paths of the global ocean currents.

## Evolution

Evolution describes how processes such as mutation, natural selection, and genetic drift change a population's gene pool over many generations. As tectonic plates move, continents are separated and populations of organisms become isolated. Geologists suggest that evolution is more pronounced when continents align along a north-south axis rather than an east-west axis. A north-south alignment alters ocean currents, producing more dramatic changes in climatic conditions. It also forces organisms to inhabit more diverse climatic zones. Over time, the organisms evolve into different species, increasing diversity on the planet.

The **supercontinent Pangaea** formed about 600 million years ago. It existed for about 400 million years. After it broke up 180 million years ago, populations of animals and plants across Pangaea became isolated on the fragments of the old continent. The isolated populations began to evolve on their own. One modern example is the flightless birds of the order Palaeognathae. These include the ostriches of Africa, emus of Australia, and tinamous of South America. All of these birds descend from a group of birds that ranged widely across Pangaea.

## Ocean Currents

One way in which plate tectonics alters climate is by altering circulation of ocean currents. As continents pull apart and come together, oceans and straits open and close. These events alter circulation patterns in Earth's hydrosphere and modify the world's climate and climate patterns.

During the last **supercontinent cycle**, Antarctica was connected to South America. When they separated, a powerful ocean current formed around Antarctica. This current isolated the continent from the warming effects of other currents. Antarctica gradually cooled and froze. North America and South America were joined together by the creation of the isthmus of Panama, about five million years ago. Before that, tropical ocean water flowed directly between the Atlantic and Pacific Oceans. When the ocean circulation patterns changed, a period of glaciation set in. These changes also impacted the distribution and evolution of marine organisms. Ocean currents are a key mechanism for distributing heat, fresh water, and nutrients in the global ocean.



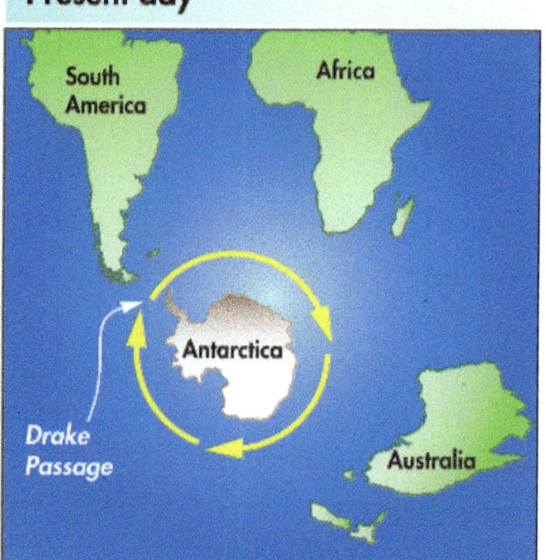


**DRAKE**

Drake Passage is a deep waterway between the tip of South America and Antarctica that opened approximately 39 million years ago. How did tectonic plates shift to create this passage?

### Climate Change

Climate conditions oscillate between two extreme scenarios. An icehouse climate describes a cold, dry planet dominated by ice. A greenhouse climate describes a planet dominated by warm, humid conditions.

The movement of plates over Earth's surface affect global climate. Plate movement alters the oceanic circulation that distributes heat around the planet. World climates are also affected when plates migrate north and south into new latitude areas. For example, as a continent moves toward a pole, its climate will respond as temperatures become gradually colder. An indirect effect of plate movements on climate comes from volcanic eruptions associated with plate movements. If a volcano erupts enough ash and sulfur dioxide into the stratosphere, these particles will reflect the sun's rays and cause Earth to cool. About 1.1 billion years ago, the formation of the supercontinent Rodinia formed a landmass that stretched from the South Pole to the Equator. Scientists describe the climate during this event as a "snowball Earth," a hypothetical scenario describing the total surface of the planet plunged into freezing conditions and covered by ice. The formation of Rodinia coincides with a series of devastating glacial events. Ice and snow covered Rodinia. Scientists even suggest that glaciers formed at the equator, and that the equator may have been as cold as present-day Antarctica.



## What Is the Supercontinent Cycle?

### Pangaea

About 300 million years ago, three continents dominated the planet's surface. Laurentia was the proto—North American continent. Baltica was the proto-European continent. **Gondwana** was the proto-African continent. Because tectonic plates were moving these continents around, Baltica eventually collided with Laurentia, forming a massive continent called Laurasia. Gondwana then collided with Laurasia to form the **supercontinent Pangaea.** During these events, the Appalachian mountain chain formed.

Pangaea began to break apart about 200 million years ago. As Pangaea broke apart, the continents that we recognize today began to form. Today, the continents continue to move apart.

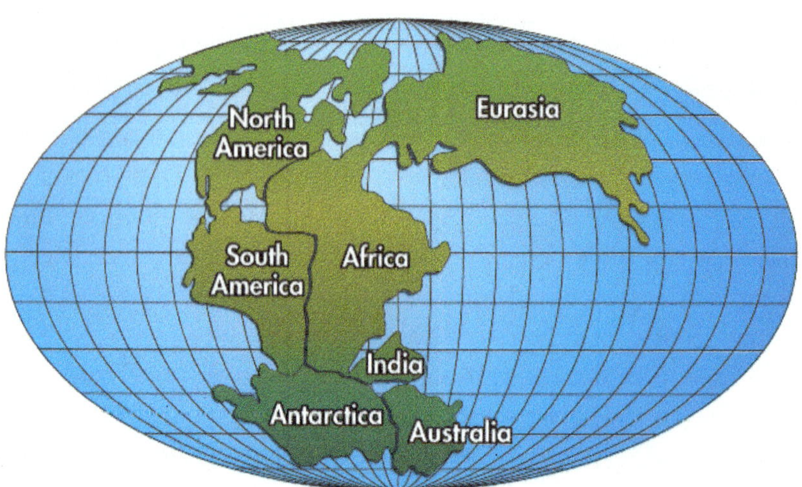


**PANGAEA**
Pangaea formed 300 million years ago. How did the shape of Earth's continents today help scientists develop the theory of continental drift?

## Supercontinent Cycle

A supercontinent forms when all of the continents on the planet's surface merge into one massive landform. Pangaea is the most recent supercontinent to form on Earth, but it is not the only supercontinent in Earth's history. Many scientists think that supercontinents follow a **supercontinent cycle**. During the cycle, supercontinents gradually form and break apart over hundreds of millions of years. Evidence on land and under the ocean of these massive landforms is often poorly preserved. The supercontinent process is important because it may have profound effects on the evolution of the planet's surface, atmosphere, climate, and organisms.

The earliest evidence of a supercontinent dates to 3.1 billion years ago. Scientists named the supercontinent Vaalbara. It began to break apart about 2.8 billion years ago. The extreme age of Vaalbara makes it difficult to reconstruct the configuration and location of the supercontinent on the planet.

The next supercontinent, Rodinia, formed about 1.1 billion years ago and broke apart over a period of 150 million years. Evidence suggests that Rodinia comprised the proto—North American continent, which was adjacent to the proto-Australian and proto-Antarctic continents. The Iapetus Ocean surrounded Rodinia. Beginning 750 million years ago, Rodinia rifted apart, opening a new ocean basin called the Panthalassic Ocean.

Around 600 million years ago, the supercontinent Pannotia formed when two pieces of the former Rodinia came together with a third landmass, enlarging the proto-African continent. Pannotia began to break apart 550 million years ago, resulting in several smaller continents, including Laurentia, Baltica, and Gondwana. About 300 million years ago the three continents merged to produce Pangaea.

### Consider the Explain Question

dlc.com/ca9026s

**What risks and benefits do humans endure as the result of plate tectonics?**

Go online to complete the scientific explanation.

### Check Your Understanding

dlc.com/ca9027s

**How do the three types of plate boundaries move relative to each other at each boundary?**

## Applying Tectonic Plate Interactions

Is there a way for scientists to monitor and warn residents of an area, such as Honshu, Japan, of an impending earthquake?

The 2011 Japanese earthquake occurred at 14:46:45 local time. The seismometer located closest to the epicenter, the spot on Earth's surface directly above the focus, detected the first tremor and sent a warning at 14:46:48, only three seconds after the earthquake began. Key populations received the warning, including TV networks, schools, factories, radio stations, and mobile phones. The city of Sendai, located 130 kilometers west of the epicenter of the earthquake, is home to one million citizens. The first S-waves reached Sendai at 14:47:17. They reached Tokyo, a much larger city, 90 seconds later. Although the system only offered a few seconds of warning, many officials believe it saved countless lives, especially when compared to previous earthquakes in the region.

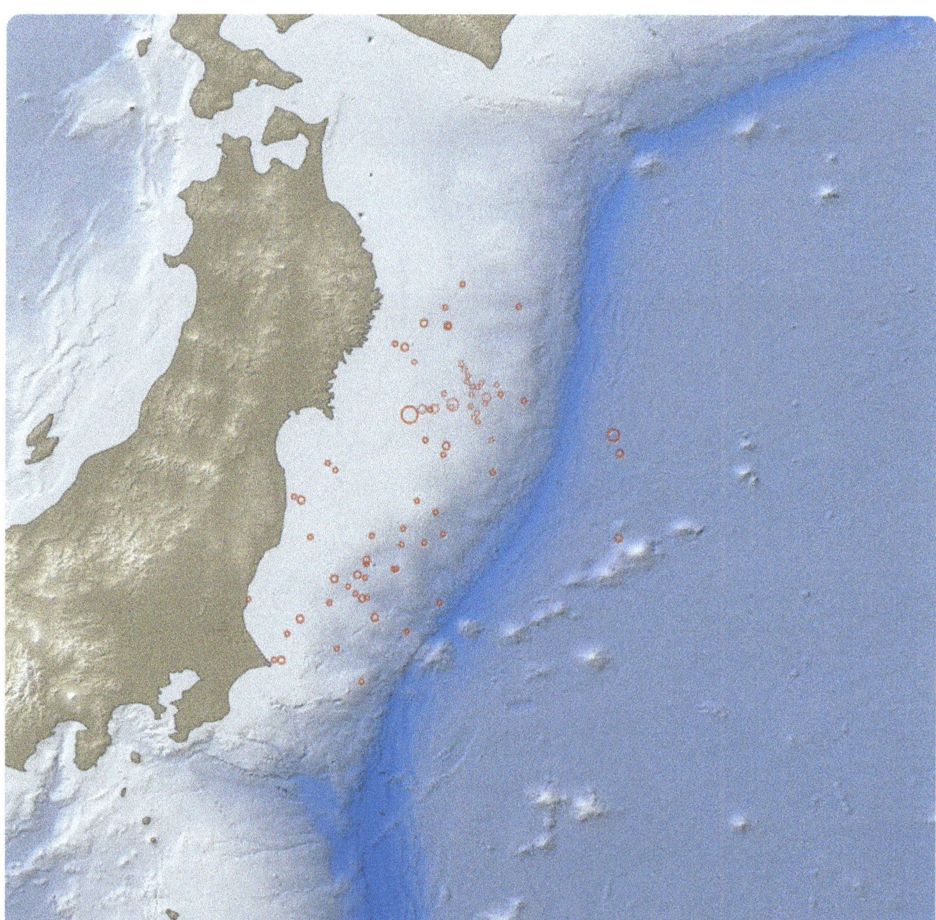

**EPICENTER OF SENDAI EARTHQUAKE**

This image shows the epicenter of the Sendai earthquake (largest red circle), along with its foreshocks and aftershocks. What can the location of the foreshocks and aftershocks tell scientists about plate movement?

The earthquake warning system that helped save lives during the 2011 earthquake was first implemented by the Japan Meteorological Society in 2007. Japanese officials recognized that because the island nation sits near a **convergent boundary**, it faced an increased risk of earthquakes. An earthquake results from the sudden rupture of rocks at the planet's surface. The rupture releases a burst of stored (potential) energy that travels through the planet's interior. Seismic waves emanate from where the rock ruptures, called the focus. The waves radiate from the focus through and around the world. The warning system used to alert citizens during the 2011 earthquake was designed to monitor those waves.

Scientists regularly monitor the warning system for the first hint of a P-wave. As soon as an earthquake starts, they send out warnings through media outlets to encourage citizens to seek protection. The announcement provides an estimate of the expected earthquake intensity and arrival time. The system also mitigates damage by undertaking measures to protect the citizenry, like stopping elevators and slowing public transportation. As a result, many lives can be saved.



Still, scientists want to be able to predict when earthquakes will happen, and the amount of damage they will cause, much earlier. After large earthquakes, teams of scientists will often drill down into the earth to take a sample from the fault line involved in the earthquake. Through lab simulations, they are able to see how shearing and other processes affect the rocks along the fault line. The more scientists know about how fault lines react to movement, the easier it will be for them to predict when major earthquakes will happen.

## STEM and Tectonic Plate Interactions

Is it possible to provide more advance warnings about when earthquakes will happen? The answer is not clear, but multiple types of scientists are working together to find out. Seismologists are scientists who study earthquakes and seismic waves, but they may not be the ones who can best predict when an earthquake will hit. While seismologists look at how seismic waves travel through the interior of Earth, other scientists focus on the initial cause of many of those seismic waves—the movement of plates along their boundaries.

Tectonophysicists, also sometimes called geodynamicists or structural geologists, study tectonic plates. They spend a lot of time in the field analyzing the movement of tectonic plates and studying the effect those movements have on Earth's continents and oceans. For example, when two faults collide, do pieces break off, or do they merge like a zipper? To collect information, Tectonophysicists use many different types of technology. They may use satellite images to get a broad view of an area and spectrometers to determine the mineral content in the rocks along plate boundaries. They also take a lot of measurements and track measurements over time to help them understand how rocks along fault lines have moved.

The information Tectonophysicists, such as Dr. Bilham, collect is often used to inform decisions made by other scientists and government officials. For example, learning how the ground reacts during an earthquake can help emergency officials develop an emergency preparedness plan or help engineers and urban planners know where to avoid constructing roads and buildings.

## The San Andreas Fault

The San Andreas fault is a real-life example of how Earth's tectonic plates are moving. One thing that technophysicists study is how much fault lines, such as the San Andreas fault, are moving.

**SAN ANDREAS FAULT**
This map of California indicates the San Andreas Fault line. Notice the two major cities that are nearby. How are they affected?

CONCEPT 1.6

# Volcanoes

## LESSON OVERVIEW

### Lesson Questions

- How do different volcanoes form?
- Where do volcanoes form?

### Lesson Objectives

By the end of the lesson, you should be able to:

- Explain the causes of volcanism.
- Identify patterns of the distribution of volcanoes.

### Key Vocabulary

Which terms do you already know?

- [ ] andesite
- [ ] hot spot
- [ ] magma
- [ ] seamount
- [ ] subduction zone
- [ ] ultramafic
- [ ] volcano

dlc.com/ca9029s

# What Causes Volcanoes?

dlc.com/ca9030s

**What are some examples of events that residents of your region must tolerate? Are they results of human activity, such as traffic jams, or are they natural phenomena?**

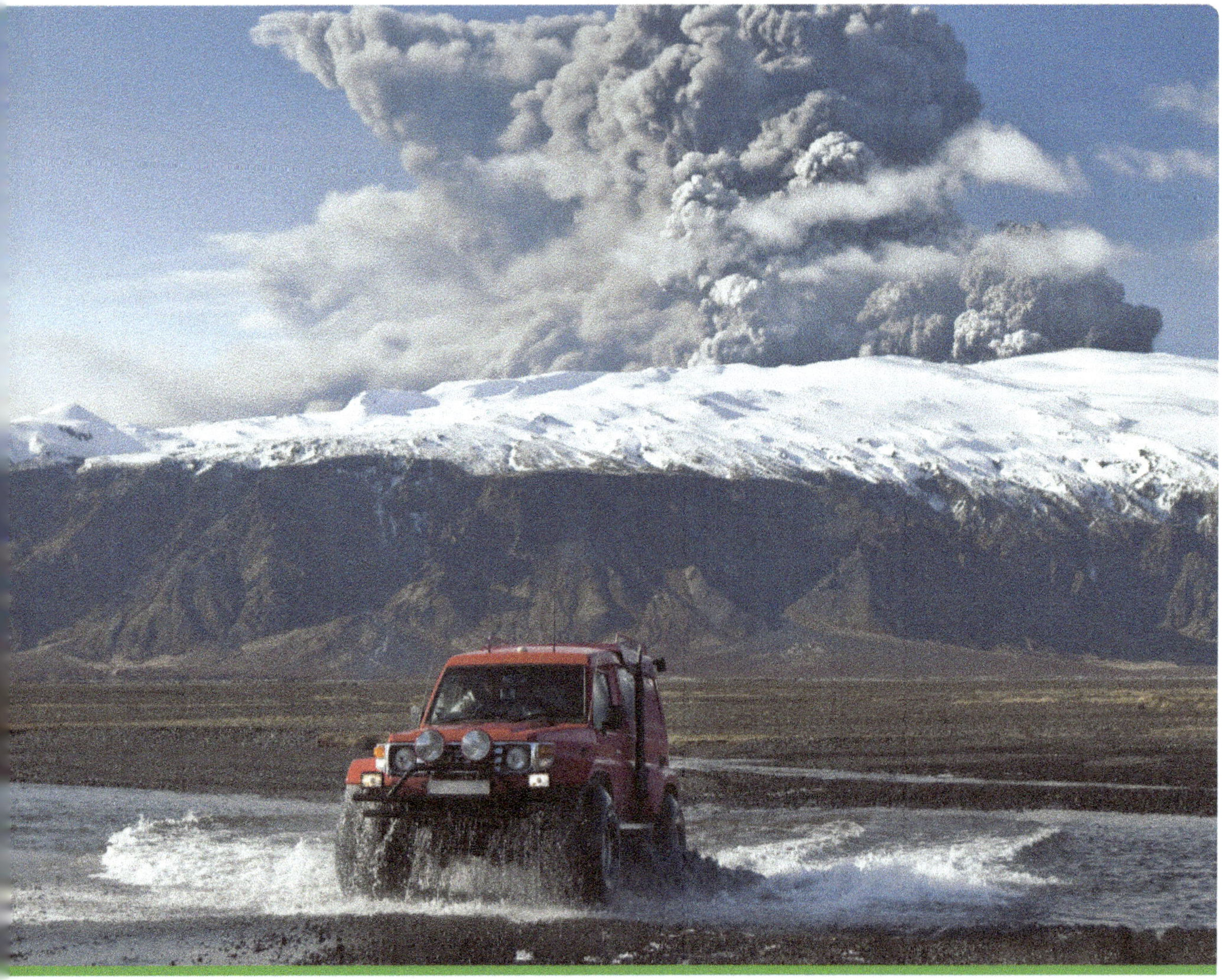

**LIVING WITH VOLCANOES**

Ash erupts from a volcano in Iceland. What adjustments do people living near volcanoes have to make in order to live with the effects of eruptions?

## EXPLAIN QUESTION

**What causes volcanoes, and why do certain types of volcanoes develop in certain places?**

# How Do Volcanoes Form?

## What Is a Volcano?

A **volcano** is a geological feature that forms around a fissure in the Earth's crust. There are a number of factors that contribute to the formation of a volcano, such as a conduit. A conduit is the means through which **magma** gets from Earth's interior to an opening or vent at the surface of Earth. It is through this opening or vent that **magma** erupts as lava. Lava, the term for magma that reaches the Earth's surface, accumulates around the vent, often forming a conical mountain. The conduit of the volcanoes also transmits steam, gases, ash, and other materials from the volcano's interior to Earth's surface and atmosphere. Conduits often branch out, carrying these materials to smaller vents off the main volcanic structure. Land is not the only place where volcanoes form; they also form in the ocean, as well as at the center of islands.



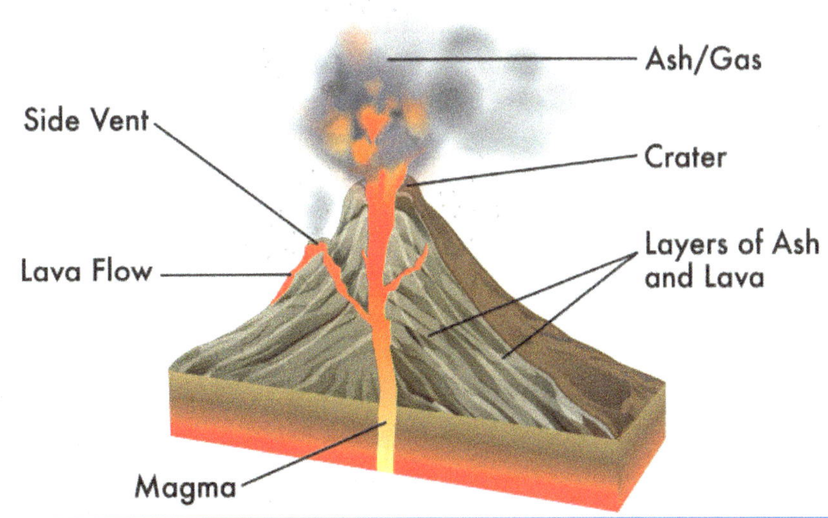


**CROSS SECTION OF A VOLCANO**

A volcano is fed by magma from Earth's upper mantle. What is the primary difference between magma and lava?

The foundation of the volcanic structure is magma: the substance that forms when rock material melts, or becomes molten, close to an Earth mantle–crust interface. This process occurs primarily by two methods: decompression melting or flux melting. Decompression melting occurs when solid, hot material rises to Earth's surface. As the pressure decreases, the rock material expands and this lowers its melting point. This allows the rock to melt at a shallow depth in Earth's interior, forming magma. Flux melting occurs when water or carbon dioxide is introduced to the rock material in the upper mantle. These materials lower the melting point of the rock material (similar to the way that salt lowers the melting point of ice). This causes the mantle rocks to melt at a shallow depth in Earth's interior, producing a pool of magma.

The gases in the molten material remain dissolved under the intense pressures of Earth's interior and, as the molten material rises to Earth's surface, the pressure decreases, and gases separate from the liquid magma. This process is similar to opening a can of soda, which releases gases from the can previously held under pressure. Common gases in magma include water vapor and carbon dioxide, as well as lower amounts of sulfur, chlorine, and fluorine. As these gases expand, they produce the explosive character of the lava at Earth's surface. The magma's composition determines the concentration of its gases.

## Where Do Volcanoes Form?

### Volcanoes at Divergent Boundaries

Volcanoes on land receive the most attention because of their proximity to human communities. However, most volcanism occurs in the ocean along divergent plate boundaries. A spreading center marks the divergent boundary where oceanic crust moves apart. A mid-ocean ridge system consists of a series of volcanoes that form along a spreading center. As the plates move apart, mantle pressure decreases and **magma** wells up from Earth's interior. The heat from the **magma** also makes the surrounding crust less dense. As a result, the crust rides higher on the asthenosphere, accentuating the height of the mid-ocean ridge system. The magma source generally has a homogenous composition that is low in silica, aluminum, and water. The lava has a low viscosity, flowing easily and producing relatively quiet eruptions. As the magma reaches Earth's surface, it sticks to the wall of the ridge and cools, forming basalt. The new ocean crust pushes the adjacent crust away from the ridge. As the ocean crust moves away from the hot magma source at the spreading center, it cools and becomes denser, riding lower on the asthenosphere.

Mid-ocean ridge systems snake around the globe. Several interconnected mid-ocean ridge systems have been identified. The East Pacific Rise (EPR) is the result of a spreading center located in the Pacific Ocean. It extends from its southernmost point near Antarctica to its northern termination at the Gulf of California. The East Pacific Rise separates the Pacific Plate on its western flank from the Nazca, Cocos, and Juan de Fuca Plates on its eastern flank. The Mid-Atlantic Ridge (MAR) bisects the Atlantic Ocean. The North American and the South American Plates are to the west of the spreading center and the Eurasian and African Plates are on the east. The Southeast Indian Ridge snakes around the Indian Ocean, dividing the Indian and Australian Plates to the north from the Antarctic Plate to the south.

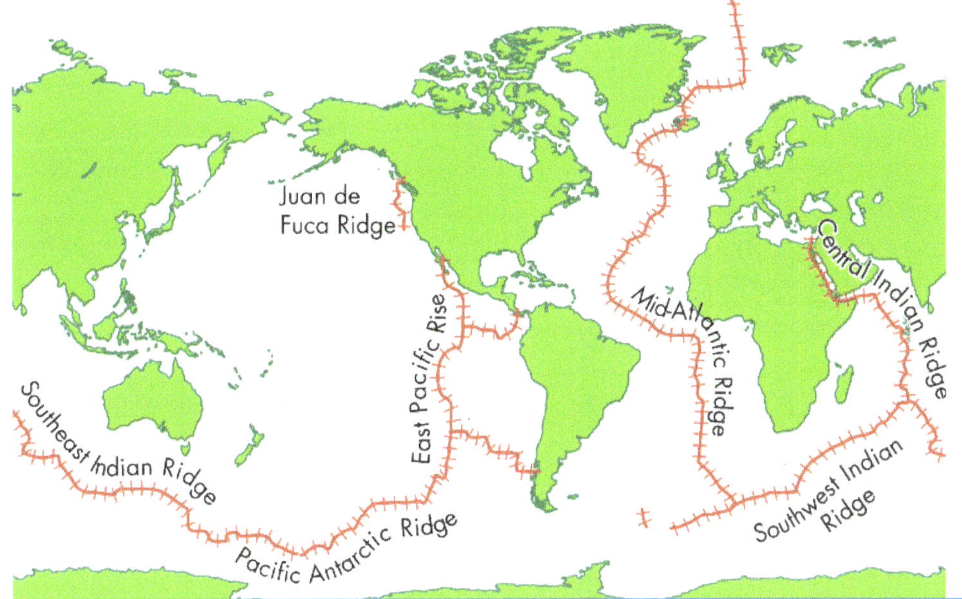


**MID-OCEAN RIDGES**

Interconnected mid-ocean ridge systems form the boundaries of many tectonic plates. Why are volcanoes often found at mid-ocean ridges?

## Volcanoes at Convergent Boundaries

Volcanoes also can form along convergent boundaries where tectonic plates come together. As the plates converge, they produce a unique feature called a **subduction zone** outlined by a trench: a deep depression in the ocean floor. As the denser plate is subducted along the trench, it is recycled back into the mantle, and, as the subducting plate enters the mantle, it begins to melt. The addition of water and ocean sediments changes the composition of the surrounding magma, producing a more buoyant plume of magma that travels upward. This plume melts through the overriding crust as it moves toward Earth's surface.

As the plume melts through the continental crust, the composition becomes even richer in silica. When the magma plume reaches Earth's surface, it feeds a series of volcanoes. The silica-rich lava is very viscous with a high percentage of gas, and, as a result, these volcanoes produce explosive eruptions that release large quantities of ash, noxious gases, and projectiles into the atmosphere.

Compared to oceanic-continental convergence, the silica content of magma produced at oceanic-oceanic convergence remains low. As the plume melts through the oceanic crust, the composition of the plume becomes only slightly enriched in silica. The lava has a low viscosity, allowing gases to escape to the atmosphere, which enables these volcanoes to produce quieter eruptions that release large volumes of lava at Earth's surface.

## Volcanoes at Hot Spots

Not every feature on the ocean floor can be explained by plate tectonics. For example, non-divergent ridges and volcanic island chains can be found outside the vicinity of plate boundaries. A **hot spot** is a relatively stationary plume of magma that scientists think originates at the mantle-core boundary. The most active hot spots are found underneath the big island of Hawaii, Reunion Island, the Galapagos Islands, and Iceland. Volcanoes over hot spots expel massive amounts of lava that contains low silica content. Conversely, hot spot volcanism on a continental crust is often violent due to the addition of silica from the continental crust.

Hot spots remain relatively stationary over long geologic periods. As the tectonic plate moves over the hot spot, the magma plume melts through the lithosphere. This action leaves a continuum of features that scar the plate. In the ocean, magma that fuels the hot spot melts through the oceanic plate and forms a submarine **volcano**, called a **seamount**, on the ocean floor. The hot magma plume supplying the hot spot makes the surrounding lithosphere more buoyant.

Eventually, a seamount may breach the surface of the water and become a volcanic island. Rising above the water's surface, volcanic islands are subject to erosional forces such as chemical and physical weathering. As the plate continues to move over the hot spot, the volcanic island is displaced from the hot magma plume and volcanism stops. The hot spot continues to melt through a new portion of the plate, forming the next seamount in the chain.



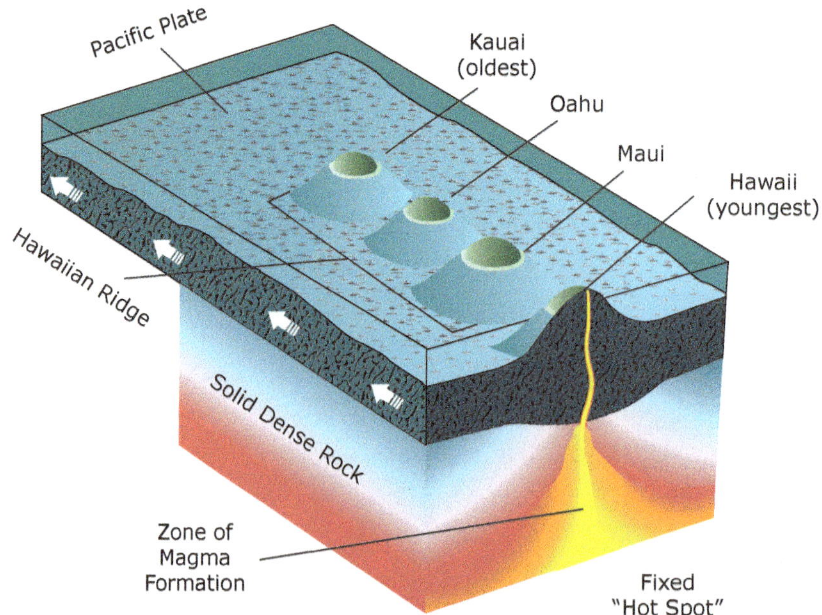


**VOLCANISM AT A HOT SPOT**

Magma from a hot spot melts through the crust, producing a series of seamounts that eventually may become volcanic islands. How does tectonic plate movement affect this process?

As the dormant volcanic island moves further from the hot magma plume, the crust begins to cool. As the distance from the hot spot increases, the island begins to subside and sinks back beneath the ocean. A guyot is a submerged volcanic island that has a flat, eroded top from the time the island was above the water. In warm, tropical waters, coral—a marine organism that builds a skeleton out of calcium carbonate—often grows around the perimeter of the sinking volcanic island. As the island slips below the surface of the water, the coral continues to build onto the reef, reaching toward the water's surface. Algae that live with the coral utilize light for photosynthesis that benefits both algae and coral. In time, the island becomes completely submerged below the water and only a ring of coral remains. The result is an atoll. The interior of an atoll is called a lagoon.

## What Are the Main Factors That Control Whether an Eruption Is Explosive or Nonexplosive?



Deep beneath Earth's surface, the mantle contains molten material known as **magma**. Magma consists of all the chemical materials needed to form rocks and minerals, but it is too hot to completely solidify. Tectonic processes allow magma to rise to the surface at locations such as subduction zones, mid-ocean ridges, and hot spots. When magma reaches Earth's surface, scientists refer to it as lava.

Magma type is defined by its chemical composition. For example, magmas that are found within the mantle are considered to be mafic. Mafic magmas are rich in iron, magnesium, silicon, and calcium, and they form basaltic lavas when they reach the surface. Mafic rocks, such as basalt, have a high density.

Magmas that are rich in silica ($SiO_2$), aluminum, and sodium are considered to be felsic. Two types of extrusive rocks, rhyolite and dacite, are formed from the eruption of felsic magmas. Granite is also formed if the felsic magmas have time to crystallize underground. Felsic rocks are generally lighter in color than mafic rocks.

Magmas that are between mafic and felsic are considered intermediate. **Andesite** is an example of rock derived from intermediate lava. Intermediate magma contains less iron and magnesium than mafic magmas. Intermediate magma also contains less silica and aluminum than felsic magmas, and its density, viscosity, and gas content are between those of mafic and felsic lavas.

## Magma Forming

*Fast Cooling Magma Forms Rock*



**FROM MAGMA TO LAVA**

Magma in the mantle may have several paths to Earth's surface, where it becomes known as lava. What are some of those paths?

Because most magma originates in the mantle, it begins as **ultramafic** magma. Mafic and felsic magma are developed by a process called magmatic differentiation. This development occurs as they rise through the crust. Mafic minerals crystallize at higher temperatures. As the magma chamber cools, mafic minerals crystallize and fall to the bottom of the magma chamber, leaving the remaining magma less mafic. If this magma moves up through continental crust, it can become more felsic as it incorporates felsic components in the surrounding rocks.

### Viscosity of Magma

Viscosity is a property of liquids that measures resistance to flow. The expression "slow as molasses" should really be "viscous as molasses," as it is the viscosity of molasses that makes it flow so slowly, relative to many other liquids.

Lava is an extremely viscous fluid, so it flows relatively slowly on Earth's surface. The viscosity of lava is largely controlled by the amount of silica it contains. Silica is a compound that forms chemical chains within a liquid. Since felsic lavas contain much more silica, they are more viscous than mafic lavas.

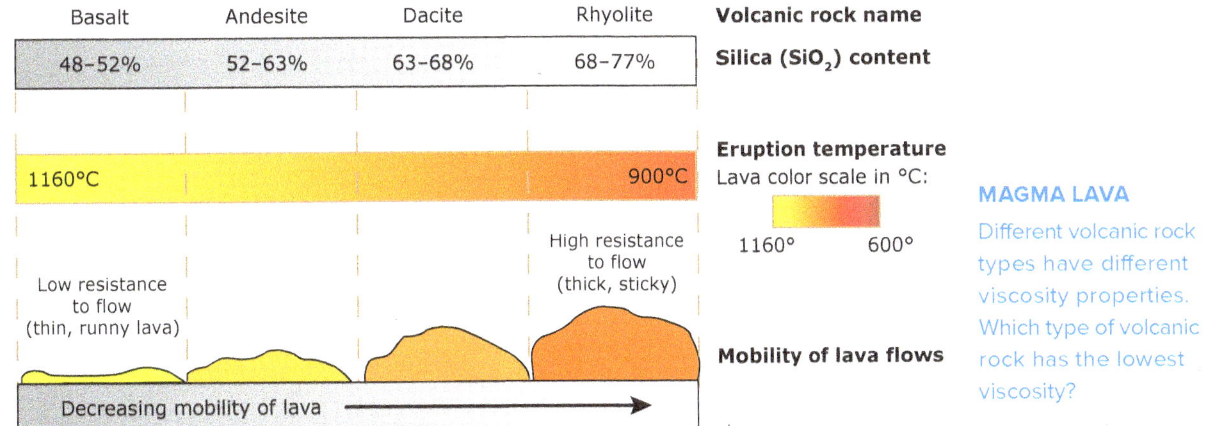


**MAGMA LAVA**

Different volcanic rock types have different viscosity properties. Which type of volcanic rock has the lowest viscosity?



### Explosiveness of Eruptions

The composition of magma also controls the explosiveness of volcanic eruptions by controlling the rate of escaping gases. Eruptions of mafic lavas are much less explosive than those of felsic lavas. Mafic lavas flow and spread out, while felsic lavas often explode violently.

An eruption's explosivity, or explosiveness, is related to the viscosity of the magma. Mafic lavas are less viscous, so gas bubbles can travel through them more easily. These bubbles escape once they reach the surface of the lava. Because felsic lavas are more viscous, it is more difficult for gas to travel through them. Some of the gas bubbles migrate toward zones of lowest pressure and accumulate to form larger bubbles. When the magma becomes exposed at the surface, the gas bubbles suddenly have a chance to escape. Because they are gathered together, the result is like opening a soda bottle after shaking it up. When all the gas tries to escape at once, the result is explosive.

## What Are Some Results of Volcanism?

### The Different Types of Lava

**Magma** reaches Earth's surface and erupts as lava while nonexplosive eruptions of mafic **magma** generate various types of lava. Pahoehoe is a basaltic lava with a relatively low viscosity. Pahoehoe flows quickly and sometimes hardens into rope-shaped bands of basalt. It usually forms where lava flows down a gradual slope.

Aa (from the Hawaiian word *a'a*) is a jagged type of basaltic lava that forms when the gas content in the lava is high. This causes increased viscosity but lower density. As aa lava cools, it solidifies as small blocks on the top of the main lava flow. Aa lava usually forms where the lava flows over steeply sloping surfaces far enough from the **volcano** vent to have cooled significantly.

Blocky lavas form from the eruption of intermediate magmas. They are more viscous and cool into larger blocks than aa, which forms into blocks less than 1 meter in length.

## Types of Pyroclastic Materials

In an explosive eruption, lava does not simply flow from a vent. Instead, exploded silica, ash, and volcanic gas erupt out of the volcano. The force of the explosion can shatter rocks around the vent and send them flying into the air. These pyroclastic materials may also cascade down the slope of the volcano as a pyroclastic flow. These mixtures of hot, toxic gas and ash cause much of the destruction associated with volcanic eruptions—including the eruptions at Mount Vesuvius and Mount St. Helens.

There are many different types of pyroclastic materials. Pyroclastic materials launched into the air are collectively known as tephra. Volcanic ash and dust are the smallest forms of tephra. They are remnants of silica from the magma that exploded when it reached the surface. Gas bubbles within the cooling lava cause it to shatter as it rapidly depressurizes. Volcanic ash and dust are all particles that are smaller than 2 millimeters in diameter.

Lapilli are small balls of volcanic glass, 2–64 millimeters in diameter. Lapilli sometimes have a teardrop shape because they cool while flying through the air. Lapilli may also be a part of a pyroclastic flow. During the flow, lapilli can become "welded" together by the heat. Lapilli tuff is an igneous rock commonly left behind after a pyroclastic flow.

Tephra larger than 64 millimeters in diameter are known as volcanic bombs or volcanic blocks. The name depends on the physical state of the materials as they are flying through the air. Volcanic bombs are molten for most of their flight, so they take on a rounded shape as they fall. Bombs that are still molten when they hit the ground will flatten; they are sometimes referred to as cow flops. Volcanic blocks are solid throughout their flight. They will usually have a more angular shape.

Pumice is a form of volcanic glass. As felsic magma travels through the air, it cools quickly enough to trap bubbles of gas inside. These trapped gas bubbles give pumice a very low density, which allows it to float on water.

## Volcano Types

There are three main types of volcanoes: shield volcanoes, composite volcanoes (stratovolcanoes), and cinder cone volcanoes. The three types differ in their shape, size, and the way that they form.

## Shield Volcanoes

Shield volcanoes are large, gently sloped volcanoes that result from the eruption of basaltic lava. The gentle slope is caused by the low viscosity of the lava, which allows it to flow great distances before cooling. The Hawaiian Islands are all made up of shield volcanoes, and there are two famous shield volcanoes on the islands. The volcanoes are Mauna Loa and Mauna Kea, each of which rises over 4,000 meters above sea level. These two volcanoes extend thousands of meters down to the ocean floor, making them among the largest mountains on Earth. They are built of many stacked flows of basaltic lava.



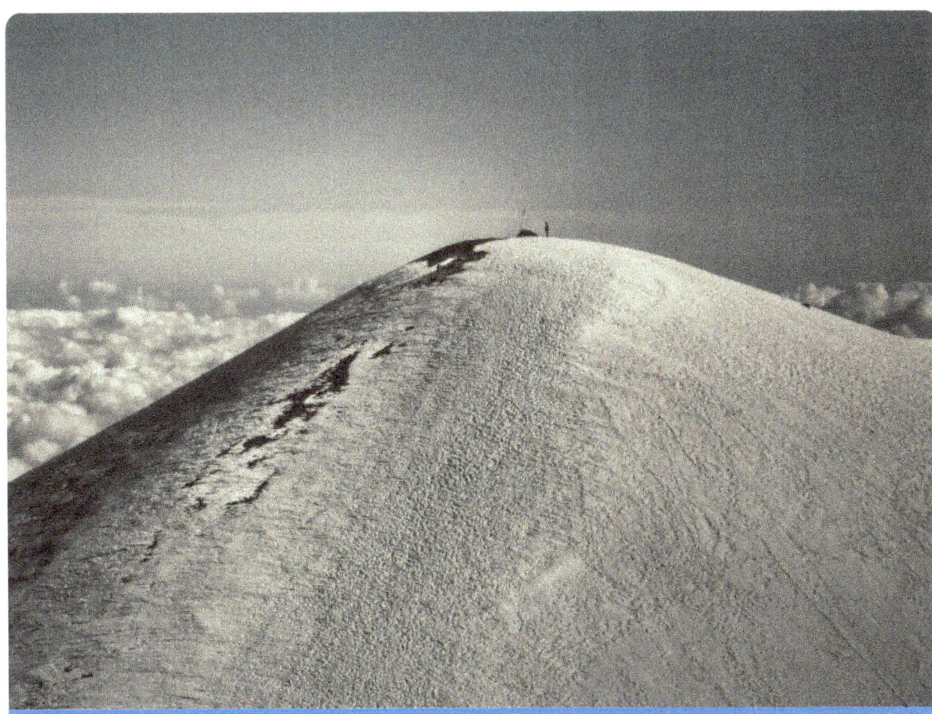

**SHIELD VOLCANO**
Mauna Kea in Hawaii is a shield volcano. How was Mauna Kea formed?

The more viscous the lava extruded by a volcano, the less distance it flows from its source. Volcanoes with viscous lava form dome-shaped structures that are called lava domes, which can reach a height of several hundred meters. The sides of these domes are composed of unstable rock debris that sometimes experience episodes of explosive eruption. If part of the dome collapses while it is still molten, it can produce pyroclastic flows.

## Composite Volcanoes (Stratovolcanoes)

Composite volcanoes are large mountains with steep sides and a distinct cone shape. They are formed by alternating layers of intermediate lavas and pyroclastic materials. Because of these alternating layers, or strata, composite volcanoes are also called stratovolcanoes. These volcanoes have steep slopes because the higher viscosity of the lava prevents it from flowing very far before it hardens. Some famous stratovolcanoes are Mount Fuji in Japan and Mount St. Helens and Mount Rainier in the United States. Because they can erupt both quietly and explosively, composite volcano eruptions are difficult to predict. They are usually considered more dangerous than shield volcanoes.

**COMPOSITE VOLCANO**

Mount Fuji is a composite volcano with steep sides formed by alternating layers of both lava and pyroclastic materials. Why is this type of volcano more dangerous than shield volcanoes?

## Cinder Cones and Spatter Cones

When compared to other volcanoes, cinder cone volcanoes are small, ranging in size from ten to hundreds of meters high. They often look like small stratovolcanoes. They are made of steep cones of pyroclastic material that are formed when lava, rich in gases, is blown violently out of a vent. A cinder cone therefore usually has a central bowl vent surrounded by volcanic ash and lapilli. Upon examination, these small stones can be seen to contain numerous gas bubbles and are often welded together. As cinder cones erupt, pyroclastic materials build up around them. The material does not fall far from the vent. Cinder cones may be found in groups forming a volcanic field, such as the San Francisco volcanic field near Flagstaff in Arizona. They are also often found on the sides of larger volcanoes. For example, Mauna Kea on Hawaii has almost 100 cinder cones associated with it.

When erupting lava contains enough gas to prevent the formation of a constant lava flow, but not enough to explode the lava into small fragments, it forms a spatter cone. As their name suggests, these cones splash out hot blobs of lava called spatter. These spatters can be up to 50 centimeters in diameter and, because they are still liquid, fuse together as they cool.

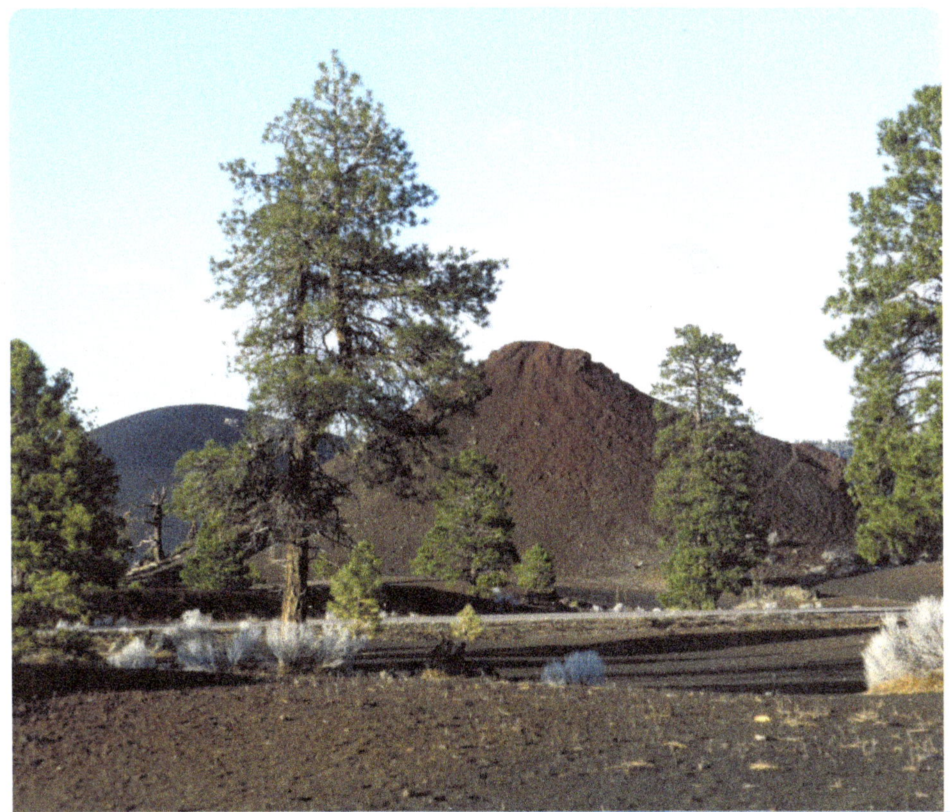

**CINDER CONE VOLCANO**

Cinder cone volcanoes are formed when ejected material falls around the volcanic vent. Where do these volcanoes usually form?

## Calderas

A caldera is formed when a volcano collapses into its magma chamber during an explosive eruption. The eruption empties the magma chamber, removing the support for the mountaintop. This collapse enlarges the crater of the volcano and fills in the crater with collapsed materials. This can be seen at Crater Lake in Oregon. A caldera formed when Mount Mazama erupted about 7,700 years ago. The former mountaintop material is now deep beneath the lake's surface.

Calderas may also be located over large magma chambers in the middle of a tectonic plate, such as the caldera at Yellowstone National Park in Wyoming. Calderas are usually formed in association with extremely explosive and powerful eruptions.

**CALDERA**
The caldera known as Crater Lake formed approximately 7,700 years ago in the Cascade Range. How are calderas formed?



## Fissure Eruptions

Sometimes nonexplosive eruptions occur along fissures or cracks in the Earth's surface. When this occurs, mafic magma flows out of the cracks. This event is common in rift zones, such as mid-ocean ridges, where the crust is splitting apart. Fissure eruptions occur in Iceland and are also common in the rift zones of shield volcanoes such as the Hawaiian volcanoes. Occasionally, lava fountains will erupt along these fissures. Fissure eruptions have the potential to be explosive, as the eruption results in the releasing of pyroclastic materials.

## Volcanism as a Constructive and Destructive Force

Volcanism has the power to build as well as to destroy. When lava flows reach the ocean, they may form new land and increase the size of islands and continents. Chains of volcanic islands formed along hot spot tracks are great examples of this process.

Volcanic eruptions can be beneficial for agriculture. Volcanic ash and soils formed from eroded lava contain many nutrients such as potassium, calcium, iron, and magnesium. These nutrients, in modest amounts, are vital for healthy plant growth. Some of the most fertile soils in the world are in the shadows of large volcanoes.

Volcanic eruptions are very powerful and can be extremely destructive. They can completely change the landscape of a mountainside, destroying the habitats of the organisms that live there. It took more than ten years for plants to begin to grow again on the slopes of Mount St. Helens after its eruption, and even more time for animals to come back to the area. Volcanic eruptions can present many different dangers, including lava flows, pyroclastic flows, lahars, earthquakes, tsunamis, volcanic bombs, and ash falls. Lahars, lava flows, and pyroclastic flows are capable of destroying almost anything in their paths. Eruptions can cause earthquakes, which can cause tsunamis, both of which can be extremely destructive. Pyroclastic materials such as volcanic bombs and blocks can be particularly damaging, as they can be hurled far away from the volcano's vents. Volcanic ash is a fine material that can damage structures as it piles up on them and can even cause them to collapse. Ash can also destroy engines if it gets caught in them. Ash can be hazardous to cars as well as aircraft. If inhaled, ash can lead to respiratory problems. When a large amount of volcanic ash is suspended in the air for a long period of time, it can block out the sun. When this happens, solar energy cannot get through to the ground. Large eruptions can block out enough solar radiation to lower global temperatures. Large ash clouds can lead to the deaths of many plants, which need the sunlight for photosynthesis. This can disrupt the food chain for long periods of time.

### Consider the Explain Question

**What causes volcanoes, and why do certain types of volcanoes develop in certain places?**

dlc.com/ca9031s

Go online to complete the scientific explanation.

### Check Your Understanding

**What causes volcanic eruptions in different types of volcanoes?**

dlc.com/ca9032s

# Applying Volcanoes

Before the 1950s, most scientists thought that the ocean floor was relatively flat and featureless. Measurements of the depth of the ocean floor were difficult to obtain. In fact, early techniques for measuring depth used a weight, often a cannonball, attached to a piece of rope. This technique hinted at the possibility of features on the ocean floor, but it was far too crude to use to create maps of bathymetric features.

In 1947, geophysicist Maurice "Doc" Ewing was determined to discover what was hidden deep in the watery realm. He enlisted the help of two young scientists: Bruce Heezen and Marie Tharp. In the 1940s and 1950s, it was unusual for a woman to be a scientist; even for those women who were scientists, they were often limited in the things they could do. So it should not be surprising that Heezen accompanied Ewing on research cruises, while Tharp stayed behind in the laboratory.

Heezen and Ewing collected depth measurements using a new device called an *echo sounder*, which measures distance from sound waves that move through the water. With this technology, they took tens of thousands of detailed measurements of the Atlantic Ocean floor.



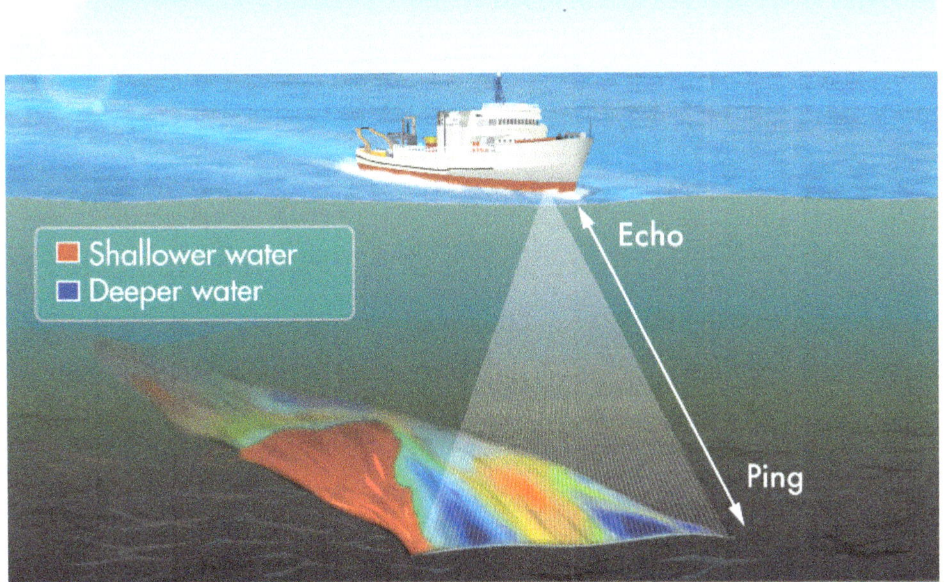


**SONAR AND BATHYMETRY**

Bathymetry is the measurement of the topography of the ocean floor. Scientists use sonar to make bathymetric measurements. The longer it takes to receive an echo from a ping, the deeper the ocean in that area. What feature would be on the ocean floor in the red area, where the water is shallower?

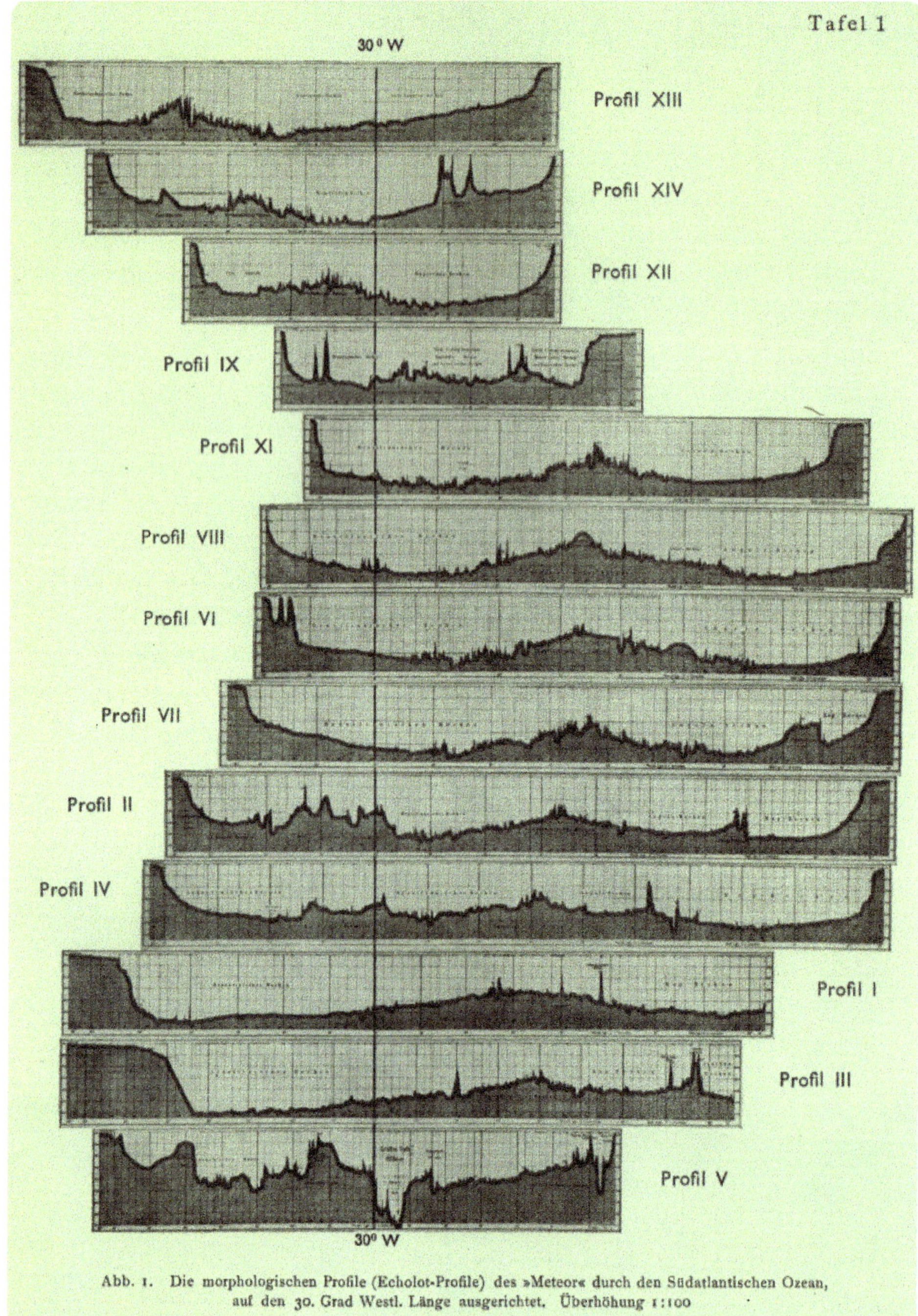


**BATHYMETRIC PROFILE OF THE MID-ATLANTIC RIDGE**

Geologic features of a section of the bottom of the Atlantic Ocean are shown here. How did Marie Tharp's analysis of the bathymetry of the Atlantic Ocean advance the theory of plate tectonics?

Back in the lab, Marie Tharp took the depth measurements collected by Heezen and Ewing and went to the drafting table to map the data. The task was tedious and laborious. Each depth measurement had to be plotted, drawn, checked, corrected, and redrawn by hand. Computers were not available to process data like this until decades later. According to Tharp, it took six weeks to assemble the hodgepodge of disjointed profiles into the proper alignment west to east across the Atlantic Ocean. She immediately noticed a distinct feature in her work—a V-shaped cleft in the center of the profiles. The individual mountains did not match up, but the cleft did, especially in the three northernmost profiles. She suspected that the "V" was actually a rift valley that cut into the crest of a long underwater **volcano**.

Tharp took her interpretation to her colleague Heezen, but he dismissed her interpretation. He thought it sounded too much like continental drift, which was not looked at favorably by the scientific community at that time. But Tharp believed in herself and her ideas. She continued looking for the rift in all of the data. She transformed the data into a 3-D model, which captured the seafloor's many textured variations. With the addition of more data, the features on the ocean floor became more prominent.

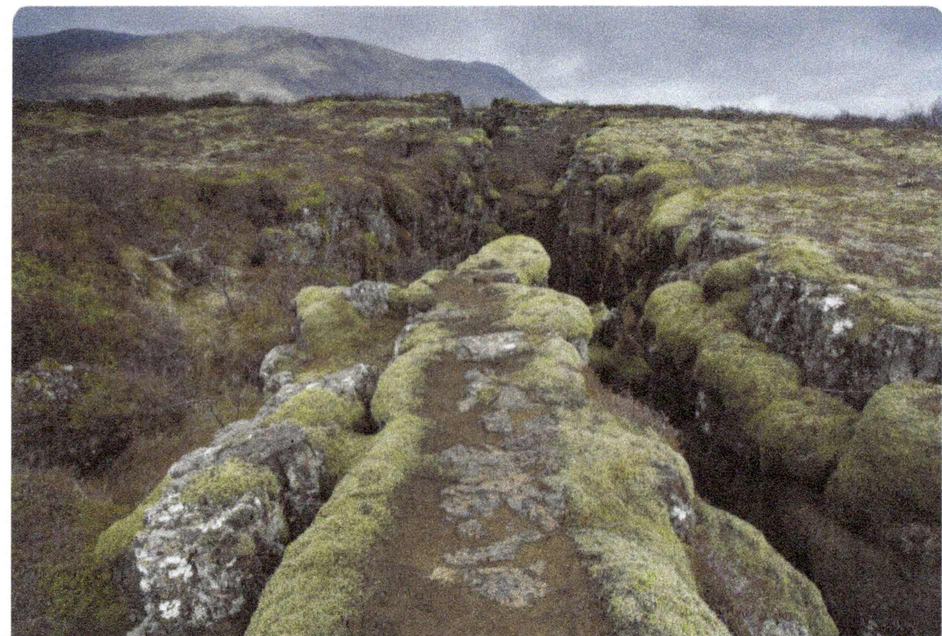

**A MID-OCEAN RIDGE ON LAND**

Once the Mid-Atlantic Ridge was discovered, geologists realized that Iceland had formed on top of it. This rift in Iceland is a continuation of the Mid-Atlantic Ridge.



By 1953, eight months after beginning the project, Tharp had convinced Heezen of the validity of her interpretation. As more maps were created, the long, sinuous, volcanic mountain chain began to snake across the Atlantic Ocean basin from the north to the south. Tharp and Heezen presented their results at the 1956 meeting of the American Geophysical Union in Toronto, Canada. The scientific community reacted with a range of emotions from amazement to skepticism to scorn. The discovery of the mid-ocean ridge system was a revelation that fit together perfectly with the new theory of seafloor spreading. The work of Tharp and Heezen constituted the first comprehensive map of the ocean floor and contributed to a revolution in the way geologists and oceanographers thought about Earth.

## STEM and Volcanoes

When aerospace engineers designed jet engines that burned hot enough to soften rock, they expected high fuel efficiency—not melted rock gumming up the works. How could rocks even bother a jet, when it flies so high in the atmosphere? Unfortunately rocks come in all sizes, from continent-sized plates that the jets take off from, to the tiny particles of dust that can melt onto their engines.

Even when volcanoes are not deadly, the materials they eject can inconvenience people living many miles away. Volcanic ash that persists in the atmosphere can be particularly troublesome, from causing global cooling to gumming up aircraft engines. Volcanic eruptions can close airports for many days, inconveniencing passengers and preventing the delivery of goods shipped by air. Delivery interruptions can affect millions of people who rely on air shipment whether or not they realize it.

While volcanologists cannot prevent these problems, they can help us to better understand the behavior of volcanic particles in the atmosphere, predict how long they will stay aloft, and advise airlines on how to keep pilots, passengers, cargo, and equipment safe by avoiding areas that may cause problems.

Controlled experiments are extremely helpful in understanding a phenomenon well enough to predict it. Volcanoes are impossible to control, however, and so is the atmosphere. That's why scientists turned to a tool developed for engineers—a wind tunnel—to test the behavior of volcanic particles and better predict their effect on aircraft.

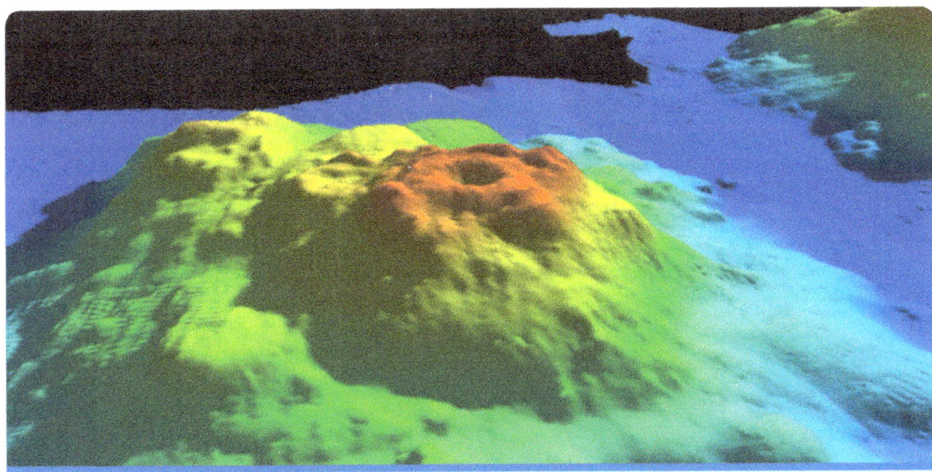

**A SUBMARINE VOLCANOE**

Scientists estimate more than a million underwater volcanoes could be found on the ocean floor worldwide. What devices could scientists use to produce an image like this?

CONCEPT

2.1

# Parts of the Atom

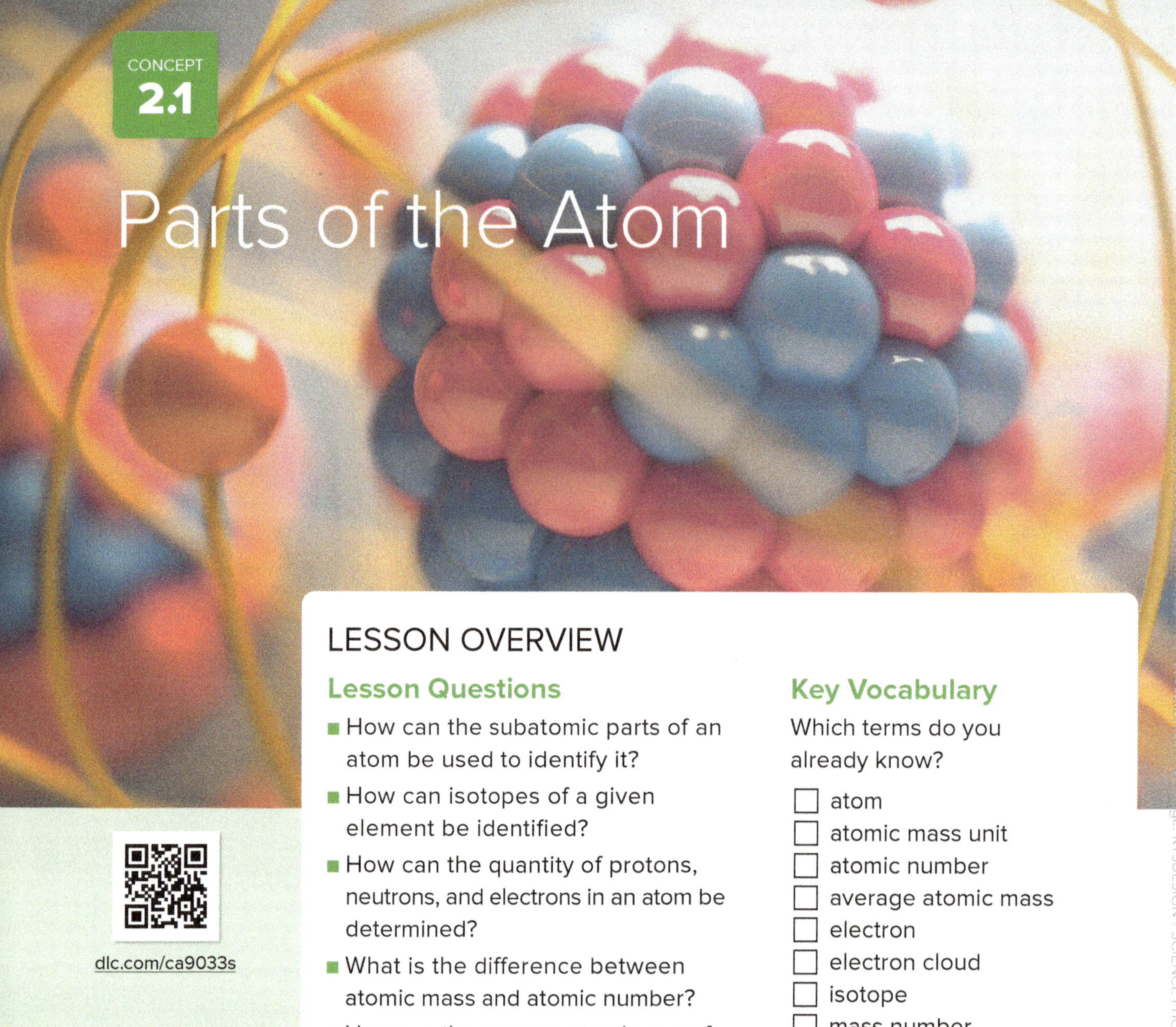

Image: KTSDESIGN / SCIENCE PHOTO LIBRARY / Getty Images

dlc.com/ca9033s

## LESSON OVERVIEW

### Lesson Questions

- How can the subatomic parts of an atom be used to identify it?
- How can isotopes of a given element be identified?
- How can the quantity of protons, neutrons, and electrons in an atom be determined?
- What is the difference between atomic mass and atomic number?
- How can the average atomic mass for an element be determined given the masses of its isotopes and their relative abundance?

### Lesson Objectives

By the end of the lesson, you should be able to:

- Determine the quantity of protons, neutrons, and electrons in an atom.
- Distinguish between atomic mass and atomic number.
- Identify isotopes of a given element.
- Given masses of isotopes and relative abundance, determine the average atomic mass for an element.

### Key Vocabulary

Which terms do you already know?

- [ ] atom
- [ ] atomic mass unit
- [ ] atomic number
- [ ] average atomic mass
- [ ] electron
- [ ] electron cloud
- [ ] isotope
- [ ] mass number
- [ ] negative charge
- [ ] neutron
- [ ] nucleus (atom)
- [ ] positive charge
- [ ] proton
- [ ] subatomic

## Discovering Energy from inside the Atom

Have you ridden in a car that suddenly ran out of gas? Have you ever experienced watching a movie only to see your TV lose power because of a thunderstorm? How does your life change when the gas tank has been refilled and the electricity has turned back on?

dlc.com/ca9034s



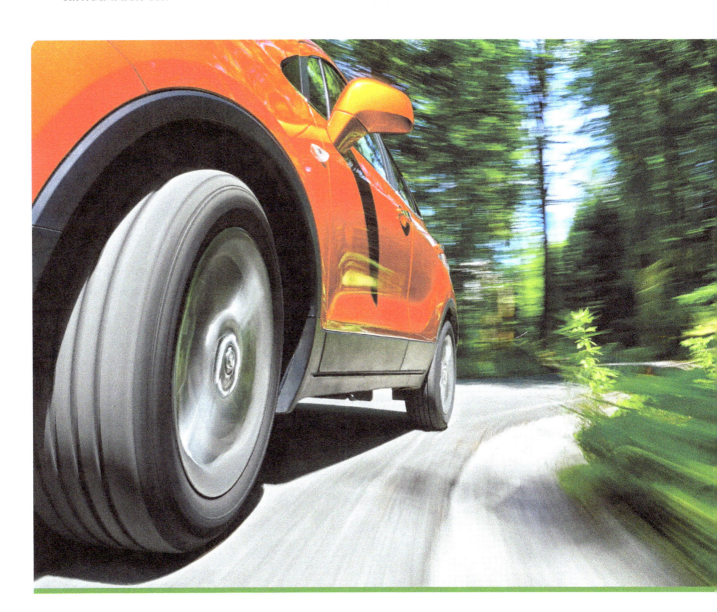

**EXPLAIN QUESTION**

What information about an atom can be determined from its parts?

**ZOOM**

The atom's nucleus is a powerhouse that supplies energy just like fuel does for a car. Where is the nucleus of an atom located?

# How Can the Subatomic Parts of an Atom Be Used to Identify It?

## Protons, Neutrons, and Electrons

Scientific thinkers long believed that atoms were the smallest particle that could be identified, as had been proposed in John Dalton's atomic theory in the early 1800s. It was not until the early 1900s that scientists discovered this to be false. We now know that atoms are made up of **subatomic** particles that have their own distinct properties.

Two kinds of subatomic particles exist in the core, or nucleus, of an **atom**—protons and neutrons. Protons are positively charged particles. Neutrons do not have a charge.

All atoms contain at least one **proton**. With the exception of the stable hydrogen atom, all atoms also contain at least one **neutron**. Protons and neutrons have the same mass, which is defined as one **atomic mass unit** (amu).

The space outside of an atom's nucleus is occupied by electrons. An **electron** is a negatively charged particle. The magnitude of the **negative charge** on an electron is equal to the amount of **positive charge** of a proton. However, the mass of an electron is extremely small; it is only 1/1,836 the mass of a proton or a neutron. Thus, almost all of the mass of an atom is found in its protons and neutrons.



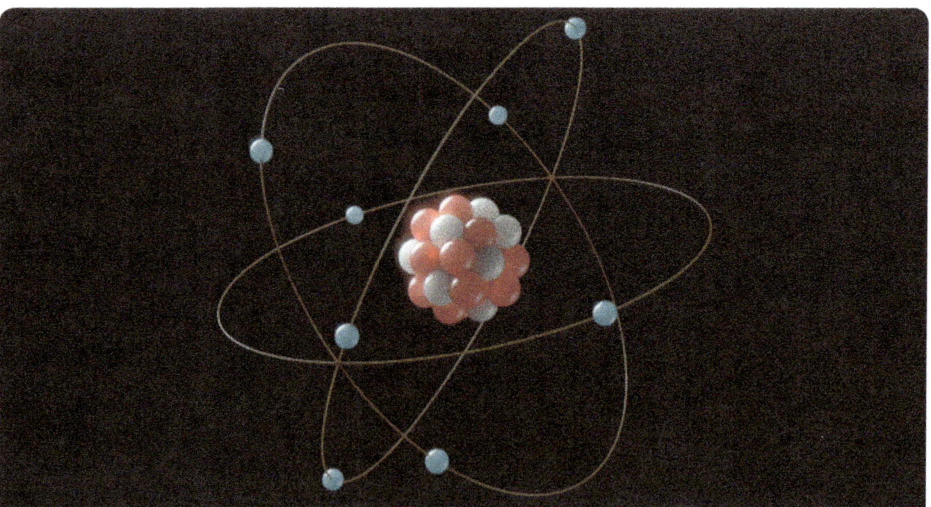

**ZIPPING AROUND**
Subatomic particles are constantly in a state of movement. What subatomic particle circles around the atom's nucleus?

Electrons orbit the nucleus in a relatively large space surrounding the nucleus. This space is called an **electron cloud**, and it accounts for most of an atom's size. The configuration of electrons within this space depends on the number of electrons present and their energy.

If an atom of any given element contains the same number of electrons as protons, its overall charge is zero. An atom with an equal number of electrons and protons is said to be in a neutral state.

However, under certain conditions an atom of an element will gain or lose electrons. In this case, the charges in the atom become unbalanced and the particle has a net charge. The charge is negative if the number of electrons is greater than the number of protons. The charge is positive if the number of electrons is less than the number of protons. Such charged particles are called ions.

## Identifying Atoms

The atoms of each element differ from one another. These differences account for the various properties that the elements exhibit. The primary detail that identifies the atoms of a specific element is the number of protons in its nucleus.

For example, an atom of the element hydrogen contains exactly one proton in its nucleus. This is true of all hydrogen atoms. An atom of the element helium contains exactly two protons in its nucleus. Again, this is true of all helium atoms.

Thus, the atoms of each known element contain a specific and unique number of protons. Put another way, the atoms of different elements contain different numbers of protons in their nuclei.



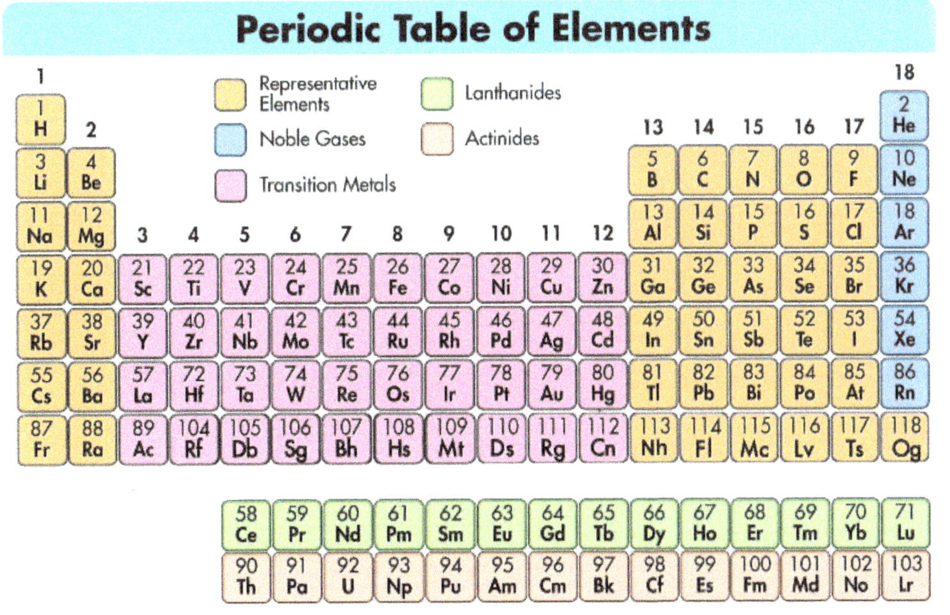


**IDENTIFYING ELEMENTS**

The numeral above the symbol of each element in the periodic table represents its atomic number. Which part of the atom determines the atomic number of an atom?

# How Can Isotopes of a Given Element Be Identified?

## Isotopes

As it turns out, not all atoms of a given element are completely identical. The most abundant hydrogen **isotope**, hydrogen-1, is also called protium. It contains one **proton** and no neutrons. The nucleus of hydrogen-2, also called deuterium or "heavy hydrogen," consists of one proton and one **neutron**. The nucleus of hydrogen-3, also called tritium, consists of one proton and two neutrons. Hydrogen-1 and hydrogen-2 are naturally occurring isotopes of hydrogen. Hydrogen-3 is an unstable radioactive isotope that is extremely rare on Earth but is created synthetically.

**The Nuclei of the Three Isotopes of Hydrogen**

| Protium | Deuterium | Tritium |
|---|---|---|
| 1 proton | 1 proton<br>1 neutron | 1 proton<br>2 neutrons |

**HYDROGEN ISOTOPES**

There are three isotopes of hydrogen. The nucleus of hydrogen-1, the most common hydrogen isotope, consists of one proton and no neutrons. What are the relative masses of these three isotopes of hydrogen?

Note that all three isotopes of hydrogen contain a single proton in their nucleus. As discussed earlier, the number of protons present is what distinguishes one element's atoms from another element's atoms.

A change in the number of neutrons in an **atom**'s nucleus does not change the atom's identity. It does, however, change the atom's mass.

# How Can the Quantity of Protons, Neutrons, and Electrons in an Atom Be Determined?

## Counting Protons, Neutrons, and Electrons

The **atomic number** of an element equals the number of protons in the nucleus of the element's atoms. Every element has a different atomic number because the elements' atoms contain different numbers of protons.

The number of electrons in a neutral **atom** equals the number of protons in the atom's nucleus. In other words, in addition to describing the number of protons in an atom, the atomic number also describes the number of electrons in a neutral atom. This only applies, however, if the atom has not lost or gained any electrons.

Recall that each **proton** and **neutron** has a mass of about 1 amu. Because the electrons in the **electron cloud** have very little mass, the entire mass of the atom can be considered to equal the mass of all the protons and neutrons in the nucleus. A neutral atom of carbon, for example, has six protons and six neutrons, each with a mass of 1 amu. The approximate mass of the stable carbon atom is therefore 12 amu. This atom has a **mass number** of 12.

The mass number expresses the total number of particles in the nucleus of an atom. Because protons and neutrons are always present in their entirety, the mass number is always a whole number. The table Atomic Number vs. Mass Number shows the atomic number and mass number of various common elements in their stable states. Note that the numbers of protons and neutrons are not always equal.

| Isotope | Atomic Number | Mass Number |
|---|---|---|
| Hydrogen-1 | 1 | 1 |
| Helium-3 | 2 | 3 |
| Carbon-12 | 6 | 12 |
| Nitrogen-14 | 7 | 14 |
| Oxygen-16 | 8 | 16 |
| Fluorine-19 | 9 | 19 |
| Sodium-23 | 11 | 23 |
| Calcium-40 | 20 | 40 |
| Iron-56 | 26 | 56 |
| Silver-108 | 47 | 108 |
| Gold-197 | 79 | 197 |
| Lead-207 | 82 | 207 |

**ATOMIC NUMBER VS. MASS NUMBER**
The atomic number is equivalent to the number of protons in the atoms of an element. The mass number is equivalent to the number of protons and the number of neutrons in the atoms. Do protons and neutrons have the same mass?

If the mass number for an element is given, the number of neutrons in its atoms can be calculated by subtracting the atomic number from the mass number.

$$\text{mass number} - \text{atomic number} = \text{number of neutrons}$$

Gold-197 has a mass number of 197 and an atomic number of 79, which means it has 118 neutrons in the nucleus of an atom.

$$197 - 79 = 118$$

Hydrogen-1 has an atomic number of 1 and a mass number of 1, so the number of neutrons in a hydrogen atom is calculated to be 0.

$$1 - 1 = 0$$

Hydrogen is the only element with atoms that contains no neutrons in the stable state.

The atomic number and the mass number are used in symbols that represent the isotopes of all the elements using the following form:

$$^{A}_{Z}X$$

Here, X represents any element. Z represents the atomic number for the **isotope**, and A represents the mass number. For example, the isotope described above, gold-197, would be represented by the following symbol:

$$^{197}_{79}Au$$

The symbols for other isotopes of gold would have the same atomic number but different mass numbers.

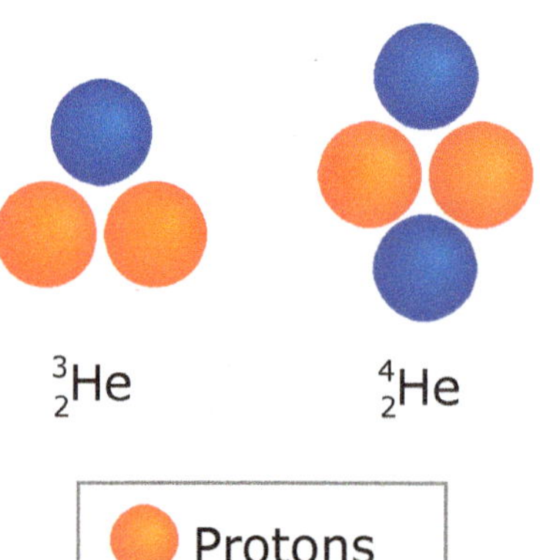


**HELIUM ISOTOPES**
Helium is the second most abundant element in the universe. The most common form is helium-4. How is it different from helium-3?

**ICEBREAKER SHIP**

The Russian mariners when navigating through the Arctic waters used these nuclear-powered ships. Why would energy from an atom be beneficial to the Russian mariners?



# What Is the Difference between Atomic Mass and Atomic Number?

## Atomic Mass vs. Atomic Number

The **atomic number** of an element equals the number of the protons in the nucleus of its atoms. The atomic number is the same for every **atom** of a particular element. For example, the atomic number of hydrogen and all its isotopes is 1.

The **mass number** will vary for different isotopes of the same element because different numbers of neutrons are present in the atoms. Mass number is usually used to describe a specific **isotope** that a scientist is working with in a lab. The isotopes of hydrogen have mass numbers of 1 and 2 for hydrogen-1 and hydrogen-2, respectively.

Modern technology has enabled scientists to calculate the exact masses of protons and neutrons. While the **atomic mass unit** (amu) is still a relative value, it has been defined as 1/12 the mass of a stable carbon atom. The precise measurement of the mass of an atom of a specific element is referred to as its atomic mass. The atomic mass of hydrogen-1 is 1.008 amu and the atomic mass of hydrogen-2 is 2.014 amu.

# How Can the Average Atomic Mass for an Element Be Determined Given the Masses of Its Isotopes and Their Relative Abundance?

## Average Atomic Mass

Often it is useful to consider the general properties of an element, even if several isotopes are present. For example, an environmental scientist might want to study the behavior of oxygen in the atmosphere. There are three naturally occurring isotopes of oxygen: oxygen-16, oxygen-17, and oxygen-18. The scientist's work will be concerned with the oxygen that actually exists in the atmosphere rather than with each **isotope** separately.

Scientists have found that they can use the **average atomic mass** of a specific element to be accurate in quantitative measurements. The average atomic mass takes into account the masses of all the naturally occurring isotopes of a particular element, weighted according to their abundance. Because it is a calculated value, it is likely to include decimals. These are the values typically listed on the periodic table of the elements. The average atomic mass of oxygen is 15.999. The whole number atomic mass of radioactive elements shown on the periodic table is the mass of the most stable isotope of that element.

The average atomic mass of hydrogen is 1.008. This is very similar to the atomic mass of hydrogen-1 because over 99.9 percent of hydrogen occurs in the stable form.

The weighted averages take into account the abundance of the naturally occurring isotopes of an element. Abundance is usually expressed as a percent of the element of each isotope. To find the atomic mass of an element, follow these steps:

1. Find the percent abundance and atomic mass of each naturally occurring isotope of the element.
2. Multiply the percent abundance by the atomic mass of each isotope.
3. Find the sum of the products.
4. Divide the sum by 100.

The sample problem that follows illustrates how weighted averages are used to calculate the atomic mass of an element.

## Average Atomic Mass: Sample Problem

Determine the average atomic mass of the three naturally occurring isotopes of magnesium given the following data:

| Isotope | Abundance (%) | Mass of Isotope (amu) |
|---|---|---|
| magnesium-24 | 79 | 23.985 |
| magnesium-25 | 10 | 24.986 |
| magnesium-26 | 11 | 25.983 |

**Solution:**

Magnesium has three isotopes, the masses of which are shown in the table. The data in the table indicates that in a sample of 100 magnesium atoms, 79 will be magnesium-24, 10 will be magnesium-25, and 11 will be magnesium-26. Thus, the average atomic mass of magnesium can be calculated as follows:

$$\text{Average atomic mass of magnesium} = \frac{(79 \times 23.985\text{ amu}) + (10 \times 24.986\text{ amu}) + (11 \times 25.983)}{100}$$

$$= \frac{2430.5\text{ amu}}{100} = 24.305\text{ amu}$$



### Consider the Explain Question

**What information about an atom can be determined from its parts?**

Go online to complete the scientific explanation.

dlc.com/ca9035s

### Check Your Understanding

**How do subatomic particles determine the chemical identity of an element?**

dlc.com/ca9036s

# Applying Parts of the Atom

When it comes to fighting cancer, doctors face two questions. Where has the cancer spread from its original location, and how can it be treated? In the case of one type of cancer, the answer to both questions may involve the **isotope** iodine-131. Iodine-131 is an artificial, radioactive isotope of iodine that releases radiation capable of killing cancer cells, especially those that may have originated in the thyroid gland. Why does iodine-131 target these particular cells?

The function of the thyroid gland is to make, store, and release thyroid hormones into your blood. Iodine is combined with an amino acid to make three different thyroid hormones. These hormones affect almost every cell in your body and help the body regulate functions such as metabolism. These hormones contain naturally occurring iodine atoms (iodine-127) that are normally absorbed by the body from foods. Iodine supplied by nutrients becomes concentrated in the thyroid gland and its cells. This is a good thing since thyroid cells need iodine to manufacture hormones that are vital to the body.

When the thyroid gland is overactive, as in hyperthyroidism, the body's processes speed up and may cause weight loss, nervousness, rapid heartbeat, tremors, or sleep problems. Endocrinologists sometimes use radioactive iodine-131 therapy to treat an overactive thyroid. A small capsule of radioactive iodine is swallowed. Once swallowed, the radioactive iodine gets into the bloodstream and is quickly taken up by the hyperactive thyroid cells. Over a period of several weeks or months, the radioactive iodine destroys the cells that have taken it up.

When treating cancer cells originating from the thyroid, oncologists take advantage of the ability of thyroid cells to attract iodine even if the cells have traveled to distant locations in the body. Radioactive iodine-131 is injected into a patient, and thyroid cancer cells throughout the body absorb it. Radiation from the iodine-131 then attacks the cancer cells. Other cells are not affected by this radiation because they do not tend to absorb significant amounts of iodine.

## STEM and Parts of the Atom

They form numbers on the faces of digital clocks and with the push of a button they light up watches. They send signals from remote controls and produce images on high-definition television screens. Electrical engineers help design and manufacture them for consumer use. What does each of these examples describe? They refer to a type of semiconductor light source called LED, or light-emitting diodes.

You might think of an LED as an extremely small light bulb. Similar to their relative the incandescent light bulb, LEDs are activated by a flow of electric current. However, the current does not flow through a metal filament as in traditional light bulbs. Instead, in LEDs the flow of an electric current causes the movement of electrons through a diode. A diode is a device that allows electricity to move in only one direction. In general, an LED is made of a material called a semiconductor, which is a type of material that regulates the flow of an electric current. Semiconductors include the metalloid elements, such as silicon and germanium.

Moving by the electric current, electrons will interact with atoms of the semiconductor, forcing them to emit packets of light called photons. The emitted light enables you to see the images on a high definition television screen and makes the numbers on a digital clock visible. It is also the light you see shining from an LED light bulb.



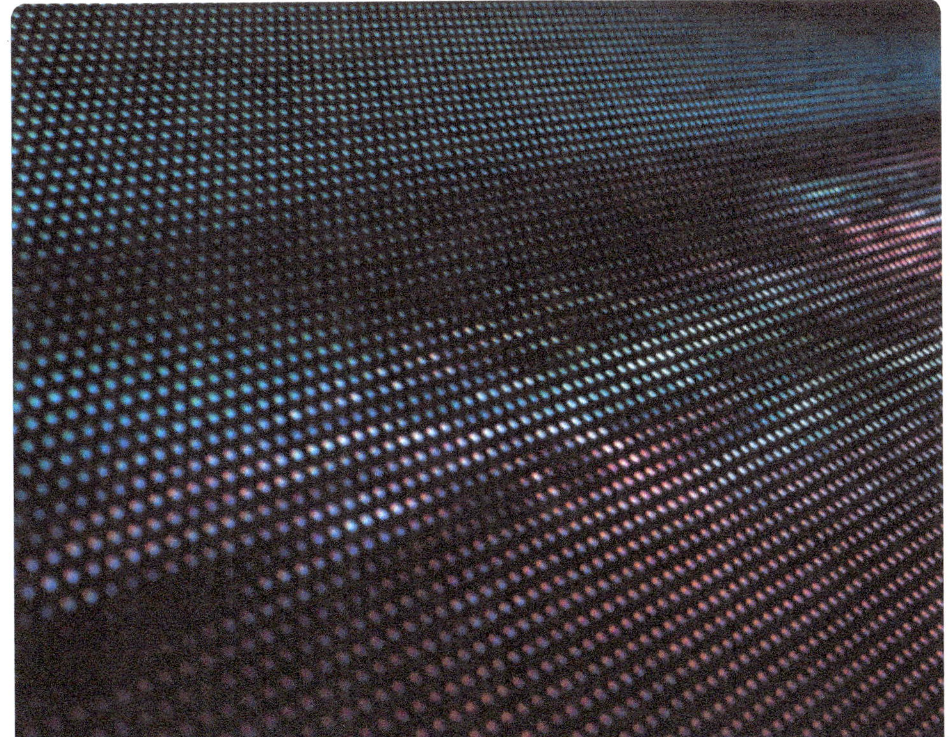

**LED LIGHT DISPLAY**

Emitted light from an LED display can produce pictures. What causes the light to be emitted?

An LCD television produces its picture by shining light produced by LEDs through color filters. Relative newcomers to the LED scene are OLEDs, organic light-emitting diodes, which use LEDs to generate both light and color. The term organic refers to the fact that the materials are carbon-based. By using electroluminescence, OLEDs are amazingly thin and can be applied to flexible surfaces. They can be bent and shaped.

An LED differs from other technological sources of light in a number of important ways. For one thing, LEDs do not produce very much heat. Your television screen does not become too warm, nor does your illuminated watch, the digital numbers on your kitchen clock, or an LED light bulb. Another important advantage of LEDs is that when compared to most other forms of lighting, they consume little energy—making them very inexpensive to operate.

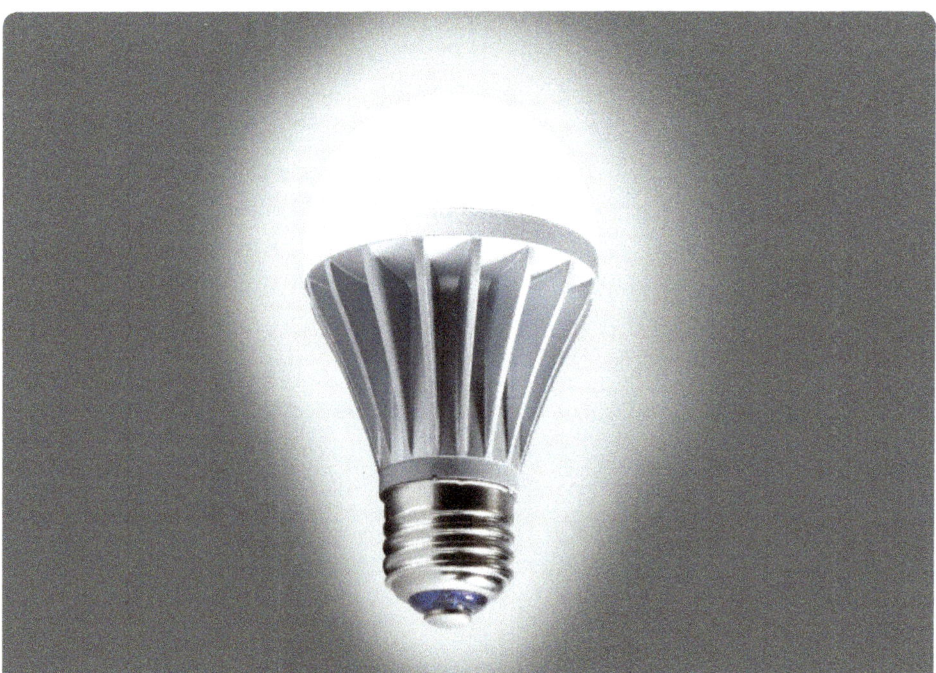

**LED LIGHT**

Light-emitting diodes, or LEDs, are widely used to light homes and our television screens. What makes an LED different from a traditional incandescent light bulb?

# Average Atomic Mass

**Use the data tables and the average atomic mass equation to solve the problems below.**

**1.** What is the average atomic mass of the isotopes shown in the table?

| Isotope | Abundance (%) | Mass of Isotope (amu) |
|---|---|---|
| Isotope-A | 68 | 36.456 |
| Isotope-B | 23 | 37.457 |
| Isotope-C | 9 | 39.459 |

**2.** What is the average atomic mass of the isotopes shown in the table?

| Isotope | Abundance (%) | Mass of Isotope (amu) |
|---|---|---|
| Isotope-A | 54 | 17.984 |
| Isotope-B | 32 | 18.985 |
| Isotope-C | 14 | 19.982 |

**3.** What is the average atomic mass of the isotopes shown in the table?

| Isotope | Abundance (%) | Mass of Isotope (amu) |
|---|---|---|
| Isotope-A | 82 | 46.113 |
| Isotope-B | 12 | 47.115 |
| Isotope-C | 6 | 48.114 |

# Using Isotopic Composition to Calculate Average Atomic Mass

**Use the data in the tables and the average atomic mass equation to fill in all empty boxes and solve the problems below.**

**1.** What is the average atomic mass of hydrogen?

| Hydrogen Isotope | Number of Protons | Number of Neutrons | Mass Number | Percent Abundance |
|---|---|---|---|---|
| A | 1 | 0 | | 99.985 |
| B | 1 | 1 | | 0.015 |
| C | 1 | 2 | | trace |

**2.** Calculate the average atomic mass of boron, which has two naturally occurring isotopes.

| Boron Isotope | Number of Protons | Number of Neutrons | Mass Number | Percent Abundance |
|---|---|---|---|---|
| A | 5 | 5 | | 20.00 |
| B | 5 | 6 | | 80.00 |

**3.** What is the average atomic mass of the noble gas neon?

| Neon Isotope | Number of Protons | Number of Neutrons | Mass Number | Percent Abundance |
|---|---|---|---|---|
| A | 10 | 10 | | 90.48 |
| B | 10 | 11 | | 0.27 |
| C | 10 | 12 | | 9.25 |

**4.** The oxygen we breathe actually exists in three isotopic forms. Find its average atomic mass.

| Oxygen Isotope | Number of Protons | Number of Neutrons | Mass Number | Percent Abundance |
|---|---|---|---|---|
| A | 8 | 8 | | 99.762 |
| B | 8 | 9 | | 0.038 |
| C | 8 | 10 | | 0.200 |

**5.** Nuclear power plants often use one isotope of uranium to produce electricity. This isotope is one of three that occur in nature. Calculate the average atomic mass for this element.

| Uranium Isotope | Number of Protons | Number of Neutrons | Mass Number | Percent Abundance |
|---|---|---|---|---|
| A | 92 | 142 | | 0.005 |
| B | 92 | 143 | | 0.720 |
| C | 92 | 146 | | 99.275 |

CONCEPT 2.2

# Arrangement of Electrons in the Atom

dlc.com/ca9037s

## LESSON OVERVIEW

### Lesson Questions

- What arrangement of electrons in an atom did Niels Bohr propose?
- How did quantum theory help establish a relationship between the energy of an electron and its position in an atom?
- What is meant by the dual nature of light?
- How are frequency and wavelength of electromagnetic radiation related?

### Lesson Objectives

By the end of the lesson, you should be able to:

- Use the Bohr model to illustrate the arrangement of electrons in atoms.
- Relate the energy of electrons to the positions of electrons in the atom.
- Cite evidence for the dual nature of light.
- Describe the relationship between frequency and wavelength for electromagnetic radiation.

### Key Vocabulary

Which terms do you already know?

- ☐ azimuthal quantum number ($\ell$)
- ☐ electromagnetic radiation
- ☐ electron
- ☐ electron cloud
- ☐ emission
- ☐ energy levels
- ☐ Ernest Rutherford
- ☐ frequency
- ☐ J.J. Thomson
- ☐ light
- ☐ magnetic orbital quantum number
- ☐ magnetic spin quantum number
- ☐ photon
- ☐ Planck's wave equation
- ☐ principal quantum number ($n$)
- ☐ proton
- ☐ quantum mechanical model of the atom
- ☐ quantum mechanics
- ☐ quantum number
- ☐ speed of light ($c$)
- ☐ wave

# Thinking about the Arrangement of Electrons in the Atom

**Thinking back, have you ever used a glowstick, the kind that you break, when you went camping or were at a party? Breaking the solids in the glowstick causes it to light up. How might the inventors of these glowsticks have been inspired?**

dlc.com/ca9038s

**EXPLAIN QUESTION**

**How is the arrangement of electrons in atoms related to their ability to emit light?**

**GLOWING JELLYFISH**

Why do these jellyfish appear to be glowing?

# What Arrangement of Electrons in an Atom Did Niels Bohr Propose?

## Rutherford's Nucleus

At the turn of the 20th century, understanding the structure of the atom was one of the most sought-after goals in physics. In 1897, **J.J. Thomson** discovered the **electron** and proposed his plum pudding model of the atom. A few years later, **Ernest Rutherford** challenged Thomson's model. Rutherford's experiments led him to conclude that the atom must have a nucleus.

Rutherford proposed that an atom's mass is concentrated at its center. Protons and neutrons had not yet been discovered, so Rutherford could not tell what the nucleus of an atom is made of.

Rutherford was also unable to determine the arrangement of electrons in the atom. He suggested that the electrons move in orbits around the nucleus, but he had no evidence to support this hypothesis. Indeed, the idea of orbiting electrons was poorly received by physicists at the time because it did not agree with the laws of classical mechanics developed by Isaac Newton. According to these laws, orbiting electrons would continually lose energy and eventually fall into the nucleus. Since this did not happen, Rutherford's hypothesis was rejected.

Niels Bohr joined Rutherford's lab in 1912 to study the arrangement of electrons in atoms. Bohr's careful research led him to propose a model of electron orbits that satisfied both experimental evidence and the laws of physics.

## Light-Emitting Atoms

When Niels Bohr began his work, several observations of atomic behavior had not yet been explained. One set of observations concerned the way that atoms emit **light**. Scientists had observed that different elements emit different wavelengths of light. These wavelengths are seen as colors of light emitted when elements are heated or when electric current is passed through them. Neon gas, for example, emits red light when an electric current passes through it. Xenon gas produces blue light. The colors emitted by elements are called **emission** spectra. Each element has a specific emission spectrum.

Each element also absorbs wavelengths of light, the pattern of which is called its absorption spectrum. The emission and absorption spectra of hydrogen were of great interest to scientists at the time Bohr began working in Rutherford's lab.

Bohr used spectroscopic observations of hydrogen and other pieces of experimental evidence to guide his research. Bohr discovered that the specific pattern of light emitted and absorbed by different elements is connected to the arrangement of electrons in their atoms.

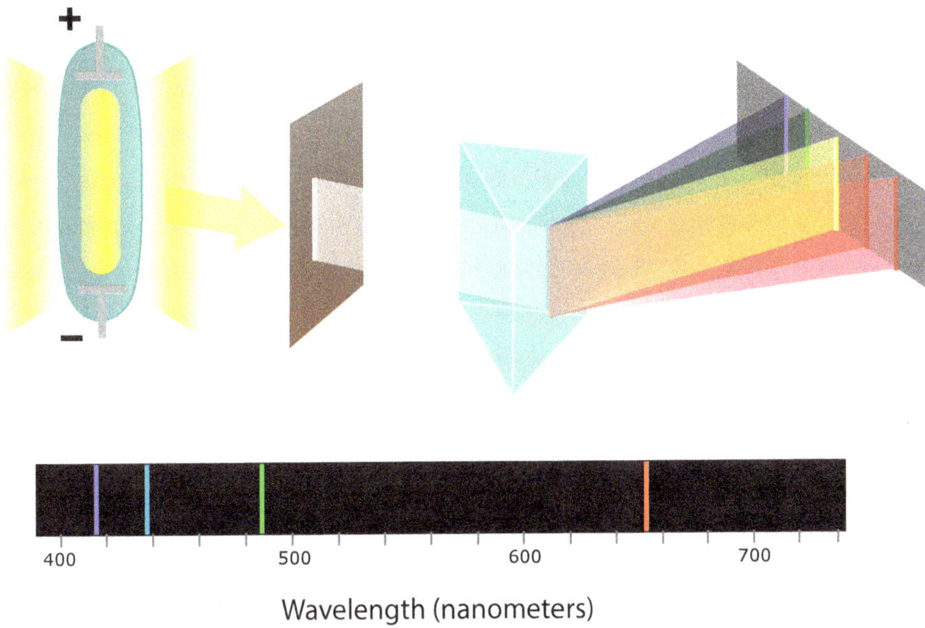


**EMISSION SPECTRUM OF HYDROGEN**

Light emitted by an excited gas is split into its component wavelengths as it passes through a spectroscope. Why is this emission spectrum different from that of other elements?



## Bohr's Model of Electron Arrangement

Bohr proposed that the light-related behavior of atoms was due to electrons traveling at specific **energy levels** in fixed paths around the nucleus. Electrons at each energy level have a fixed amount of energy. Electrons remain in their energy levels unless something causes them to move to a different energy level.

Bohr showed that when an electron moves from a higher energy level to a lower one, it emits energy. The specific amount is equal to the difference in energy between the two levels. This energy is emitted in the form of light. The color or wavelength of the light emitted is related to its amount of energy.

Conversely, for an electron to move from a lower energy level to a higher one, it must absorb energy. The amount it absorbs is equal to the energy difference between the levels. The wavelength that is absorbed is related to this amount of energy.

Bohr called the specific amount of energy emitted or absorbed by electrons moving between energy levels a quantum of energy.

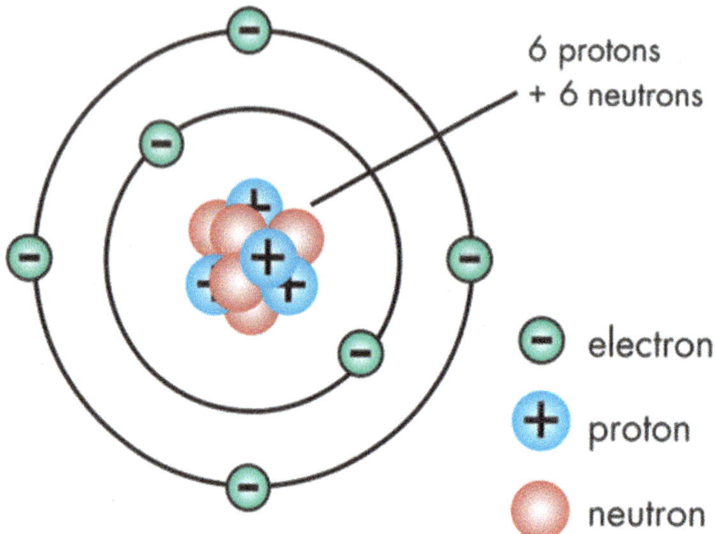


**CARBON ATOM**
In the Bohr model, electrons travel in orbits around the nucleus. What do the different levels of the orbits mean?

## Bohr's Calculations

Bohr used mathematics to show how his model explained the emission spectrum of hydrogen. He calculated the energy of a hydrogen electron using the equation below.

$$E = (-2.178 \times 10^{-18}\ \text{J})\left(\frac{1}{n^2}\right)$$

Here, $E$ is the energy of an electron and $n$ is an integer that corresponds to the energy level where the electron is located.

When Bohr applied this equation to determine the energy levels in the hydrogen atom, he found that the values matched those predicted by hydrogen's emission spectrum. This meant that his model could explain the observed behavior of hydrogen.

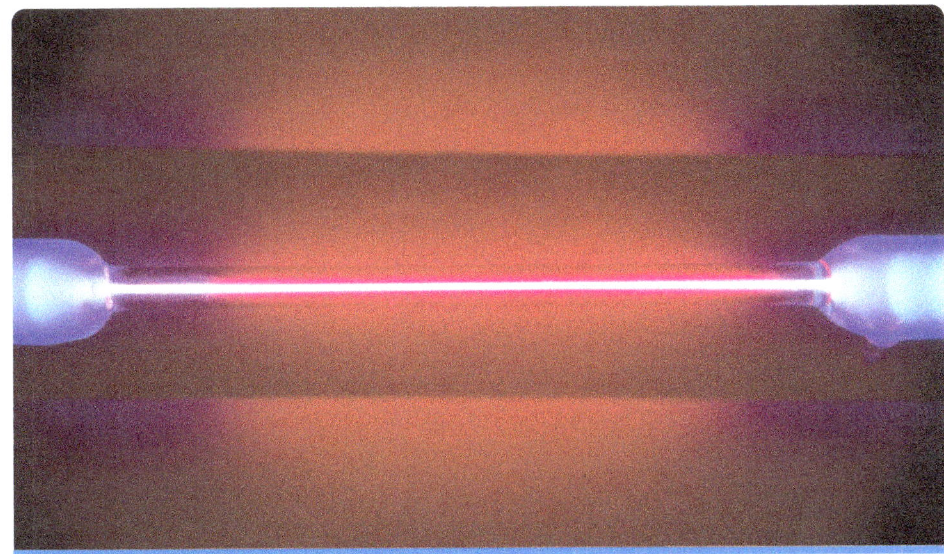

**HYDROGEN EMISSION TUBE**

The hydrogen gas in this glass tube emits colored light when an electric current is passed through it. How does the color of the light relate to the energy level of electrons in the hydrogen gas?



## Bohr's Calculations: Sample Problem

Calculate the energy of an electron in the fourth energy level ($n = 4$) of hydrogen.

**Solution:**

The equation relating energy and energy level is

$$E = (-2.178 \times 10^{-18}\ \text{J})\left(\frac{1}{n^2}\right)$$

To calculate $E$, substitute the value of 4 for $n$ and solve:

$$E = (-2.178 \times 10^{-18}\ \text{J})\left(\frac{1}{n^2}\right)$$

$$E = (-2.178 \times 10^{-18}\ \text{J})\left(\frac{1}{4^2}\right)$$

$$E = -1.361 \times 10^{-19}\ \text{J}$$

Note that the energy has a negative sign associated with it. This is because $E$ approaches zero as $n$ gets larger. Thus, an electron far away from the nucleus has zero energy. Its energy becomes more negative as the electron moves toward the nucleus.

### The Limitations of Bohr's Model

Bohr's model fit only data related to hydrogen atoms. It could not be used to calculate energy levels in other elements. This shortcoming prompted other physicists to continue unraveling the mysteries of atomic structure.

Although the Bohr model was useful only for the hydrogen atom, it marked an important milestone in atomic physics. The Bohr model made it clear that electron shells in atoms were quantized, or related to specific amounts (or quantities) of energy. The emission and absorption of light by atoms could be explained by assuming that electrons gain or lose specific quantities of energy when they move from one energy level to another.

## How Did Quantum Theory Help Establish a Relationship between the Energy of an Electron and Its Position in an Atom?

### From Bohr's Model to Quantum Theory

Bohr's model of **electron** arrangements in atoms had serious flaws. It applied only to one element, hydrogen. It also incorrectly described electrons as moving in fixed circular orbits in an atom.

Although Bohr's work showed that electrons had discrete energies, each electron's movement could not be easily defined. A scientist named Werner Heisenberg discovered that it was not possible to describe both the velocity and location of an electron at the same time. This concept became known as the Heisenberg uncertainty principle.

According to the Heisenberg uncertainty principle, electrons cannot be described by precisely defined movement. Instead, it is only possible to describe regions surrounding the atom where that electron is most likely to be found.

A new theory of atomic structure began to develop as a replacement for the Bohr model. This theory is called quantum theory or **quantum mechanics**.

In this theory, electrons occupy discrete **energy levels** as Bohr proposed. However, instead of traveling in fixed orbits around the nucleus, electrons are found around the nucleus in regions of space called atomic orbitals. An orbital's size and shape is defined by the probability of finding an electron in it. Thus, electrons are not pinpointed to exact locations. Instead, they have only a probability of being found in certain regions. The term *electron cloud* is often used to refer to such probabilities.

## Quantum Numbers

Quantum theory uses quantum numbers to describe the electrons in an atom. There are four quantum numbers.

| Quantum Number | Letter Designation | Description |
|---|---|---|
| principal quantum number | $n$ | energy level |
| azimuthal quantum number | $\ell$ | orbital shape |
| magnetic orbital quantum number | $m_l$ | orbital orientation in space |
| magnetic spin quantum number | $m_s$ | direction of spin |

**QUANTUM NUMBERS**

Each electron in an atom has a unique set of quantum numbers. What does each quantum number correspond to physically?

Each electron in an atom is assigned its own set of quantum numbers. The first three quantum numbers describe the probable location of the electron. The fourth **quantum number** describes the spin of the electron.

Taken together, the set of quantum numbers for any electron describes the amount of energy found in that electron.

## The Principal Quantum Number

The principal quantum number ($n$) indicates the energy level in which an electron is found. Values for $n$ are integers starting with one and continuing upward.

$$n = 1, 2, 3, 4, 5 \ldots$$

The principal quantum number corresponds to the periods on the periodic table. Each period from top to bottom on the periodic table represents a new energy level and an increase in principal quantum number.

### The Azimuthal Quantum Number

In general, the azimuthal quantum number ($\ell$) indicates the shape of the orbital in which an electron moves. You might sometimes see it called the angular momentum quantum number or the orbital quantum number. Values for $\ell$ are integers starting with one and continuing upward. Instead of using a number, a letter is used to represent the azimuthal quantum number:

$$\ell = 0, 1, 2, 3, 4 \ldots$$
$$\ell = s, p, d, f, g \ldots$$

The shape of an orbital depends on whether it is an *s*, *p*, *d*, *f*, or *g* orbital. In this model, *s* orbitals are spherical and *p* orbitals are shaped like dumbbells. The shapes become more complex for orbitals beyond *d*.

For any given electron, the value for the azimuthal quantum number is always dependent on the principal quantum number, $n$. The value of $\ell$ ranges from 0 to $(n - 1)$. The table below shows the distribution of $\ell$ values for some values of $n$.

**AZIMUTHAL QUANTUM NUMBER**
The azimuthal quantum number is bounded according to its principal quantum number. What relationships are there between the other quantum numbers?

| Principal Quantum Number ($n$) | Azimuthal Quantum Number ($\ell$) |
|---|---|
| 1 | 0 |
| 2 | 0 |
| | 1 |
| 3 | 0 |
| | 1 |
| | 2 |

### The Magnetic Orbital Quantum Number

The **magnetic orbital quantum number** ($m_l$) provides the orientation of the orbital shape along the *x*, *y*, and *z* axes. Values for $m_l$ are positive and negative integers that depend on the azimuthal quantum number. The value of $m_l$ ranges from $-\ell$ to $\ell$. The following table shows the distribution of $m_l$ values for each $\ell$ value shown in the previous table.

| Principal Quantum Number ($n$) | Azimuthal Quantum Number ($\ell$) | Magnetic Orbital Quantum Number ($m_l$) |
|---|---|---|
| 1 | 0 | 0 |
| 2 | 0 | 0 |
| | 1 | −1 |
| | | 0 |
| | | 1 |
| 3 | 0 | 0 |
| | 1 | −1 |
| | | 0 |
| | | 1 |
| | 2 | −2 |
| | | −1 |
| | | 0 |
| | | 1 |
| | | 2 |

**MAGNETIC ORBITAL QUANTUM NUMBER**

The magnetic orbital quantum number is bounded by the absolute value of its azimuthal quantum number. What relationships are there between the other quantum numbers?

## The Magnetic Spin Quantum Number

So far, three quantum numbers describe the orbital in which an electron is found. These numbers designate the energy level, shape, and orientation of an orbital. The three numbers do not completely define each electron in an atom. This is because each electron spins. An electron can spin in a clockwise direction or in a counterclockwise direction.

A maximum of two electrons can occupy one orbital. The physicist Wolfgang Pauli postulated that the two electrons spin in opposite directions. One electron spins clockwise and the other spins counterclockwise.

The **magnetic spin quantum number** ($m_s$) indicates the direction of spin of each electron. Here, there are only two possibilities for any given electron. Either the electron is spinning in a clockwise direction or it is spinning in a counterclockwise direction.

This condition leads to the only possible values for $m_s$: $-½$ and $+½$.

Thus, according to the Pauli exclusion principle, no two electrons in a given atom have the same set of quantum numbers.

# What Is Meant by the Dual Nature of Light?

## Using Photons to Describe Light

As quantum theory was developing, it became clear that nature appears to behave differently at the atomic level than it does at the macro level, which we experience in our daily lives. This conflict forced physicists to find new ways of describing natural phenomena, especially when classical descriptions and explanations did not agree with the models of quantum theory. The phenomenon of **light**, which is fundamental to the study of atomic structure, is one example.

By the early 1900s, scientists understood that light is a form of **electromagnetic radiation** that has no mass and travels in waves. This description served well because the behavior of light generally matched this description. However, there were two exceptions—blackbody radiation and the photoelectric effect. Neither of these phenomena was explained by the **wave** model of light generally accepted at the time.

Blackbody radiation occurs when a metal emits different colors of light depending on the temperature to which it is heated. The photoelectric effect results when a metal emits electrons when exposed to light at certain frequencies. The intensity of light does not alter the effect. Physicists in the 1900s were puzzled by these observations because they could not be explained by the classical description of light.

In 1905, Albert Einstein used quantum concepts to describe the behavior of light. He expanded the description of light as a wave, as traditionally understood. He also hypothesized that light behaves like a stream of particles, each made up of a packet of energy. He used the term **photon** to name these packets of light energy.

## The Dual Nature of Light

By combining the concept of quanta and Einstein's description of the particle behavior of light, physicists now describe light as having a dual nature. Light has properties and behaviors that make it act like both a particle and wave.

**BIOLUMINESCENT DINOFLAGELLATES**
Organisms that shine are said to be bioluminescent. These dinoflagellates emit light and make the ocean glow. What molecule could be involved?



# How Are Frequency and Wavelength of Electromagnetic Radiation Related?

## The Electromagnetic Spectrum

The electromagnetic spectrum is the full range of electromagnetic radiation, from lowest to highest energy. It includes visible light as well as invisible forms of radiation such as radio waves and ultraviolet radiation.

Energy and frequency are directly proportional to each other. At lower energies on the electromagnetic spectrum, light has a lower frequency. At higher energies, the frequency is higher.

Energy and wavelength are inversely proportional to each other. Thus, at lower energies, light has a longer wavelength. At higher energies, the wavelength is shorter.

Wavelengths and frequencies can vary greatly. Consider the following examples:

- Radio waves emitted by the antennas from broadcasting stations have wavelengths as long as a football field.
- Under certain conditions, elements emit light in or near the visible light range. This type of radiation has a wavelength about the size of a single bacterium.
- Gamma rays are emitted during nuclear changes, such as radioactive decay. This is an extremely high-energy form of radiation. Its wavelengths are shorter than the radius of a hydrogen atom.

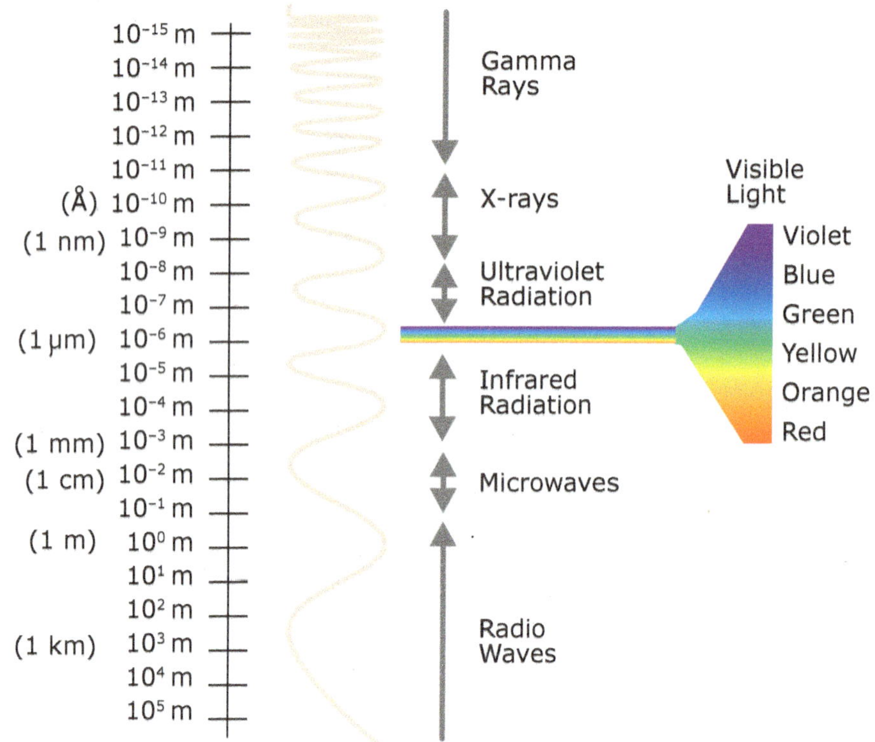


**ELECTROMAGNETIC SPECTRUM**

Energy decreases as wavelength increases in the electromagnetic spectrum. Do X-rays or microwaves have more energy?

## Wavelength and Frequency

As a form of **electromagnetic radiation**, **light** has properties that can be described as both waves and as particles. The dual nature of light is counterintuitive when viewed in terms of classical mechanics, but this duality is the best way to describe the behavior of light. In many situations, light's **wave**-like properties provide the most convenient description.

A wave has a repeating pattern of peaks and troughs. Two characteristics define a wave:

- A wave travels in a straight line, so an observer can count the number of peaks in the wave that pass by in a given time period. This characteristic is known as the wave **frequency** and is given the Greek symbol nu, $\nu$. The SI unit of **frequency** is the hertz (Hz). One hertz is one cycle per second, $s^{-1}$.
- The distance from peak to peak in any given wave is constant. This characteristic is known as the wavelength and is given the Greek symbol lambda, $\lambda$.

These two characteristics are inversely proportional. As frequency increases, wavelength decreases, and vice versa. The inverse of one is equal to the other multiplied by a constant:

$$\nu = c\left(\frac{1}{\lambda}\right)$$

where $c$ is the speed of light, $\nu$ is wave frequency, and $\lambda$ is wavelength.

When the above equation is rearranged to isolate $c$, we get the following equation:

$$c = \lambda\nu$$

The constant, $c$, is equal to the speed of light in a vacuum, which is the same for all wavelengths of light ($2.9979 \times 10^8$ m/s).

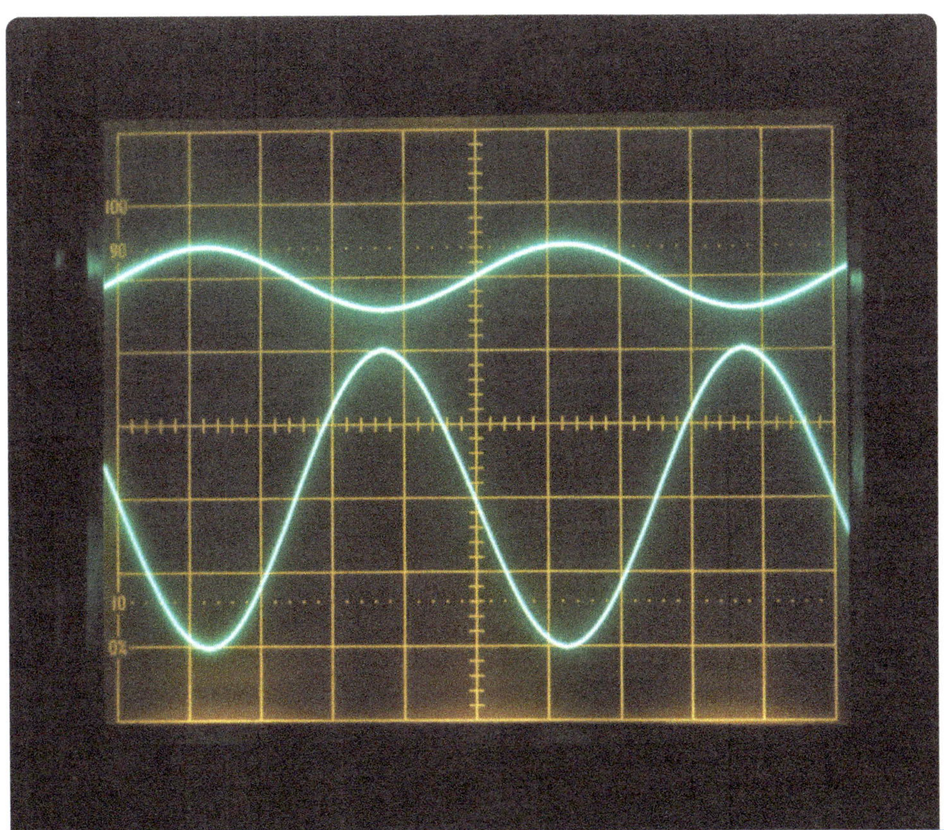

**WAVES**

Waves can take many forms. How are the wavelength, frequency, and energy of a wave related?

## Calculating the Frequency of Light: Sample Problem

Calculate the frequency of light that has a wavelength of $3.3 \times 10^{-7}$ m.

**Solution:**

Use the equation:

$$c = \lambda\nu$$

Solving for frequency ($\nu$), we get

$$\nu = \frac{c}{\lambda}$$

$$\nu = \frac{2.9976 \times 10^{8} \text{ m/s}}{3.3 \times 10^{-7} \text{ m}}$$

$$\nu = 9.1 \times 10^{14} \text{ s}^{-1}$$

## Calculating the Wavelength of Light: Sample Problem

Calculate the wavelength of light that has a frequency of $3.8 \times 10^{13}$ $s^{-1}$.

**Solution:**

Use the equation:

$$c = \lambda\nu$$

Solving for wavelength ($\lambda$), we get

$$\lambda = \frac{c}{\nu}$$

$$\lambda = \frac{2.9979 \times 10^{8} \text{ m/s}}{3.8 \times 10^{13} \text{ s}^{-1}}$$

$$\lambda = 7.9 \times 10^{-6} \text{ m}$$

## Planck's Wave Equation

Observations have shown that many elements can emit light under certain conditions. For example, neon gas emits red light when an electric current is passed through it. Likewise, xenon emits blue light.

Many elements emit light when brought to high temperatures. German physicist Max Planck found that there was a mathematical relationship between the amount of energy absorbed by an element and the frequency of light emitted by that element. The relationship is called Planck's wave equation and is shown below:

$$E = h\nu$$

where $E$ is the amount of energy absorbed, $\nu$ is the frequency of light emitted, and $h$ is a constant called Planck's constant.

This agrees well with the portion of Bohr's model that proposed the existence of discrete **energy levels** for electrons in an atom. With Planck's equation, we can calculate energy changes that occur during light **emission**. All that is needed is the frequency of the light that is emitted.

Since frequency is related to wavelength through the following equation, we can rewrite Planck's equation in terms of wavelength:

$$\nu = \frac{c}{\lambda}$$

$$E = h\nu = \frac{hc}{\lambda}$$

Note that frequency and energy are directly proportional to each other. Wavelength and energy are inversely proportional to each other.



## Planck's Wave Equation: Sample Problems

**Problem 1:** Calculate the energy of light with frequency $3.5 \times 10^{13}\ s^{-1}$. Planck's constant is $6.63 \times 10^{-34}\ J \cdot s$.

**Solution:**

The equation relating energy and frequency is Planck's law:

$$E = h\nu$$

where $E$ is the amount of energy absorbed, $\nu$ is the frequency of light emitted, and $h$ is Planck's constant.

To find energy, substitute the values for Planck's constant and frequency into the equation and solve:

$$E = h\nu$$

$$E = (6.63 \times 10^{-34} J \cdot s)(3.5 \times 10^{13} s^{-1})$$

$$E = 2.3 \times 10^{-20} J$$

**Problem 2:** Calculate the energy of light with wavelength of $5.7 \times 10^{-6}$ m.

**Solution:**

Use the frequency wavelength equation and substitute it into Planck's wave equation for frequency.

Rearrange the frequency-wavelength equation to solve for frequency:

$$c = \lambda\nu$$

$$\nu = \frac{c}{\lambda}$$

Substitute the value for frequency from the equation above into the energy equation:

$$E = h\nu$$

$$E = h\left(\frac{c}{\nu}\right)$$

$$E = 6.63 \times 10^{-34}\ \text{J} \cdot \text{s}\left(\frac{2.9979 \times 10^{8}\ \text{m/s}}{5.7 \times 10^{-6}\ \text{m}}\right)$$

$$E = 3.487 \times 10^{-20}\ \text{J} = 3.5 \times 10^{-20}\ \text{J}$$

**Consider the Explain Question**

**How is the arrangement of electrons in atoms related to their ability to emit light?**

dlc.com/ca9039s

Go online to complete the scientific explanation.

**Check Your Understanding**

**Go online to check your understanding of this concept's key ideas.**

dlc.com/ca9040s

# STEM in Action

## Applying the Arrangement of Electrons in the Atom

Fluorescence is a form of **light emission** that is found throughout the natural world. Some examples include pigments made by living organisms, minerals, and laundry detergents. The fluorescence that you are probably most familiar with is that of lights. You probably know that fluorescent lights in homes and stores emit white light, but did you know that the glow of neon signs is also a form of fluorescence?

The ideas developed by Bohr, Heisenberg, Einstein, and Planck can be used to explain fluorescence. In fluorescent lighting, a glass tube is filled with an inert gas such as neon. Light is produced by exciting electrons within the glass tube, usually by heating a filament or applying an electric charge. The electrons in the gas absorb this electromagnetic energy, and the excited electrons in the gas jump to a higher energy level. As the electrons return to their original energy level, they give off energy in the form of light.

**FLUORESCENT LIGHTS**
How do the arrangement of electrons in an atom relate to these fluorescent lights?

## STEM and Arrangement of Electrons in the Atom

Lasers are light sources. They produce light that is limited to one or just a few wavelengths. Lasers also amplify the light being produced. This means that the light beam is very intense. The amplification process takes advantage of the light absorption and emission properties of elements. A chain of absorption and emission events occurs inside the laser that ends in the final output of high-intensity light across a narrow band of wavelengths—or even just one wavelength. The photons emitted by the electrons in a laser as they return to their unexcited state are all at the same wavelength and are "coherent," meaning the crests and troughs of the light waves are all in lockstep. In contrast, ordinary visible light comprises multiple wavelengths and is not "coherent."

Technology applies lasers in many functions. Everyday examples include barcode scanning at store checkout stations, and scanning of compact discs (CDs) and digital video discs (DVDs). Because they can produce such high-intensity light, surgeons use lasers to cut soft tissues. Lasers are also used for cutting hard materials in manufacturing processes.

Forensic experts use lasers to help visualize human fingerprints. The laser uses electric current to excite atoms of argon. As the excited electrons return to their original state, they emit light of a specific wavelength. When directed onto a fingerprint, this light causes pigments in the fingerprint to fluoresce.

## Lasers in Flow Cytometry

A common use of lasers in biology and medicine is a technique called flow cytometry. Flow cytometry can be used to count cells in a fluid. A mixture of cells of different types, for example healthy and diseased blood cells, are suspended in a fluid. A fluorescent compound is selected that tags one group of these cells but not the other. This suspension of tagged and non-tagged cells are then forced down a narrow tube. A laser is shone on them. Those that are tagged fluoresce. A detector tuned to the wavelength of light given off when this fluorescence occurs is used to count the individual tagged cells. Another detector uses the laser light scattered from all the cells to count the total number of cells in the sample. By using multiple types of fluorescent tags and multiple detectors, each sensitive to different wavelengths, it is possible to scan for a wide range of cell characteristics at once.

Flow cytometry is used in the diagnosis of diseases such as blood cancers, the detection of antigens, and a wide variety of research.

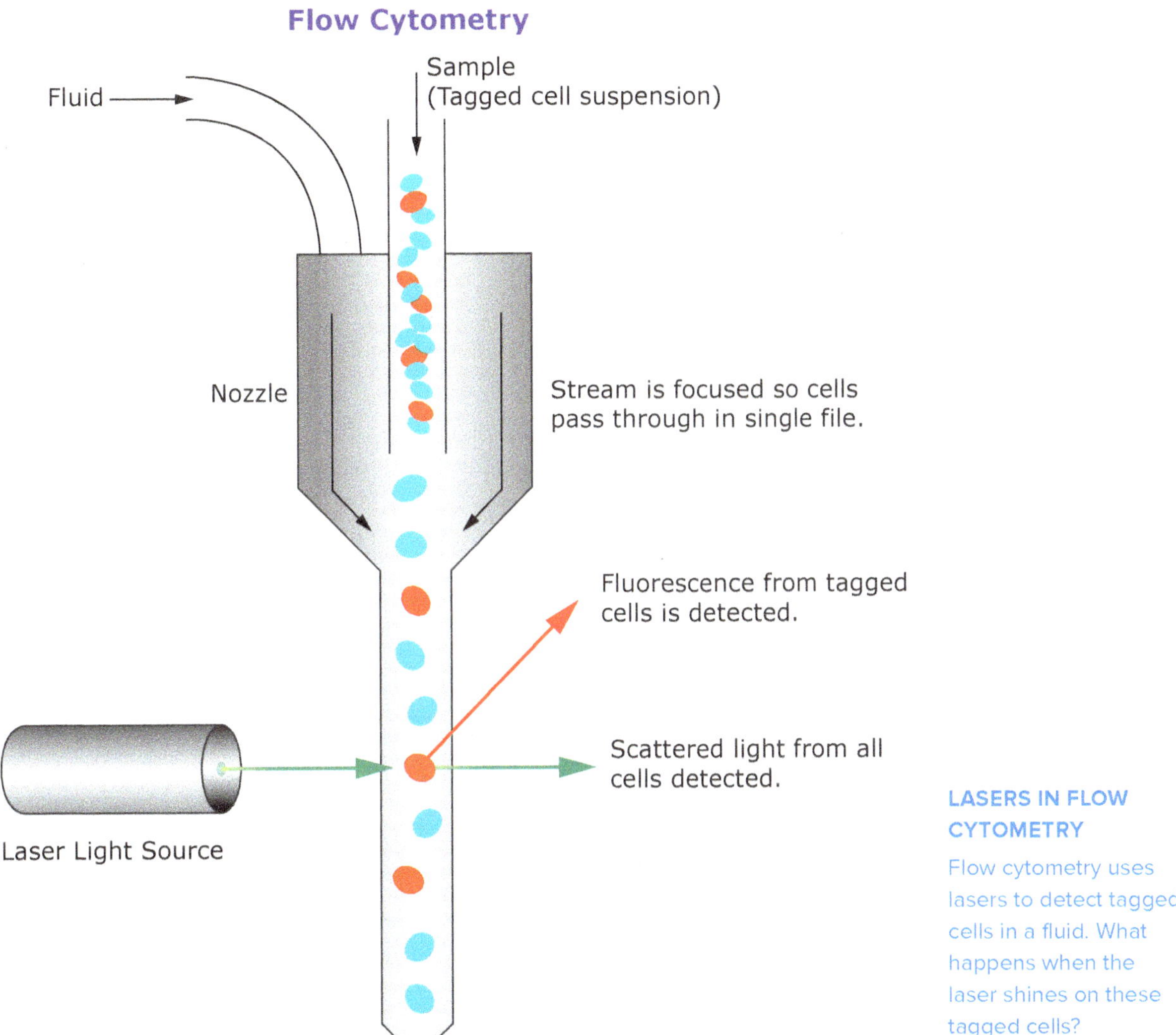
Fluid
Sample
(Tagged cell suspension)
Nozzle
Stream is focused so cells pass through in single file.
Fluorescence from tagged cells is detected.
Scattered light from all cells detected.
Laser Light Source

**Flow Cytometry**

**LASERS IN FLOW CYTOMETRY**

Flow cytometry uses lasers to detect tagged cells in a fluid. What happens when the laser shines on these tagged cells?

## Bohr's Calculations

**Use Bohr's equation to calculate the energy of an electron as it occupies different energy levels within the hydrogen atom.**

1. What is the energy of an electron in the first energy level of hydrogen?

2. What is the energy of an electron in the second energy level of hydrogen?

3. What is the energy of an electron in the third energy level of hydrogen?

4. What is the energy of an electron in the eighth energy level of hydrogen

5. Fill in the table below with your answers from above. How would you calculate the $\Delta E$ (change in energy) of an electron as it moves between any two levels? Give an example.

| Energy Level | Energy (J) |
|---|---|
| 1 | |
| 2 | |
| 3 | |
| 8 | |

# Light Wavelength, Frequency, and Energy

**Use the wavelength frequency equation or/and Planck's Wave Equation to solve the problems below.**

1. Calculate the wavelength of light that has a frequency of $2.2 \times 10^{12}\ s^{-1}$.

2. An ultraviolet ray of light has a wavelength of $1.5 \times 10^{-8}$ m. What is the ray's frequency?

3. Calculate the energy of light that has a frequency of $8.8 \times 10^{16}\ s^{-1}$.

4. An infrared beam of light has a wavelength of 7.7 millimeters. What is its energy?

# Planck's Wave Equation

**Use Planck's wave equation to solve the problems below. Some constants you may need include: $h$ = 6.63 $\times$ $10^{-34}$ Js, c = 3.00 $\times$ $10^{8}$ m/s**

1. A line in the emission spectrum produced by a sample of excited lithium atoms has a frequency of 4.47 $\times$ $10^{14}$ $s^{-1}$. What is the energy associated with this emission line?

2. Calculate the frequency of radio waves having a wavelength of 222 m.

3. What is the wavelength associated with light with frequency of 6.28 $\times$ $10^{14}$ $s^{-1}$? Express the result in nanometers.

4. Photons corresponding to radiation in the microwave range have an approximate energy of 8.34 $\times$ $10^{-24}$ J. What is the wavelength in millimeters that corresponds to these photons?

CONCEPT 2.3

# Electron Representations



## LESSON OVERVIEW

dlc.com/ca9041s

### Lesson Questions

- What is the configuration of electrons within an atom?
- How is an orbital diagram created?
- How are valence electrons represented within a Lewis structure?
- How does the octet rule describe the arrangement of electrons in noble gases?

### Lesson Objectives

By the end of the lesson, you should be able to:

- Explain the configuration of electrons in an atom.
- Demonstrate how to use orbital diagrams.
- Use Lewis structures to illustrate the valence electrons in atoms.
- Explain the stability of the noble gases based on the octet rule.

### Key Vocabulary

Which terms do you already know?

- ☐ atomic radius
- ☐ chemical symbol
- ☐ coefficient
- ☐ electron configuration
- ☐ inert gas
- ☐ Lewis (electron) dot structure
- ☐ magnetic orbital quantum number
- ☐ magnetic spin quantum number
- ☐ noble gas
- ☐ noble gas configuration
- ☐ octet rule
- ☐ orbital diagram
- ☐ Pauli Exclusion Principle
- ☐ periodic table
- ☐ principal quantum number (*n*)
- ☐ quantum number
- ☐ valence electron

## Thinking about Electron Representations

**Why are some roses bright red, while other roses are yellow or a soft pink color? What makes some lilacs have a deep purple color, while others are pale lavender? The answers to these questions lie in the elements that make up each object.**

dlc.com/ca9042s



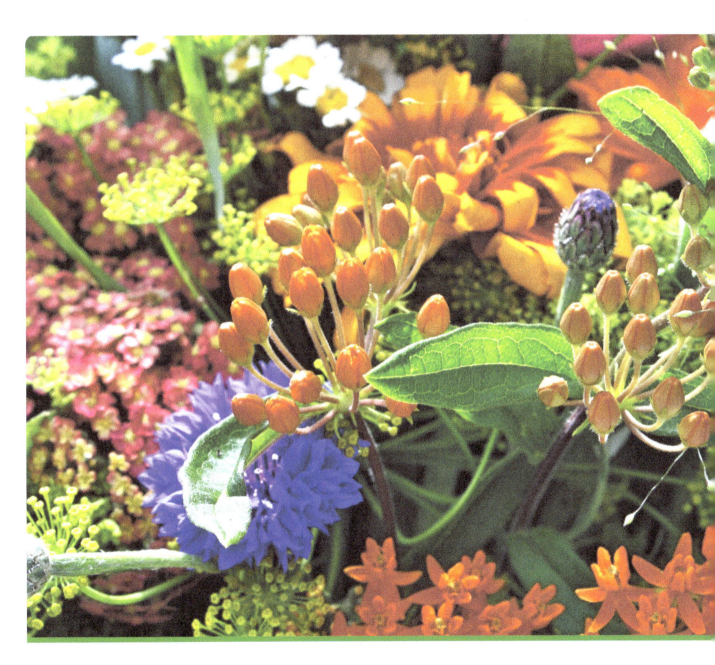

**EXPLAIN QUESTION**

**How are electron arrangements described by Lewis structures, orbital diagrams, and electron configurations?**

**COLOR**

What role do electrons play in determining the color of these flowers?

# What Is the Configuration of Electrons within an Atom?

## Electron Configuration Notation

To study the properties of an element, scientists consider where the electrons are located within an atom. Quantum theory uses specific rules to describe the locations of the electrons. Scientists must know how to write electron configurations to apply these rules. They use a shorthand method to describe the configurations, using a system of quantum numbers. The system uses the principal **quantum number** (*n*) and the azimuthal quantum number ($\ell$) to describe the configuration of electrons in an atom.

For example, a helium atom has two electrons orbiting its nucleus. Helium has the following **electron configuration**:

$$1s^2$$

This system of notation includes three pieces of information. The first is the energy level, or the principal quantum number of the electrons. In the case of helium, the atom's two electrons occupy the energy level nearest the nucleus. The numeral 1 represents this energy level.

The second piece of information is the orbital type. Think of the orbital as the pattern of electron density surrounding an atom's nucleus. The letters *s*, *p*, *d*, and *f* represent different types of orbitals. In a helium atom, the electrons are most likely to occupy the *s* orbital.

Finally, the number of electrons in each orbital is shown by a superscript after each orbital-type letter. A helium atom contains two electrons in the 1*s* orbital.

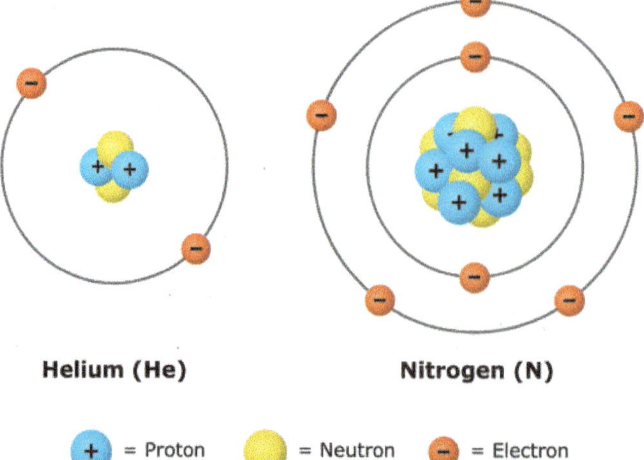


**HELIUM AND NITROGEN**

Helium has two electrons that orbit its nucleus in the 1s shell. How does helium differ from nitrogen?

The following list shows the electron configurations of a selection of elements:

| | |
|---|---|
| H | $1s^1$ |
| He | $1s^2$ |
| Li | $1s^22s^1$ |
| B | $1s^22s^22p^1$ |
| N | $1s^22s^22p^3$ |
| Ne | $1s^22s^22p^6$ |
| Na | $1s^22s^22p^63s^1$ |

## Writing Electron Configurations

Each orbital in an atom can hold up to two electrons. According to the **Pauli exclusion principle**, no two electrons can have exactly the same quantum number. In other words, no two electrons can be in the same place at the same time. Two electrons can occupy the same orbital, but each electron is assigned a "spin" so that it can be identified individually.

In electron configuration notation, the first electron added to an orbital is denoted with the up arrow (↑), and the second has a down arrow (↓). The direction of the arrows reflects the spin of each electron. The spins of the two electrons in an orbital are always opposite one another.

Some other basic rules apply to electron configuration notation. First, as the atomic number increases, the number of orbitals required for the electrons increases. Also, all the orbitals of one type (*s*, *p*, *d*, or *f*) within an energy level (1, 2, 3, and so on) must be completely filled before an electron can occupy the next highest orbital. Finally, Hund's rule states that all the orbitals of the same energy will be filled with one electron before any of the orbitals take on a second electron. This rule means that electrons in nitrogen, for example, are placed in each of the three *p* orbitals before any two electrons can share an orbital.

**ELECTON CONFIGURATIONS**

In electron configuration notation, the first electron added to an orbital is represented by the up arrow. Why is the next electron in the orbital shown as a down arrow?

| | 1s | 2s | $2p_x$ | $2p_y$ | $2p_z$ | 3s | Electron configuration |
|---|---|---|---|---|---|---|---|
| H | ↑ | | | | | | $1s^1$ |
| He | ↑↓ | | | | | | $1s^2$ |
| Li | ↑↓ | ↑ | | | | | $1s^22s^1$ |
| B | ↑↓ | ↑↓ | ↑ | | | | $1s^22s^22p^1$ |
| N | ↑↓ | ↑↓ | ↑ | ↑ | ↑ | | $1s^22s^22p^3$ |
| Ne | ↑↓ | ↑↓ | ↑↓ | ↑↓ | ↑↓ | | $1s^22s^22p^6$ |
| Na | ↑↓ | ↑↓ | ↑↓ | ↑↓ | ↑↓ | ↑ | $1s^22s^22p^63s^1$ |

## Using the Periodic Table

The electron configurations of elements closely follow trends in the periods of the **periodic table**. Elements within the same period, group, or region have similar electron configurations.

Across a period, for instance, the number of valence electrons in the last sublevel increases as each orbital is filled. In Period 2, the number of electrons in the 2*p* orbitals increases across the period.

B $1s^22s^22p^1$

C $1s^22s^22p^2$

N $1s^22s^22p^3$

O $1s^22s^22p^4$

F $1s^22s^22p^5$

Ne $1s^22s^22p^6$

Na $1s^22s^22p^23s^1$

The electron configuration shows a pattern within groups, too. The number of principal energy levels used increases as you go down through a group. The electron configurations of Group 1 elements always end with one electron in the highest energy orbital.

| | |
|---|---|
| H | $1s^1$ |
| Li | $1s^2 2s^1$ |
| Na | $1s^2 2s^2 2p^6 3s^1$ |
| K | $1s^2 2s^2 2p^6 3s^2 3p^6 4s^1$ |

The standard way to organize the periodic table is in blocks that signify the types of orbitals occupied by the atoms' valence electrons. The elements in Group 1 and 2 are often called the "*s* block." Their electron configurations end with one or two electrons occupying the highest energy *s* orbital available. The "*p* block" consists of elements that have unfilled *p* orbitals. The "*d* block" has elements with unfilled *d* orbitals. Elements in a given block have electrons in some of the orbitals of that type, but the orbitals are not necessarily filled.



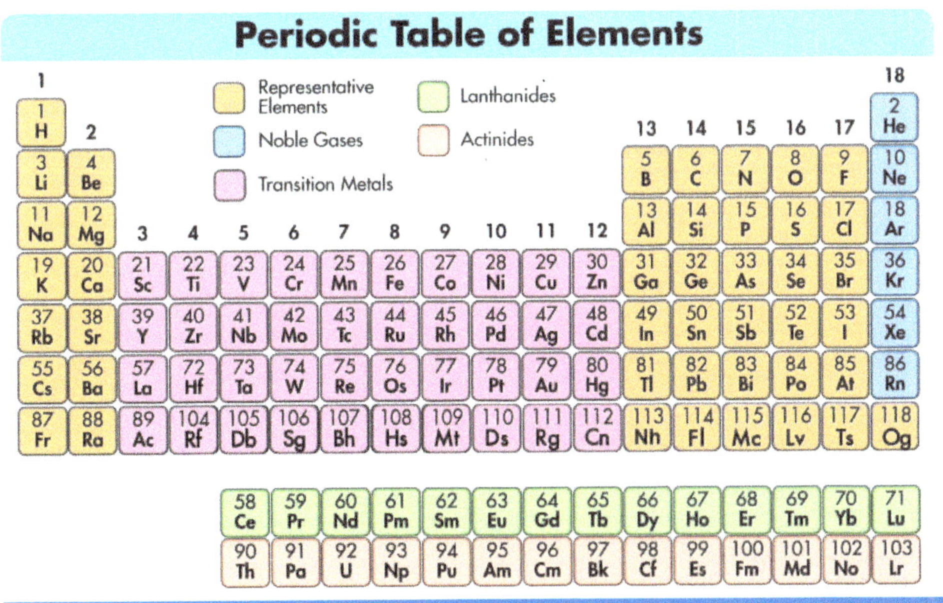


**THE PERIODIC TABLE OF THE ELEMENTS**

Elements in the periodic table follow trends. How do patterns in the arrangement of elements relate to the electron configurations of their atoms?

## Noble Gas Notation

The electron configuration of elements that have high atomic numbers becomes complex. There is a shortcut for these elements to make their electron configurations easier to write. This method is called **noble gas** notation.

In noble gases, electrons completely fill the orbitals at a particular energy level. For instance, neon's electron configuration is $1s^22s^22p^6$. Each of the four orbitals in the second energy level contains two electrons. Sodium, which has one more proton, has the same electron configuration as neon plus an added electron in the lowest orbital of the third energy level (3*s*). All of the elements in Period 3, in fact, have the same electron configuration as neon in the $n = 1$ and $n = 2$ energy levels. Always having to write "$1s^22s^22p^6$" for these elements would be tiresome. To simplify notation, the "base" electron configuration can be shown by writing the symbol for neon, Ne, with square brackets. The notation describing the electrons in the third energy level is simply added to the "base." For example, the noble gas notation for sodium's electron configuration is:

$$[\text{Ne}]3s^1$$

Noble gas notation allows the electron configuration of any element with a large atomic number to be written in shorthand. It is as simple as writing the bracketed atomic symbol of the noble gas followed by the additional orbitals that are filled.

| Atom | Electron configuration | Previous noble gas | Valence configuration | Noble gas notation |
|---|---|---|---|---|
| Na | $1s^22s^22p^6$ $3s^1$ | Ne | $3s^1$ | $[\text{Ne}]3s^1$ |
| Cl | $1s^22s^22p^6$ $3s^23s^5$ | Ne | $3p^5$ | [Ne] $3s^23p^5$ |
| Ca | $1s^22s^22p^6$ $3s^23s^64s^2$ | Ar | $4s^2$ | $[\text{Ar}]4s^2$ |

**NOBLE GAS NOTATION**

Noble gas notation simplifies electron configurations. What is the relationship between noble gas notation and electron configuration?

# How Is an Orbital Diagram Created?

## Visualizing Electron Configurations

When creating electron configurations, consider that the 1*s* orbital is always filled with electrons before the 2*s* orbital, which is always filled before the 2*p* orbital. This sequence is followed because the 1*s* orbital has less energy than the 2*s* orbital, which has less energy than the 2*p* orbital, and so on. In atoms that have electrons in more than two energy levels, the order in which orbitals are filled is more complex. In these cases, an **orbital diagram** is helpful.

An orbital diagram shows the orbitals used in a particular atom and the orbitals' relative energies. Usually, empty squares are used to indicate the orbitals. These squares are arranged to determine which orbitals are at higher energy levels than others.

The closer an orbital is to the atomic nucleus, the lower its energy. The 1*s* orbital is closest to the nucleus and has the lowest energy. This energy level is always the first to be filled.

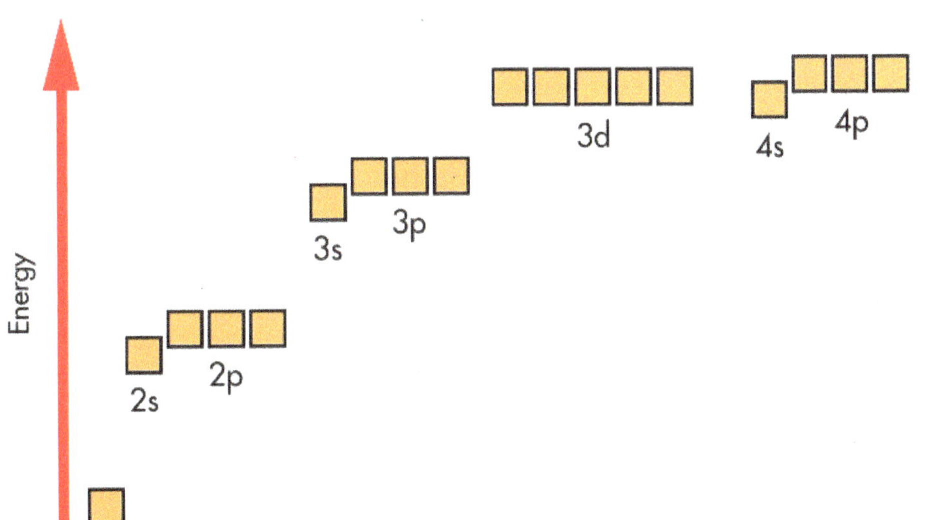


**ORBITAL DIAGRAM**

When diagramming an atom's electron configuration, orbitals are filled in order of their energies. Are higher or lower energy electrons nearest the nucleus of the atom?

Within an energy level, the higher the azimuthal **quantum number**, the more energy in the orbital. The 2*p* orbitals (azimuthal number 1) are slightly higher in energy than the 2*s* orbitals (azimuthal number 0).

The differences between the energy levels generally decrease further from the atomic nucleus. The energy gap between the 1*s* and 2*s* orbitals, for example, is larger than the gap between the 2*s* and 3*s* orbitals. In some cases, the orbitals in an energy level further from the nucleus have lower energy than those in the next energy level down. For example, the 4*s* orbital has slightly lower energy than the 3*d* orbitals. In the case of potassium, the **electron configuration** is [Ar]$4s^1$. In an orbital diagram for potassium, electrons are placed in the 4*s* orbital before they are placed in the 3*d* orbitals.

## Filling an Orbital Diagram

Once the relationship between energy levels and electron configurations is clear, electrons can be added to the orbital diagram.

When filling an orbital diagram, remember the following:

1. Fill the lowest energy orbital that is available.
2. Place one arrow in each orbital of the same energy level before putting pairs of arrows in any orbital at that level. All the single electrons should be shown as having the same spin. In other words, all of the single arrows representing electrons in different orbitals at the same energy level should point up.
3. If two electrons share an orbital, they must have opposite spins. No more than two electrons can share an orbital.

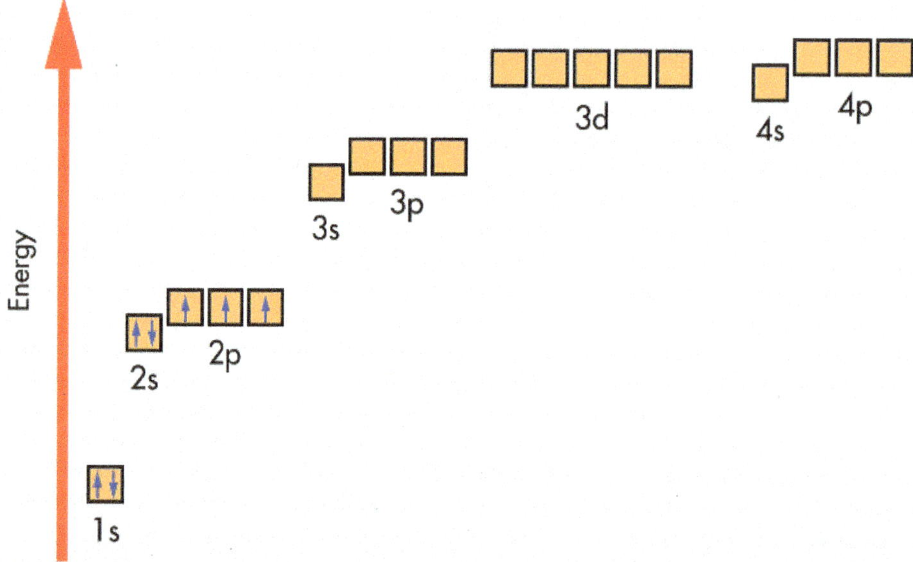


**COMPLETED ORBITAL DIAGRAM**

The completed orbital diagram for nitrogen shows pairs of electrons with opposite spin in the lower energy levels. Why are electrons paired with opposite spin?

# How Are Valence Electrons Represented within a Lewis Structure?

## Lewis Structures

In most elements, electrons in the outermost energy level of an atom are valence electrons. They reside in the highest occupied energy level of the atom. Many properties of an element are related to these valence electrons. Such properties include how the elements form ionic and covalent bonds.

The number of valence electrons in an atom can be determined from the **electron configuration** of the element. The total number of electrons at the highest energy level dictates the number of valence electrons. For example, carbon (C) has four electrons in its outermost energy level: two in the 2*s* orbital and two in the 2*p* orbital. Carbon, therefore, has four valence electrons.

In 1916, Gilbert Lewis developed an easy method for representing atoms and their valence electrons. These simple diagrams are called Lewis structures or electron dot structures. A Lewis structure includes chemical symbols for the elements, surrounded by dots that represent the valence electrons.

Lewis structures follow trends in the **periodic table**. Moving across a period, a dot is added for each additional electron in an atom of the element. The number and pattern of dots is the same moving down a group, making Lewis structures convenient representations of both the elements and their valence electrons.

To draw a Lewis structure, begin by writing the atomic symbol of the element. Next, identify the number of valence electrons. The number of valence electrons determines how many dots to use in the structure. Then, place the dots around the **chemical symbol**, making sure that only one dot is placed on each side before any are paired up.

| 1 | 2 | 13 | 14 | 15 | 16 | 17 | 18 |
|---|---|---|---|---|---|---|---|
| H | | | | | | | He |
| Li | Be | B | C | N | O | F | Ne |
| Na | Mg | Al | Si | P | S | Cl | Ar |
| K | Ca | | | | Se | Br | Kr |
| Rb | Sr | | | | Te | I | Xe |
| Cs | Ba | | | | | | |

**LEWIS STRUCTURES IN PERIODIC TABLE**

This cropped version of the periodic table shows the dot structure of some representative elements. Why does helium have its two electron dots on the same side?

Lewis structures can also be used to represent molecules. In molecules, valence electrons involved in covalent bonds are shown as lines. Valence electrons that are not involved in bonding, however, are still represented as dots. These dots are often referred to as lone pairs of electrons, to differentiate them from bonding electrons. The Lewis structure of a water molecule ($H_2O$) is:

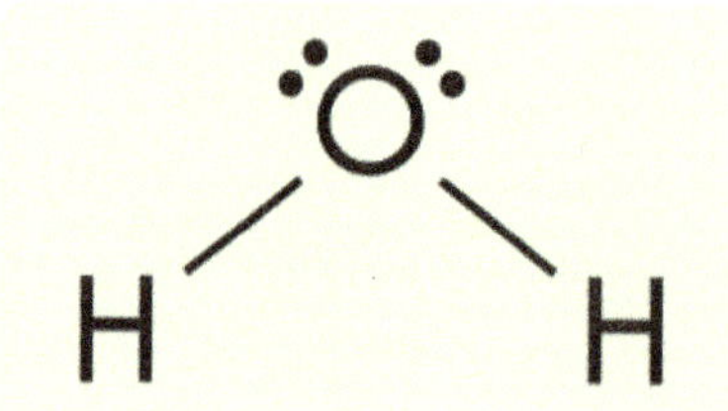


**WATER MOLECULE**

This illustration shows the Lewis structure of a water molecule. What do the line segments represent?

Each hydrogen atom has one **valence electron** and the oxygen atom has six valence electrons. Each of hydrogen's valence electrons forms a covalent bond with oxygen. These bonds are shown as solid lines. Oxygen's remaining four valence electrons are represented with two sets of dots placed on the other two sides of the chemical symbol.

| | Valence electrons | Electron configuration |
|---|---|---|
| H | 1 | $1s^1$ |
| Mg | 2 | $1s^22s^2$ |
| C | 4 | $1s^22s^22p^2$ |
| F | 7 | $1s^22s^22p^5$ |
| Na | 1 | $1s^22s^22p^63s^1$ |

**VALENCE ELECTORONS AND ELECTRON CONFIGURATION**

An element's electron configuration can be used to determine the number of valence electrons in an atom. Which energy level are valence electrons located in?



## How Does the Octet Rule Describe the Arrangement of Electrons in Noble Gases?

### The Octet Rule

Gilbert Lewis studied the electron configurations of the various elements in the **periodic table** and derived the **octet rule**. The octet rule states that atoms react with one another in ways that make their electron configurations the same as that of a **noble gas**. In a noble gas, orbitals within an energy level are completely filled. In other elements, filling an energy level stabilizes the arrangement of the electrons.

Consider the halogens, the elements in Group 17. They have seven valence electrons and are one short of the **noble gas configuration**. Halogens readily form ions by acquiring an electron from other atoms or molecules. The ions formed in this way, known as halides, have a charge of –1. This charge is indicated by a superscript minus sign next to an atomic symbol, such as $F^-$.

# Periodic Table of Elements

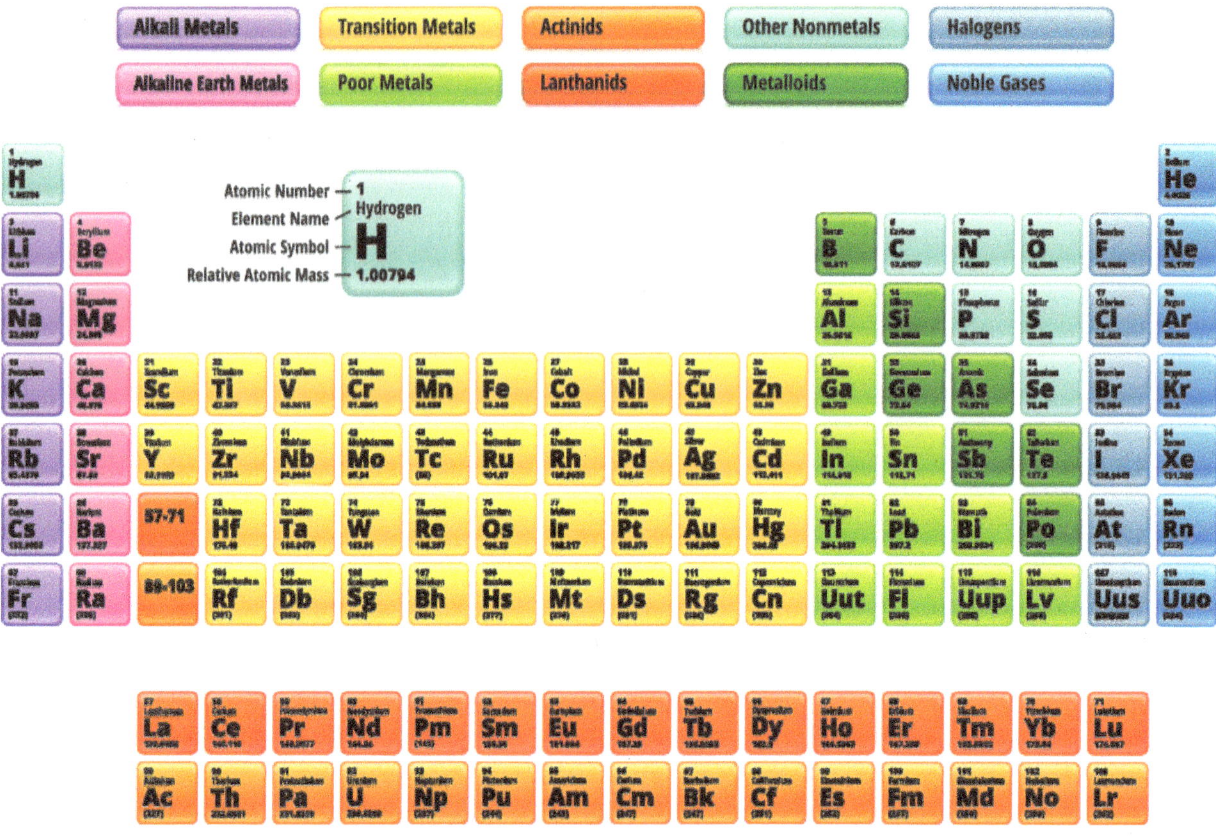


**PERIODIC TABLE OF ELEMENTS**

Groups such as the alkali metals, halogens, and noble gases have similar characteristics. Which characteristic is used when predicting the ion that will form from a neutral atom?

Similarly, the Group 1 alkali metals have one **valence electron**. If an atom of sodium loses its valence electron, it forms a sodium ion, $Na^+$, with a +1 charge. The sodium ion has the **electron configuration** of the noble gas neon.

The outer shell electron configurations of $F^-$ and $Na^+$ are the same as Ne. Together, these two ions can form the ionic molecule sodium fluoride (NaF). Each of the atoms has the same outer shell electron configuration, but their nuclei have different numbers of protons.

The octet rule is a powerful tool for predicting the ions that will form from neutral atoms. It is very consistent for elements in the *s* and *p* blocks.

Noble gases have completely filled valence shells. This configuration makes them chemically stable, which is why they are also known as inert gases. These elements are inert because they very rarely react with other elements.

To form a compound, atoms have to gain, lose, or share valence electrons. For noble gases, gaining or losing an electron would violate the octet rule. Some noble gases will share their electrons; however, such compounds are unstable, so this occurs very rarely and only at extremely low temperatures.

| | 1s | 2s | $2p_x$ | $2p_y$ | $2p_z$ | 3s | Electron configuration |
|---|---|---|---|---|---|---|---|
| F | ↑↓ | ↑↓ | ↑↓ | ↑↓ | ↑ | | $1s^2 2s^2 2p^5$ |
| $F^-$ | ↑↓ | ↑↓ | ↑↓ | ↑↓ | ↑↓ | | $1s^2 2s^2 2p^6$ |
| Ne | ↑↓ | ↑↓ | ↑↓ | ↑↓ | ↑↓ | | $1s^2 2s^2 2p^6$ |
| $Na^+$ | ↑↓ | ↑↓ | ↑↓ | ↑↓ | ↑↓ | | $1s^2 2s^2 2p^6$ |
| Na | ↑↓ | ↑↓ | ↑↓ | ↑↓ | ↑↓ | ↑ | $1s^2 2s^2 2p^6 3s^1$ |

**THE OCTET RULE**

The octet rule says that atoms react in a manner that will give them electron configurations like those of the noble gases. Why do atoms want to achieve a full octet?



## Consider the Explain Question

**How are electron arrangements described by Lewis structures, orbital diagrams, and electron configurations?**

dlc.com/ca9043s

Go online to complete the scientific explanation.

## Check Your Understanding

**Can you model what happens to electrons during the formation of covalent bonds?**

dlc.com/ca9044s

## Applying Electron Representations

Orbital diagrams represent orbitals that are occupied by electrons in an atom. How are the orbitals represented when atoms form covalent bonds in molecules? The electrons in these molecules are shared between atoms. Orbital diagrams cannot easily depict these bonded electrons. Another different kind of **orbital diagram** and **electron configuration** is used to depict the outer shell electrons in the central atom of the molecule.

When atoms, like carbon, have electrons in different orbitals in the same energy level, the orbitals combine to form equal energy orbitals when bonding to another atom. This process is hybridization. The most energetically favorable arrangement of the molecule determines which orbitals hybridize.

Covalent bonds form when atomic orbitals overlap. These shared, overlapping orbitals allow for the electrons to be located around either nucleus in the bond at any given time.

Carbon nanotubes are one allotropic form of carbon that has a cylindrical shape. Their applications are vast and include uses in medical and dental fields, batteries, electronics, and fuel cells. Their size and shape make them ideal candidates for engineers to use in their fields.



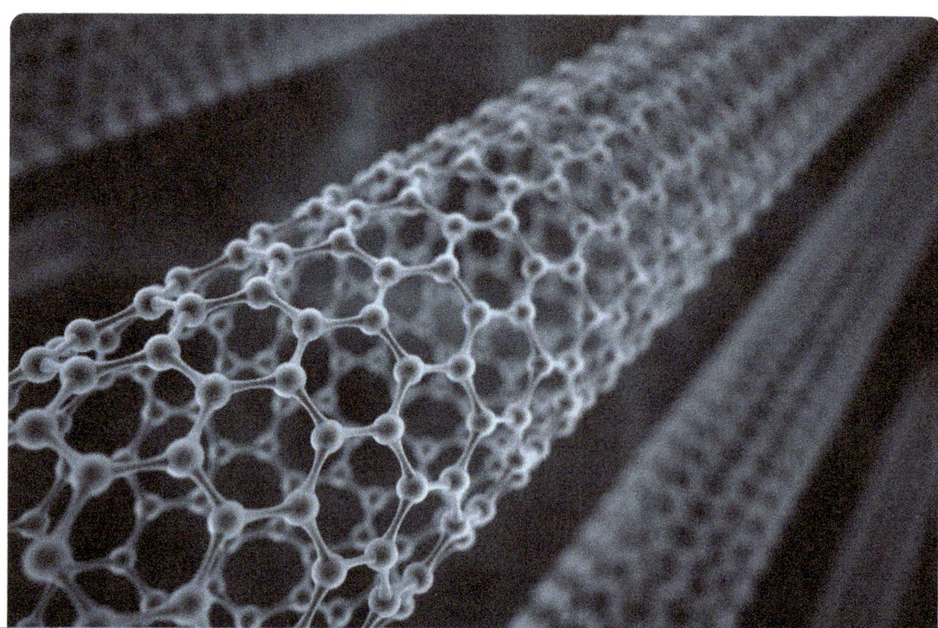

**CARBON NANOTUBES**

Carbon nanotubes consist of carbon atoms covalently bonded to form a tube-like shape. How does the bonding in nanotubes make this form of carbon so strong?

Orbital hybridization best describes the chemical bonding in nanotubes. They are made up entirely of $sp^2$ bonds, similar to those of graphite. These bonds, which are stronger than the $sp^3$ bonds found in diamonds, provide nanotubes with their unique strength.

## STEM and Electron Representations

Certain materials can act as conductors, semiconductors, or insulators. A materials scientist is someone who is interested in understanding how a material is processed, its structure, and its properties and performance. Material scientists have developed a theory, called band theory, to explain conductors, semiconductors, and insulators.

Many metals are conductors. As solids, they form a lattice structure of atoms. The orbitals of these atoms are so close together that they overlap. The overlapping makes it easier for valence electrons to move to higher energy levels, which means that they are able to move around more freely. The band gap is the energy difference between the ground state and excited states of these valence electrons. Molecular orbital diagrams, which describe the electronic energy levels of individual molecules, are used to represent this theory. They can be useful in determining the energy differences.

An electrical current can be thought of as adding an electron to one end of a metal and removing an electron from the other. The ease with which the electrons flow is related to the band gap. The rule of thumb is the smaller the band gap, the better the conductor. Copper has a small band gap. It is one of the best conductors, which is why it is widely used in electrical devices.

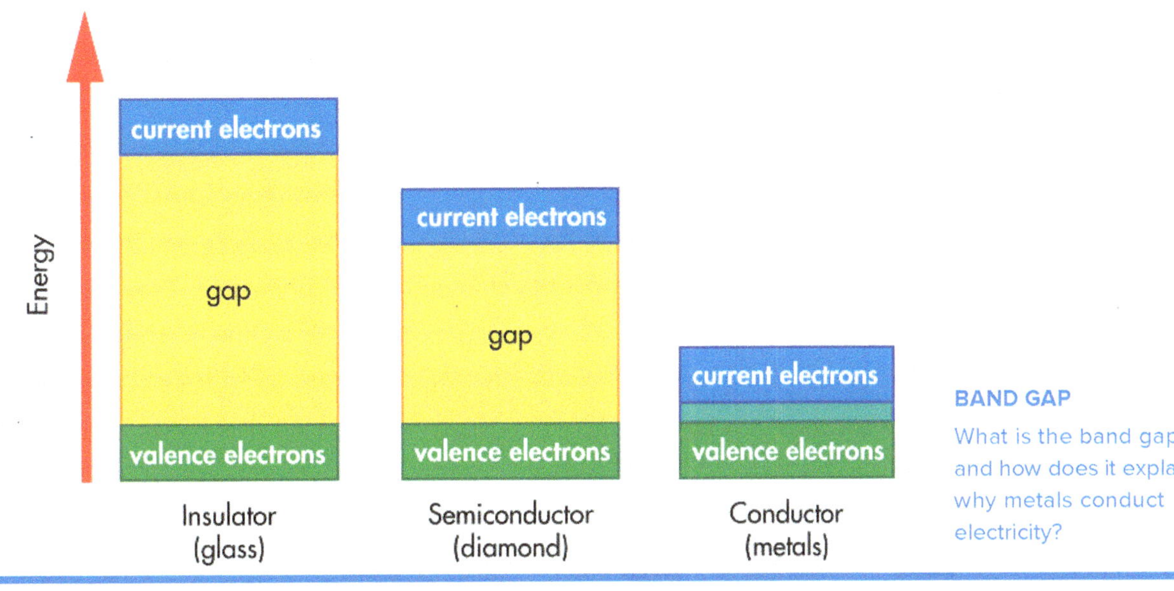


**BAND GAP**

What is the band gap, and how does it explain why metals conduct electricity?

CONCEPT
2.4

# Structure of the Periodic Table

dlc.com/ca9045s

## LESSON OVERVIEW

### Lesson Questions

- How do scientists distinguish between metals, metalloids, and nonmetals?
- How are the columns of the periodic table organized, and what do the elements in each group have in common?

### Lesson Objectives

By the end of the lesson, you should be able to:

- Distinguish the structures and properties of metals, metalloids, and nonmetals.
- Interpret the columns of the periodic table as groups or families that contain elements with similar properties.

### Key Vocabulary

Which terms do you already know?

- ☐ actinide series
- ☐ alkali metals
- ☐ alkaline earth metals
- ☐ atomic radius
- ☐ conductor
- ☐ ductility
- ☐ electron configuration
- ☐ electronegativity
- ☐ families
- ☐ groups
- ☐ halogens
- ☐ ionic bond
- ☐ ionization energy
- ☐ lanthanide series
- ☐ malleability
- ☐ melting point
- ☐ metalloids

## Lesson Objectives continued

- Relate the location of an element in the periodic table to its general properties for: the alkali metals, the alkaline earth metals, the halogens, transition metals, and the noble gases.
- Show how the result of a chemical reaction is related to atomic structure, trends in the periodic table, and patterns of chemical properties.
- Cite evidence to show how the structure of substances relates to the strength of electrical forces between particles.
- Show how the functioning of designed materials depends on molecular-level structure.

## Key Vocabulary continued

- ☐ metals
- ☐ noble gas
- ☐ nonmetals
- ☐ Pauli Exclusion Principle
- ☐ period
- ☐ periodic table of the elements
- ☐ periodic trend
- ☐ transition metals

## Understanding the Structure of the Periodic Table

dlc.com/ca9046s

**How do you organize things in your life? Do you file your documents in alphabetical order? Perhaps you color code certain items for easy identification. Just as we sort and organize things in a logical manner the same concept was applied when structuring the periodic table.**

**FRUITS AND VEGETABLES IN A GROCERY STORE**

Items are organized in a grocery store according to their characteristics. How does organization make shopping easier?

**EXPLAIN QUESTION**

**How can general properties influence element location on a periodic table?**

# How Do Scientists Distinguish between Metals, Metalloids, and Nonmetals?

## The Periodic Table

The periodic table includes information about all of the elements that have been discovered. Elements with similar chemical and physical properties are placed together within **groups** or **families** (vertical columns). There are also patterns in the way elements appear within periods (horizontal rows). The **period** (or row) number indicates the number of electron energy levels in an atom of that element.

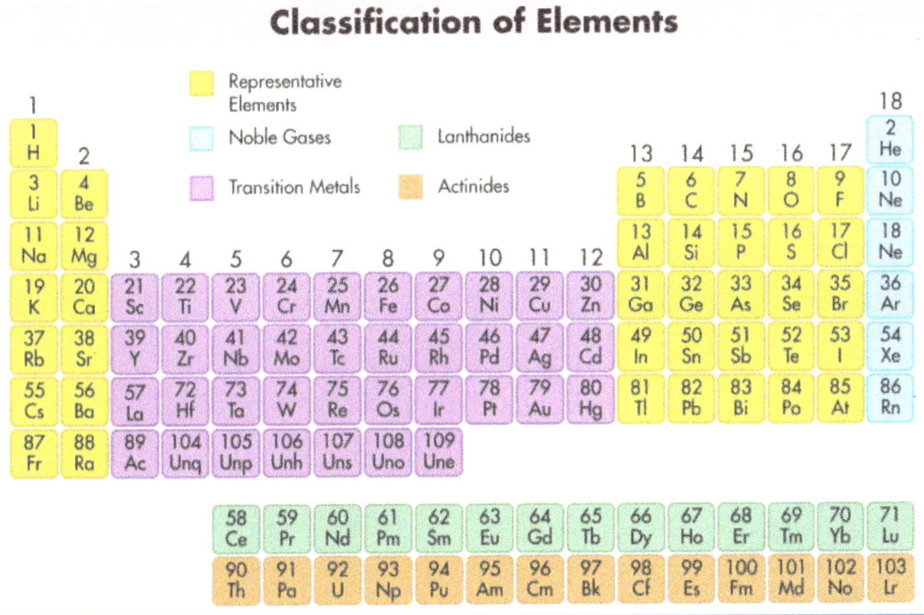


**ORGANIZATION OF THE PERIODIC TABLE.** The periodic table is divided into groups of elements with similar electron configurations and properties. What patterns in the periodic table would allow you to predict the chemical properties of specific elements?



There are many versions of the periodic table that include slightly different information. The designations for the groups, for example, are presented in three main styles. The European and U.S. conventions use "A" and "B" labels for different parts of the periodic table. In the U.S. model, the "A" groups include the "main group elements" in columns 1, 2, and 13–18. These are often called the "representative elements." The "B" groups are the **transition metals**. The International Union of Pure and Applied Chemistry (IUPAC), a worldwide organization working to standardize chemistry conventions, simply numbers the groups 1–18.

In all versions of the table, elements are placed in the periodic table according to increasing atomic number, which is the number of protons in an atom's nucleus. The configuration of the electrons in an atom can also be predicted using the periodic table. These details about atomic structure affect the chemical and physical properties of an element.

Elements that appear in certain places on the periodic table exhibit similar properties. For example, elements can be classified as **metals**, **nonmetals**, or **metalloids** depending on their properties.

## Metals

Metals are located on the left-hand side of the periodic table. Metals are primarily in Groups 1–12. Some elements in Groups 13, 14, and 15 are also metals.

Metals exhibit similar properties. They have high melting points. **Melting point** is the temperature at which a solid substance transforms into a liquid. Metals are usually solid at room temperature because of this property. Mercury is a highly toxic metal. It is one of only two elements that exist in the liquid state at room temperature.

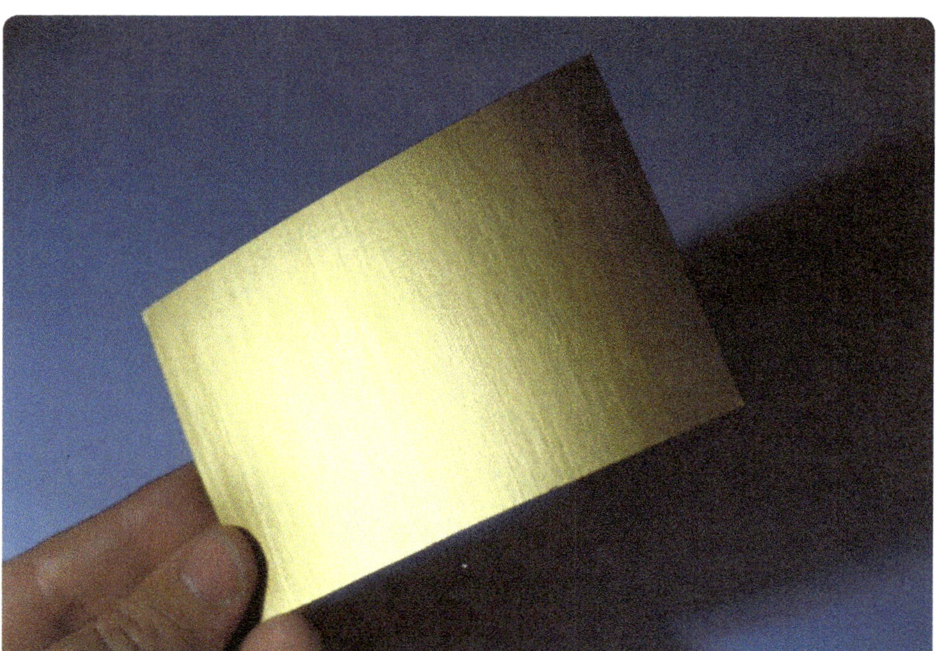

**METALS**
Metals are also typically shiny and have high melting points. How could you use a metal's position on the periodic table to predict these properties compared with other elements?



Valence electrons are an important factor in determining the properties of an element. Valence electrons are those that atoms tend to gain, lose, or share during a chemical reaction. The group number corresponds with the number of valence electrons in the atoms of many elements. For representative elements, they are electrons in the outer energy level. Elements in Group 1 contain one valence electron. Elements in Group 2 contain two valence electrons. For elements in Group 13–18, the number of valence electrons is 10 less than the group number. An element in Group 15, for example, has 5 valence electrons. The elements in Groups 3–12 have varying numbers of valence electrons.

Valence electrons produce many of the familiar properties of metals. Most metals are quite reactive because they easily gain, lose, or share valence electrons. Metals are typically conductive, malleable, and ductile, and they have shiny surfaces. A metal is a good **conductor** of heat and electricity. In solid metals, the valence electrons are shared equally among positive metal ions. These electrons flow freely between the nuclei of the metal atoms. They form a "sea of electrons" and move easily through the outer shells of metallic atoms, which explains why metals are good conductors. However, the attraction between the positive metal ions and the sea of electrons is strong. This explains why most metals have high melting points.

**Malleability**, **ductility**, and luster also are physical properties exhibited by metals. Malleability is the ability of an element to be molded into another shape without breaking or cracking. Ductility is the ability of an element to be pulled into a long, thin wire. The sea of electrons can flow with metal ions when they move. As a result, the metal ions slide past each other without breaking the bonds that hold them together. Luster refers to the shine reflected from an element's surface. Physical and chemical properties are important to consider when choosing a metal for a specific project.



## Nonmetals

The nonmetal elements are located on the upper right-hand side of the periodic table. Atoms of most nonmetals tend to gain valence electrons during chemical reactions. The noble gases are found in the far right column. They are stable and tend not to react with other substances.

**NONMETALS**

Nonmetals are stable and not very reactive. How do these features relate to the number of their valence electrons?

Nonmetals tend to have low melting points. This means that nonmetals are more likely to exist as gases at room temperature. When solid, nonmetals often exhibit properties such as brittleness, poor conductivity, and dull surfaces. The term *brittleness* is used to describe how easily something crumbles or breaks. Glass is brittle and will often break if dropped. Steel is not brittle and will not break if dropped.

The properties of nonmetals are also explained by attractions between the valence electrons and the nuclei of the atoms. Unlike metals, nonmetals hold their valence electrons tightly. Nonmetals share electrons between a few atoms to make molecules. However, the molecules do not share electrons with each other. They are held together by weaker forces, which is why nonmetals often have low melting and boiling points. Nonmetals are brittle because the electrons do not flow from atom to atom. Bonds break when the atoms move.

## Metalloids

There are only seven elements that are considered metalloids. The metalloids appear in Groups 13–16. They form a jagged line between the metals and nonmetals.

The metalloids exhibit properties of both metals and nonmetals. Metalloids are solid at room temperature and have luster, like metals. However, metalloids are usually brittle and behave like nonmetals.

The physicals properties of metalloids are related directly to the attractions between the valence electrons and nuclei. Metalloids hold on to their electrons more tightly than metals. But they hold them less tightly than nonmetals. Metalloids tend to share electrons the way nonmetals do. The attraction between the electrons and the nuclei holds the electrons between atoms. Because electrons cannot flow freely from atom to atom, metalloids tend to be brittle.

Yet, under certain conditions, electrons can move between atoms, which is why a metalloid typically acts as a semiconductor. Semiconductors conduct heat and electricity more easily than a nonmetal (insulator), but not as well as a metal (**conductor**). Semiconductors are very important in the production of modern electronic devices.

# How Are the Columns of the Periodic Table Organized, and What Do the Elements in Each Group Have in Common?

## Groups or Families on the Periodic Table

There are 18 **groups** (vertical columns) in the periodic table, which may also be called **families**. The table includes seven periods (horizontal rows). Two additional rows appear separately from the table. These elements are part of the **lanthanide series** and the **actinide series**. These **metals** are not assigned group numbers and they have unique characteristics that make them different from the other **transition metals**.

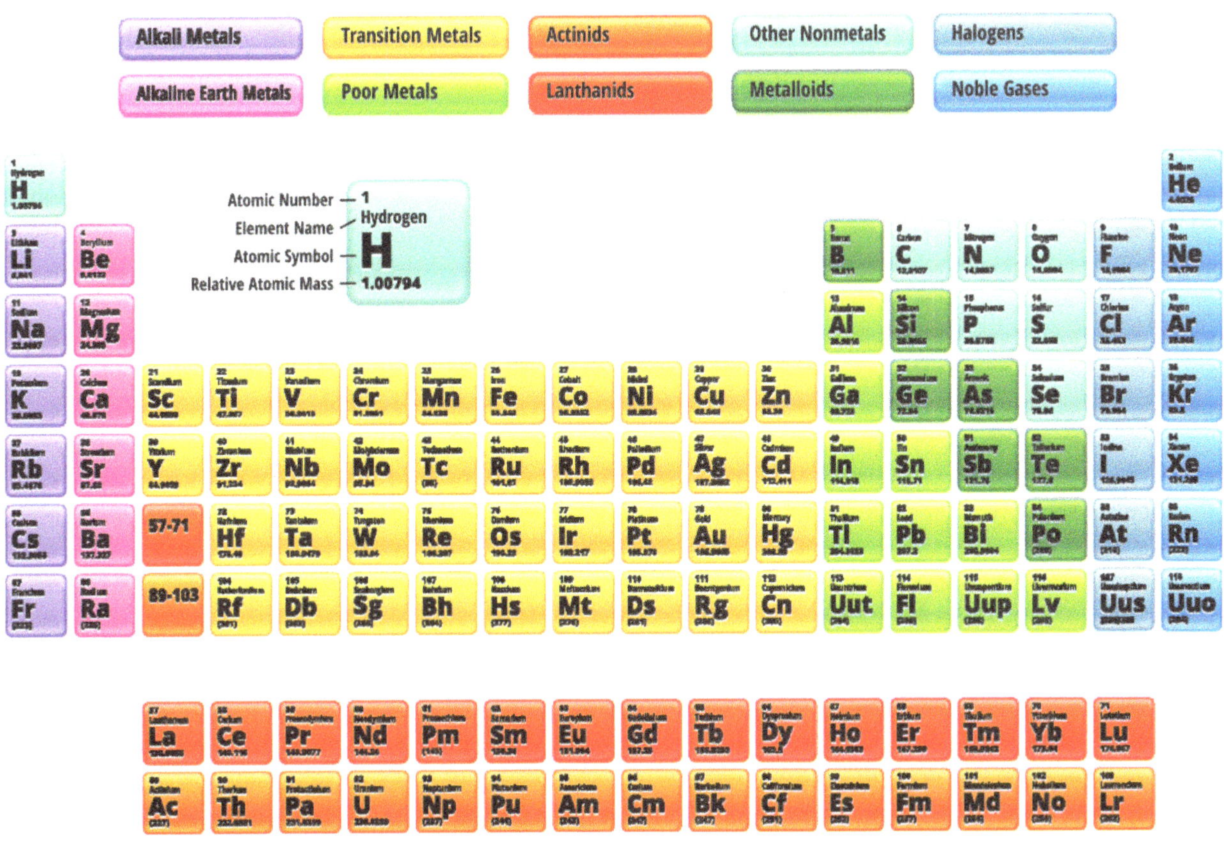


**GROUPS IN THE PERIODIC TABLE**

Elements in each column of the periodic table are grouped together because they share particular characteristics. Which group contains elements that are unreactive gases?

## Alkali Metals

The **alkali metals** make up Group 1 of the periodic table.

The atoms of these elements each contain a single valence electron, which makes them quite reactive. They tend to lose this electron during chemical reactions. Alkali metals easily react with Group 17 elements, such as chlorine or bromine. The element sodium (Na), for example, is an alkali metal. It combines with chlorine (Cl) to form sodium chloride (NaCl). Each sodium atom gives its valence electron to a chlorine atom during the reaction.

Alkali metals also quickly combine with oxygen when exposed to air. They react vigorously when placed in contact with water and form strong bases such as sodium hydroxide (NaOH). In fact, the term "alkali," for which this group of elements was named, means "basic."

## Alkaline Earth Metals

**Alkaline earth metals** are found in Group 2 of the periodic table. The atoms of these elements contain two valence electrons, which makes them fairly reactive. These elements tend to lose electrons during chemical reactions.

The behavior of alkaline earth metals is fairly predictable, though the properties of beryllium and magnesium vary a bit compared to the other metals in the group. These metals also tend to form weak bases when placed in water, and the salts that contain alkaline earth metals tend to display a distinct color when placed in a flame.

When exposed to oxygen, the alkaline earth metals tend to form oxides with high melting points. (An oxide is a compound made of oxygen and another element.) They often exist in very hot environments such as volcanoes.

## The Halogens

The halogen group, Group 17, includes five nonmetallic elements. This group is located on the right-hand side of the periodic table. Because the atoms of these elements contain seven valence electrons, they are very reactive. The atoms of these elements usually need to gain one valence electron to make them stable. Since elements from Group 1 contain one valence electron, **halogens** easily bond with these metals.

The word *halogen* means "salt-former." This refers to the fact that when halogens react with metals, salts are produced. Sodium chloride (NaCl) is a common example of this, as shown by its chemical equation below:

$$Na^{+} + Cl^{-} \rightarrow NaCl$$

The halogen group includes elements that exist in different forms at standard temperature and pressure. Astatine and iodine most commonly occur in solid form, while bromine is usually a liquid, and chlorine and fluorine are usually gases. This is the only group that includes elements commonly found in each of three states of matter.

## The Noble Gases

The six noble gases in Group 18 are in the far right column of the periodic table. These elements are generally unreactive because they each contain a stable set of valence electrons. The valence shell of a helium atom holds two electrons. The valence shells of the other noble gases hold eight.

Because the noble gases tend not to bond with other elements, they typically appear as monatomic elements (as a single atom). As a result, noble gases have low boiling points and usually exist in the gas state. However, the noble gases make up only a small percentage of Earth's atmosphere.

**HELIUM BALLOONS**

Hydrogen was first used in dirigibles and balloons, before helium was used. What properties of helium make it a suitable alternative to hydrogen?

## The Transition Metals

There are many elements in Groups 3 through 12, found in the middle block of the periodic table. These are referred to as the transition metals. Like other metals, the transition elements are ductile, malleable, and conductive. Some transition metals (for example, iron, cobalt, and nickel) are able to produce a magnetic field.

The atomic structure of the transition metals allows not only electrons from the valence shell but also some from the next inner shell to participate in bonding. This makes it more difficult to predict an element's properties.

## The Lanthanide Series

Two rows of elements appear in a separate section below the main body of the periodic table. This is done only to keep the table from being too wide. These elements are sometimes referred to as the inner transition metals, because the atomic numbers for these elements fall within the transition metals.

The lanthanide series is part of **Period** 6. It contains 15 elements with atomic numbers from that of lanthanum (La, 57) to lutetium (Lu, 71). These metals have properties similar to those of other metals.

One of the lanthanide metals, promethium (Pm), was synthesized in a laboratory. The rest commonly occur in nature, usually mixed together. This series is therefore sometimes called the "rare earth metals." The name likely came about because the properties of these metals are so similar, and it is difficult to isolate the inner transition metals in their pure elemental forms.

## The Actinide Series

The actinide series is part of Period 7 on the periodic table. It contains 15 elements with atomic numbers from that of actinium (Ac, 89) to lawrencium (Lr, 103). Only three of these metal elements occur naturally. The others must be synthesized in a laboratory, and most are quite unstable. All of the actinide series metals are radioactive.

Uranium and thorium are widely used in nuclear reactors because of their radioactive properties. Americium-241 is a common radioactive isotope that is used in smoke detectors and in medical diagnostic equipment.

## Consider the Explain Question

**How can general properties influence element location on a periodic table?**

dlc.com/ca9047s

Go online to complete the scientific explanation.

## Check Your Understanding

**How are groups or families organized on the periodic table of the elements?**

dlc.com/ca9048s

# STEM in Action

## Applying the Structure of the Periodic Table

What else can the periodic table be used for besides locating an element? Are there other elements that exist in nature? Scientists are constantly exploring ways to understand the different characteristics of matter and how this relates to the elements and structure of the periodic table.

All throughout history, scientists have tried to classify elements into a system that is easy to understand and easy to use. Nowadays, math and technology allow us to solve hard science problems. Technology helps us generate much needed information on the properties of molecules. It allows us to make a chemical prediction and possibly find new elements to add to the periodic table. Advancements in techniques through the use of computers along with science concepts and math models have really paved the way for a very important and useful branch of study called computational chemistry.

Computational chemistry allows scientists to study the chemical properties of all kinds of matter. Advanced software based on the Aufbau principle is able to theoretically determine the structures of atoms. According to this principle, electrons fill the space that makes the atom most stable. This software can also be used to predict the properties of molecules and solids, including atom location, reactivity, and much more.

**ANTIBODY 1GT**

Models used by computational chemists include the properties of atoms to predict the structure of complex molecules such as antibodies. How could these visualizations help scientists to discover new drugs?

How does the world exactly function? Why are the characteristics of this newly discovered drug what they are? Can we simulate how a molecule would behave in different scenarios? Computational chemists are faced with many questions such as these. Whether it is understanding existing physical properties of different elements such as their corrosive nature or ability to be a superconductor, this field does a great job of taking the knowledge used to develop the periodic table to create useful tools for the 21st century and beyond.

## STEM and Structure of the Periodic Table

Can we take what we know about an element, based on its location on a periodic table, and use it for our technological benefit? This is what manufacturers think about as they design many of the items we use in products such as electronics. Let's take semiconductors for example.

Semiconductors are chemical compounds used in electronics. These compounds are special because they conduct electricity only when needed. One type of semiconductor, silicon, is very good at gaining or losing up to four electrons. In fact, most modern technology is highly dependent on purified silicon including the production of computer chips, computer circuits, and electronics. One property that contributes to the use of silicon for a variety of purposes is its abundance on Earth. Silicon is extremely common, and compounds containing a high proportion of silicon (such as sand) make up more than 90 percent of the planet's crust. Silicon is similar to boron (another metalloid) and carbon (a nonmetal) in its chemical and physical properties. Now that you are an expert about all things silicon, where is this element located on the periodic table?

How would life be without the semiconductor? There is a large positive impact of semiconductors on our everyday living. Computer engineers use computer science and electronic engineering to develop a wide range of computer systems or technological devices. Their research work on semiconductors has paved the way for a lot of the modern technology we see today. Years ago portable radios were the hot item. Now music through a portable headphone is the way to go. Back then there was only a desktop computer available for use. Now there are tablets, smartphones, and smart watches that function similarly to the desktop computer. Leaders in pioneering the future of computer hardware, such as semiconductors, computer engineers make sure we stay on the cusp of modern technology.

CONCEPT 2.5

# Periodic Trends

## LESSON OVERVIEW

dlc.com/ca9049s

### Lesson Questions

- What is the relationship between the observed chemical reactivity of an element and its position on the periodic table?
- How can the observed periodic trends in reactivity of the elements be explained?
- What other periodic trends besides reactivity occur, and how are they explained?

### Key Vocabulary

Which terms do you already know?

- [ ] atomic radius
- [ ] chemical reactivity
- [ ] electron
- [ ] electron configuration
- [ ] electronegativity
- [ ] energy levels
- [ ] families
- [ ] groups
- [ ] ion
- [ ] ionic bond
- [ ] ionic radius
- [ ] ionic size
- [ ] ionization energy
- [ ] metallic character
- [ ] nucleus (atom)
- [ ] orbital model of the atom
- [ ] Pauli Exclusion Principle

## Lesson Objectives

By the end of the lesson, you should be able to:

- Relate an element's chemical reactivity to its position on the periodic table.
- Explain the basis for the observed periodic trends in reactivity of the elements.
- Describe and explain the basis for trends in atomic radius, ionic size, metallic character, ionization energy, and electronegativity of the elements across groups and periods within the periodic table.

## Key Vocabulary continued

- ☐ periodic table of the elements
- ☐ periodic trend
- ☐ shielding effect
- ☐ valence electron

## Understanding Periodic Trends

dlc.com/ca9050s

**Have you ever noticed that some of your family members share the same characteristics? Whether it is height, facial features, hair color, or athletic ability, certain characteristics often seem to occur in family groups. Similarly, scientists have observed that atoms in families share some of the same characteristics.**



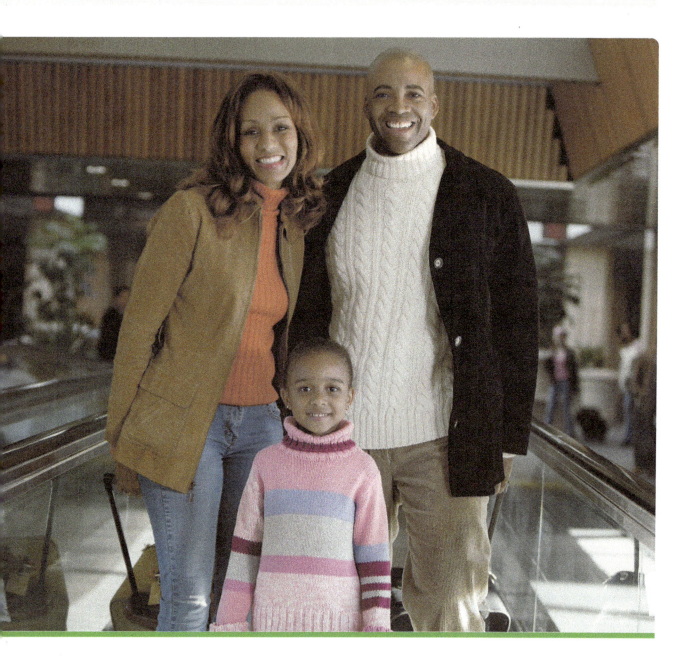

**FAMILY PORTRAIT**

What characteristics do children inherit from their parents?

### EXPLAIN QUESTION

**How do trends in the periodic table help predict the properties of an element?**

# What Is the Relationship between the Observed Chemical Reactivity of an Element and Its Position on the Periodic Table?

## Reactivity of Metals: The Alkali Metals

The periodic table that we use today was developed after a great deal of experimentation and observation. A fundamental observation was that elements in the same group had similar physical and chemical properties.

A good example is found in Group 1. The elements in Group 1 are known as the alkali metals and include lithium, sodium, potassium, rubidium, cesium, and francium. These metals have similar physical properties such as low melting points and low densities. These properties make the alkali metals very soft and easy to cut with a knife. The alkali metals are also highly reactive when placed in water.

Not all alkali metals have the same **chemical reactivity**, however. Sodium reacts more vigorously with water than lithium. Potassium is even more reactive than sodium. Going down the group, the alkali metals become more and more reactive when placed in water. The most reactive metal is francium (Fr), in the bottom left corner of the periodic table.



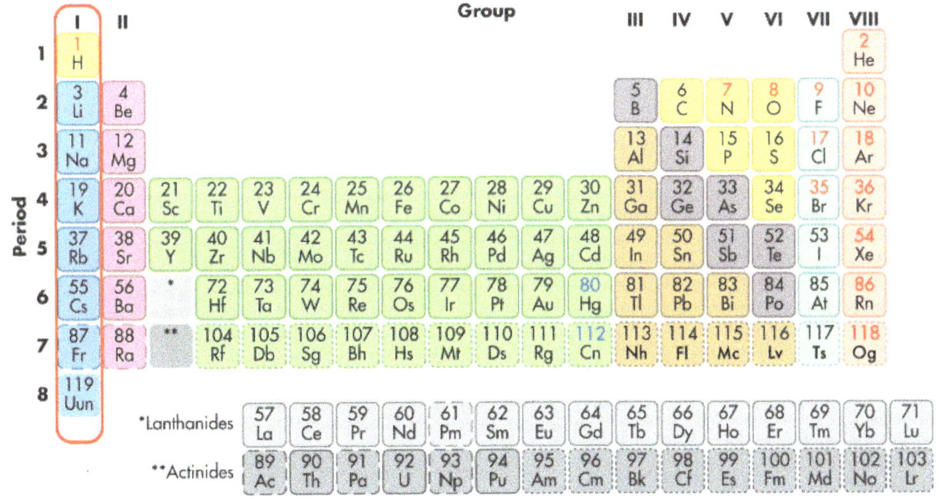


**ALKALI METALS**

The alkali metals (shown circled in red) are the most reactive of all metals. Why do alkali metals have such high reactivity?

## Reactivity of Metals: The Alkaline Earth Metals

Group 2 is known as the alkaline earth metals. These elements are nearly as reactive as the alkali metals. However, compared to the alkali metals, these elements are harder and have higher melting points.

We see a similar pattern of reactivity as we saw with the alkali metals as we go down within Group 2. The farther down within the group, the more reactive the alkaline earth metal.

**ALKALINE EARTH METALS**

The alkaline earth metals (shown circled in red) are nearly as reactive as the alkali metals, but are much harder and denser. How do the melting points of alkaline earth metals compare to those of the alkali metals?

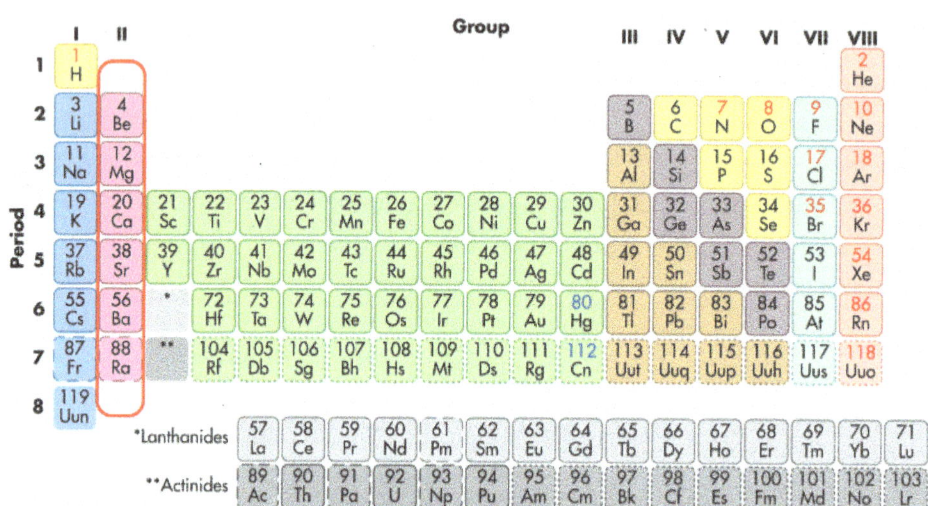


## Reactivity of Metals: The Transition Metals

Moving further to the right on the periodic table, we encounter the transition metals, which include the lanthanide and actinide metals at the very bottom.

**TRANSITION METALS**

The transition metals, including the lanthanides and actinides, are much less reactive than the alkaline earth metals and the alkali metals. What other properties do they possess?

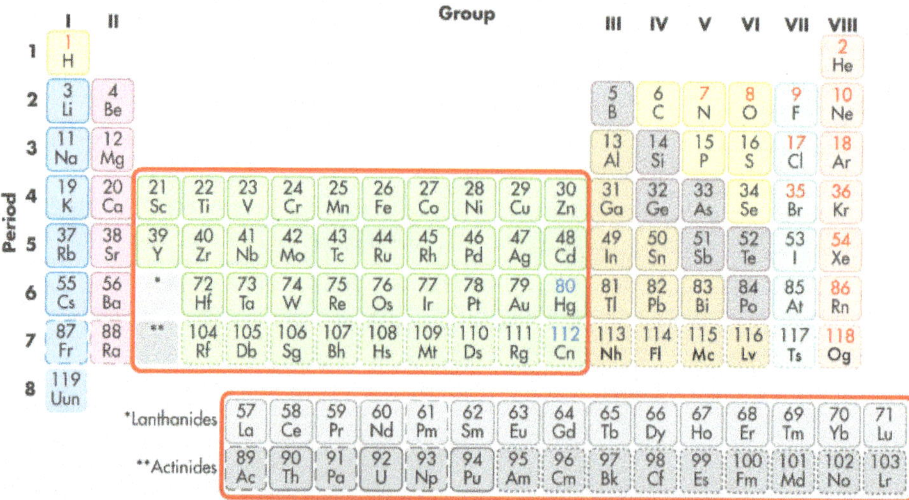

Transition metals tend to be very hard and have high densities. They are also much less reactive than either the alkaline earth metals or alkali metals. This low reactivity makes them suitable for making products that must resist change over long time periods.

## Reactivity of Metals: Periodic Trends

In summary, the reactivity of metals tends to decrease from left to right across the periodic table. The reactivity of metals also tends to increase as you move down within a group. The most reactive metal is francium (Fr) in the bottom, left corner of the periodic table.

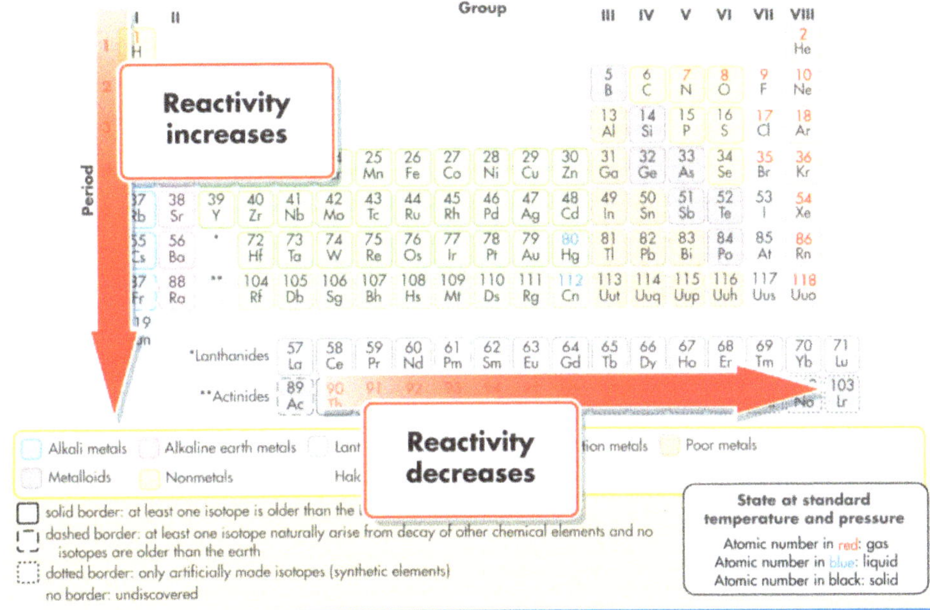


**REACTIVITY OF METALS**

Reactivity of metals tends to increase down a group and decrease from left to right across a period. Which metal is the most reactive?

## Summary of Metal Reactivity with Water, Hydrochloric Acid, and Air

| Metal | Reaction with Water | Reaction with Dilute Hydrochloric Acid | Reaction with Air |
|---|---|---|---|
| Potassium | Reacts violently with cold water. Hydrogen ignites. | Very violent dangerous reaction producing hydrogen gas that ignites | Rapidly forms a white oxide, burns violently with a lilac flame |
| Sodium | Reacts violently with cold water. Hydrogen may ignite. | Very violent dangerous reaction producing hydrogen gas that ignites | Rapidly forms a white oxide, burns violently with a yellow flame |
| Calcium | Reacts slowly producing bubbles of hydrogen gas | Reacts vigorously producing hydrogen gas | Slowly forms a white oxide, burns with a red flame |
| Magnesium | Reacts slowly producing bubbles of hydrogen gas. Burns in steam. | Reacts fairly vigorously producing hydrogen gas | Burns rapidly with a bright white flame |
| Aluminum | No reaction* | Reacts slowly* | No reaction* |
| Zinc | Reacts slowly with steam | Reacts fairly vigorously producing hydrogen gas | Zinc wool burns with a blue green flame. Forms an oxide very slowly in air |
| Iron | Reacts slowly with steam | Reacts slowly | Forms a reddish oxide but iron wool and iron filings burn quickly with sparks. Forms an oxide in the presence of air and water. |
| Lead | No reaction | No reaction | Slowly forms an oxide film, oxidizes readily on heating |
| Copper | No reaction | No reaction | Slowly forms an oxide film, oxidizes slowly on heating |
| Silver | No reaction | No reaction | No reaction |

**Aluminum is protected by a thin film of aluminum oxide. If the film is removed it behaves in a manner intermediate between magnesium and zinc.*

## Reactivity of Nonmetals: Groups 14 and 15

How do the nonmetals vary in reactivity? Do they follow the same **periodic trend** as the metals? Again, we can turn to observations of the elements to find answers to these questions.

Carbon is representative of Group 14. This element is not highly reactive, although in some forms it can easily form bonds with other elements. Two familiar forms of carbon are graphite, used to make pencils, and diamond, a crystalline form of carbon. Neither of these carbon forms is very reactive.

Nitrogen is in Group 15. In its elemental form, nitrogen atoms bond to one another to form diatomic nitrogen, $N_2$. This gas is found in abundance in Earth's atmosphere and tends to be nearly inert, or nonreactive.

**CARBON GRAPHITE**

Graphite is one of the forms of carbon. What other forms does carbon take?

## Reactivity of Nonmetals: Groups 16 and 17

As we move to the right in the periodic table, we find that reactivity of the nonmetals increases. Whereas carbon in Group 14 and nitrogen in Group 15 are relatively nonreactive, oxygen in Group 16 is more reactive.

Moving still further to the right are the halogens in Group 17. These elements are the most reactive nonmetals.

## Reactivity of Nonmetals: Noble Gases

The least reactive of all the elements are the noble gases that make up Group 18. These elements exist as gases and are colorless, odorless, and inert.

## Reactivity of Nonmetals

If we exclude the noble gases, which are the most inert of all the elements, the reactivity of nonmetals tends to decrease from right to left across the periodic table. The reactivity of nonmetals also tends to increase as you move up within a group. These trends are opposite to those observed for the metals. The element fluorine (F) at the top of Group 17 is the most reactive of all the elements.

**REACTIVITY OF NONMETALS**

Excluding the noble gases, the reactivity of nonmetals tends to increase moving up a group and decrease from right to left across a period. Why is the periodic table organized in this manner?

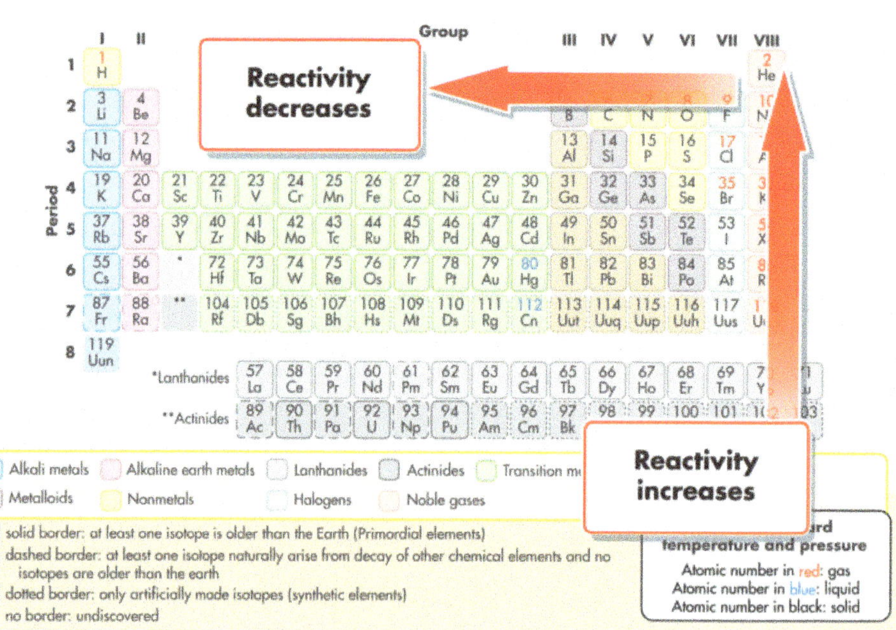


## Periodic Trends in Reactivity

In summary, we can say that metals and nonmetals both show a periodic trend for reactivity, although these trends are opposing. Metal reactivity increases down within a group and decreases from left to right across a period. Reactivity for nonmetals other than noble gases varies in the opposite way. It increases as you move up within a group and decreases from right to left across a period.

# How Can the Observed Periodic Trends in Reactivity of the Elements Be Explained?

## Electron Configurations and the Periodic Table

The periodic table was originally developed as a way to organize elements according to their properties. As chemists learned more about atomic structure, they began to understand that the arrangement of the periodic table also reflected the arrangement of electrons in atoms.

Periods correspond to the **energy levels** occupied by electrons. Elements in Period 1 have electrons in only one energy level, whereas elements in Period 2 have electrons in two energy levels, elements in Period 3 have electrons in three energy levels, and so on.

Elements in each group have the same number of electrons in their outermost energy level. For example, all alkali metals have one **electron** in their outermost energy level, and all alkaline earth elements have two electrons in their outermost energy level.



## Valence Electrons and Reactivity

The outermost electrons of an atom are called the valence electrons. These are the electrons involved in chemical reactions. Elements tend to react with one another to either donate, take, or share electrons in order to achieve a filled outer energy level. This makes them more stable.

Metals tend to lose electrons and nonmetals tend to gain electrons in order to achieve a full outer level. It is this tendency to gain or lose electrons that determines how reactive an element is. The greater the tendency to gain or lose electrons, the more reactive the element.

## Explaining Trends in Chemical Reactivity from Group to Group

Elements within the same group have the same number of valence electrons in their outermost energy level. Therefore, the general pattern of reactivity seen between elements within the same group can be explained by this similarity in valence electrons.

For example, all alkali metals have one electron in their outermost energy level. We can explain their highly reactive nature from this feature. Because they have only one *s* orbital electron to lose to achieve a filled outer energy level, these elements lose this electron readily. This explains their high reactivity.

In the case of transition metals, the tendency to react is much less. Most of these elements have filled *s* and *p* orbitals. Only their *d* orbitals are partially filled. It is energetically less favorable for atoms with partially filled *d* orbitals to react. As a result these metals are comparatively stable.

With respect to nonmetals, the most reactive elements are the halogens, which gain one electron to form a filled outer energy level. Nonmetals in Group 15 must gain three electrons, and those in Group 16 must gain two electrons, which is less likely to happen. That makes these atoms less reactive.

The pattern that emerges from these observations is that, for many atoms, the closer the atom is to being able to achieve eight valence electrons in its *s* and *p* orbitals, the more reactive it is.

Of course, at the very far extreme are the noble gases. These elements have filled *s* and *p* orbitals. Their inert character results from this feature because they have a tendency to neither lose nor gain electrons. As a result, noble gases are very nonreactive.



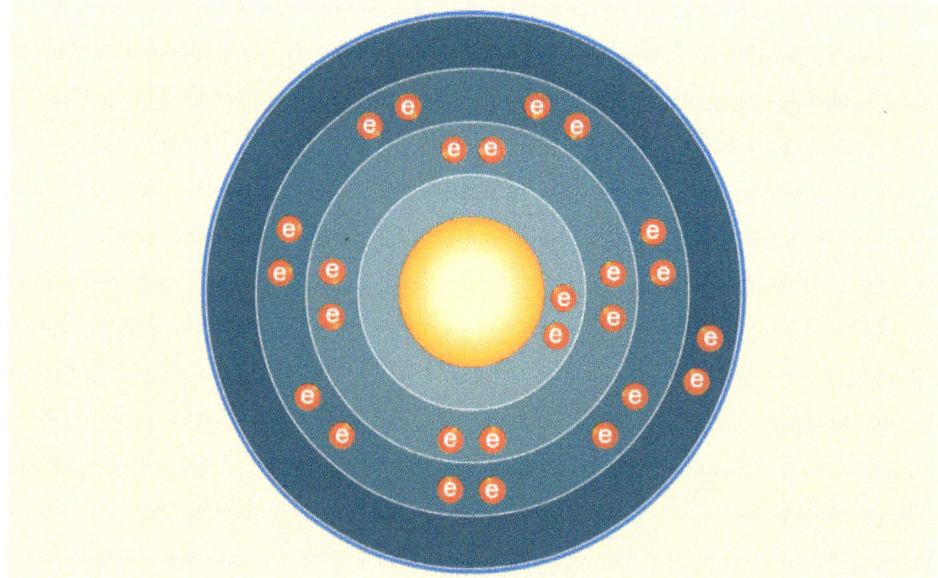


**THE VALENCE SHELL: IRON**

The image shows the arrangement of electrons in an iron atom. How does iron's electron configuration impact its reactivity?

### Explaining Trends in Chemical Reactivity from Period to Period

We saw that reactivity increases as we go down in a group of metals and decreases as we go down in a group of nonmetals. Why is that?

The answer has to do with the ease of gain or loss of valence electrons by an element.

We can take the alkali metal group as an example. As you move down this group, the number of energy levels increases. As these energy levels add layers to the structure of the atom, the valence electrons are located farther away from the positively charged nucleus. More electrons are present in the inner energy levels. Those inner electrons shield the outer electrons from the positive nuclear charge. With so much shielding, the valence electrons have less pull from the nucleus and can move away with greater ease. Because the alkali metals lose electrons to achieve stability, greater shielding results in greater reactivity.

If we examine the halogens as an example of the nonmetals, we observe something different. As we move down the group, the number of energy levels increases just as with the metals. This again increases the shielding of the outermost energy levels from the nuclear charge. The nonmetals, however, gain electrons to achieve stability instead of losing them, and the increase in shielding decreases reactivity in the lower periods of a group because the pull of the nucleus is limited.



## What Other Periodic Trends besides Reactivity Occur, and How Are They Explained?

### Atomic Radius: Observed Periodic Trends

One measure of atomic size is the distance from the center of the nucleus to the outermost electrons. This distance is defined as the **atomic radius**. The size of the radius is hard to describe because the **electron** cloud does not have a well-defined surface. A common way to describe it is half the distance between the nuclei of neighboring atoms. These might be identical atoms that are bonded together or atoms next to each other in a crystal.

The atomic radius does not vary much for transition elements, but it does for main group (representative) elements. The chart, Atomic Radii, shows the changes.

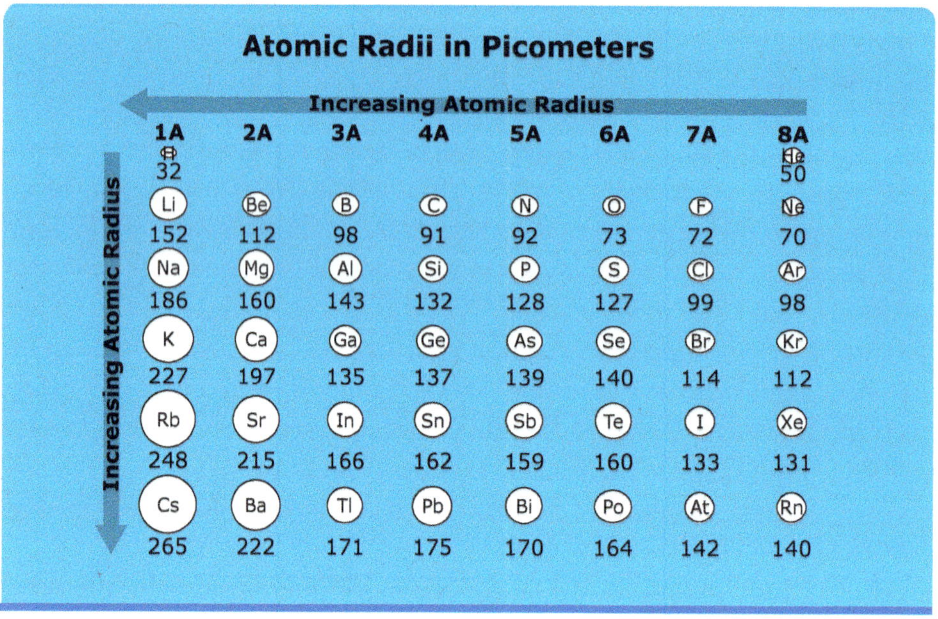
Atomic Radii in Picometers
Increasing Atomic Radius
Increasing Atomic Radius
1A 2A 3A 4A 5A 6A 7A 8A
H 32
He 50
Li 152 Be 112 B 98 C 91 N 92 O 73 F 72 Ne 70
Na 186 Mg 160 Al 143 Si 132 P 128 S 127 Cl 99 Ar 98
K 227 Ca 197 Ga 135 Ge 137 As 139 Se 140 Br 114 Kr 112
Rb 248 Sr 215 In 166 Sn 162 Sb 159 Te 160 I 133 Xe 131
Cs 265 Ba 222 Tl 171 Pb 175 Bi 170 Po 164 At 142 Rn 140

**ATOMIC RADII**

For main group elements, atomic radius increases as you move down within a group but decreases as you move from left to right across a period. What does this pattern reveal about the properties of the elements?

For main group elements, the atomic radius increases as you go down a group. This increase results because a new energy level is added as you move from one period down to the next. The outermost electrons occupy **energy levels** that are farther and farther from the nucleus with each shift to a new period.

Look at the diagram to see what happens to the atomic radius of main group element atoms as you move from left to right within the same period. With each move to the right, one more proton and one more electron are added. You might think that the addition of these particles would result in an atom of larger size. However, the diagram shows that for most main group elements, atomic radius decreases from left to right across a period. Something else besides a physical increase in numbers of subatomic particles affects the radius. What might explain this observation?

### Atomic Size: The Effect of Nuclear Charge

The decrease in atomic size across a period can be explained by the effect of increasing nuclear charge. As you move from left to right within a period, the numbers of protons and electrons increase. This increases the amount of positive charge in the nucleus. Since the electrons are being added to one energy level, they occupy the same region. As a result, they are held more tightly by the increasing positive charge of the nucleus.

### Factors Underlying Periodic Trends

The factors that explain the observed trends in **chemical reactivity** and atomic radius apply to other properties of the elements. Along with chemical reactivity and atomic radius, the important factors are:

- total number of electrons in an atom
- the size of the nuclear charge
- the distance between the nucleus and the outermost valence electrons
- amount of shielding between the positively charged nucleus and negatively charged electrons in the outermost energy level

All other periodic trends can be explained using these four factors.

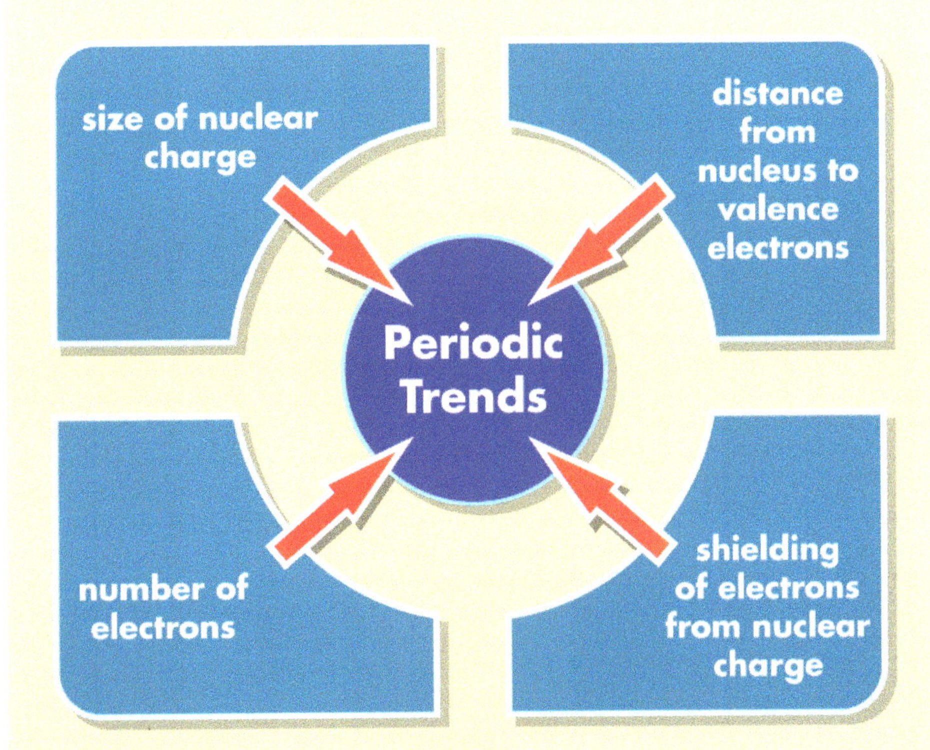


**FACTORS AFFECTING PERIODIC TRENDS**

Four main factors contribute to the trends exhibited in the periodic table. How do these factors impact the organization of elements?

## Periodic Trends in Ionic Radius

An **ion** is a charged particle that results when an atom loses or gains electrons. The charge results from an imbalance in the numbers of protons and electrons present.

**Ionic radius** is the distance from the center of the nucleus to the outermost electrons of an ion. The ionic radius, like the atomic radius, is hard to describe because the electron cloud does not have a well-defined surface.

Metal atoms become ions by losing electrons. In these ions, the net charge is positive because protons outnumber electrons. These ions are smaller than their neutral atoms because there is more pull by the nucleus on the electrons in the outer energy level. Nonmetal atoms become ions by gaining electrons. The net charge is negative because electrons outnumber protons. These ions are larger than their neutral atoms because the pull of the nucleus on the electrons is weaker. Among metals and among nonmetals, ionic radii tend to decrease as you move left to right across a period. The nonmetal ions, however, tend to be larger than the metal ions.



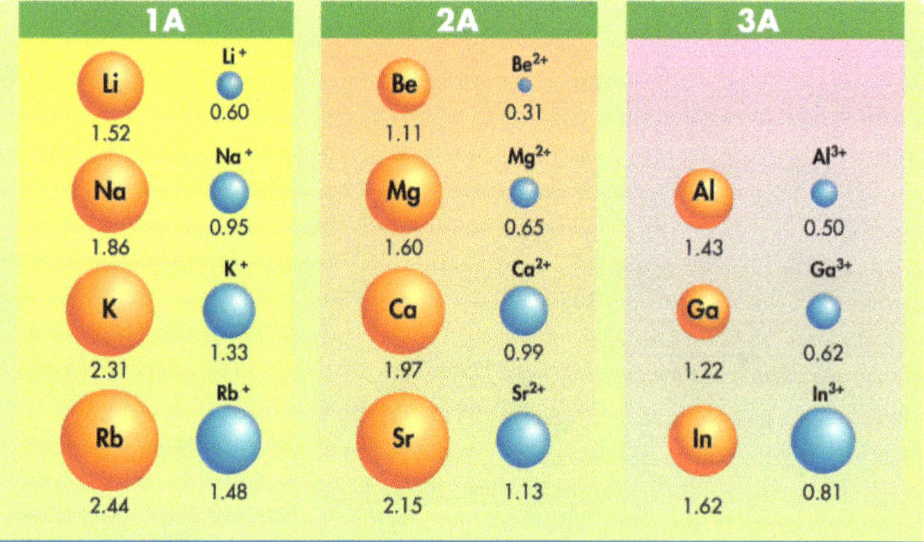


**IONIC RADII**

Ionic radii differ from atomic radii but follow the same general trends within a group and across a period, as shown by this chart for metals. What are the ionic radii for other elemental categories likely to reveal?

## Periodic Trends in Metallic Character

**Metallic character** is the measure of how easily an atom loses an electron.

The diagram summarizes the periodic trends in this property. Metallic character decreases across a period, and increases down a group. The four factors noted earlier can explain these trends.

The addition of new energy levels as you move down a group has two effects. The distance between the nucleus and the outermost electrons increases, and greater numbers of energy levels occupy this distance. The **shielding effect** of the inner electrons and the increasing distance reduce the effective nuclear charge experienced by the outermost electrons. As a result, less energy is needed to remove one outer electron, and so metallic character increases.

Within a period, however, only the nuclear charge factor is at work. The number of protons increases as you move from left to right but no new energy levels are introduced. Therefore, as you move from left to right, the outermost electrons experience a greater and greater positive charge from the nucleus, and it becomes more difficult for an electron to be removed. Metallic character decreases as a result.



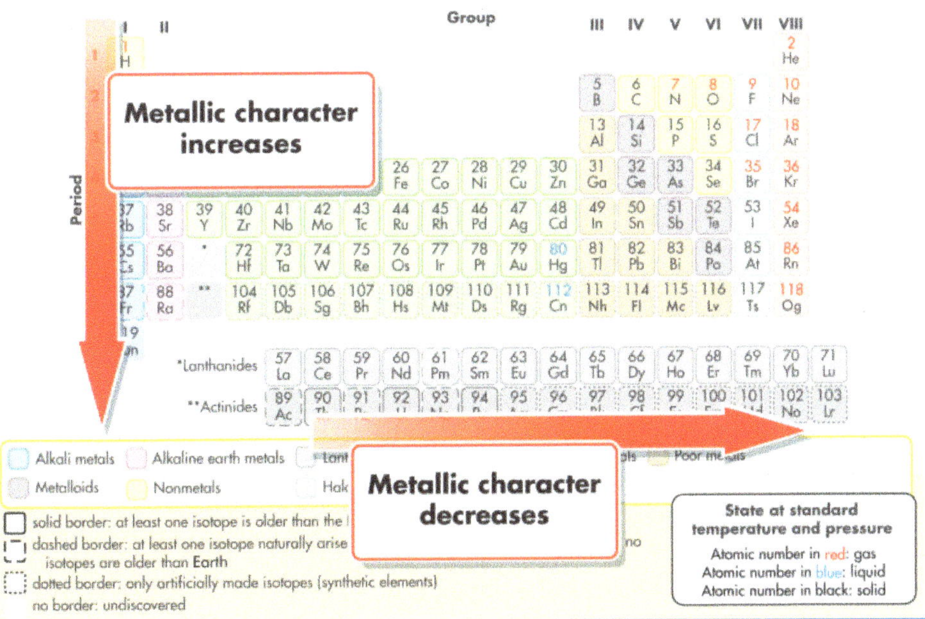


**METALLIC CHARACTER**

Metallic character increases down a group and decreases across a period. How do variations in the energy levels of the elements explain these changes?

## Periodic Trends in Ionization Energy

**Ionization energy** is the energy needed to remove one electron from the outermost energy level of a gaseous neutral atom to form an ion having +1 charge.

$$Na\ (g) \rightarrow Na^{+}\ (g) + e^{-}$$

The energy required to remove the outermost electron from the gaseous +1 ion is called the second ionization energy. It is larger than the first ionization energy and the third ionization energy is higher than the second. This is because fewer electrons remain to shield the nucleus from the electrons and its pull is stronger on the remaining electrons. Thus, each successive electron removed from the atom feels an increasingly stronger effective nuclear charge.

As shown in the Ionization Energy Chart, ionization energy decreases as you go down within a group and increases as you go from left to right across a period. Ionization energy behaves opposite to metallic character.

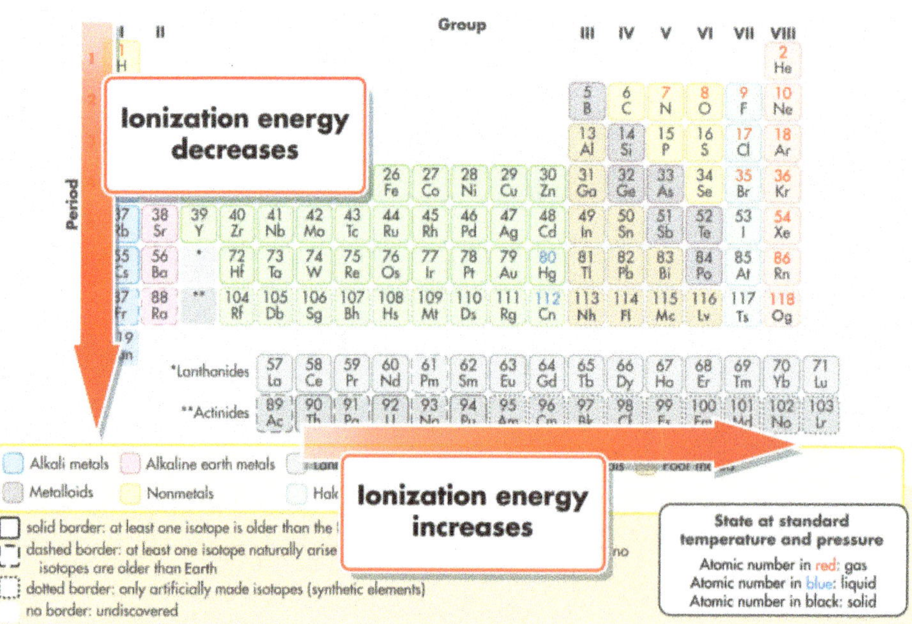


**IONIZATION ENERGY CHART**

Ionization energy decreases down a group and increases across a period. How does this pattern compare to trends in metallic character on the periodic table?

The addition of new energy levels as you move down a group has two effects. The distance between the nucleus and the outermost electrons increases. Greater numbers of energy levels occupy this distance between them as well. The **shielding effect** of the inner electrons and the increasing distance reduce the effective nuclear charge experienced by the outermost electrons. Less energy is required to remove an outer electron as a result. This increases metallic character, but decreases ionization energy.

Within a period, only the nuclear charge factor affects metallic character. The number of protons increases as you move from left to right but no new energy levels are introduced. Therefore, as you move from left to right, the outermost electrons experience a greater positive charge from the nucleus. More energy is required to overcome this pull. Therefore, metallic character decreases, but ionization energy increases.

## Periodic Trends in Electronegativity

**Electronegativity** is defined as the tendency of an atom to attract electrons.

The Electronegativity Chart shows the periodic trends observed for this property. Electronegativity decreases as you go down within a group and increases from left to right across a period. Once again, the factors discussed earlier become important when explaining this trend.

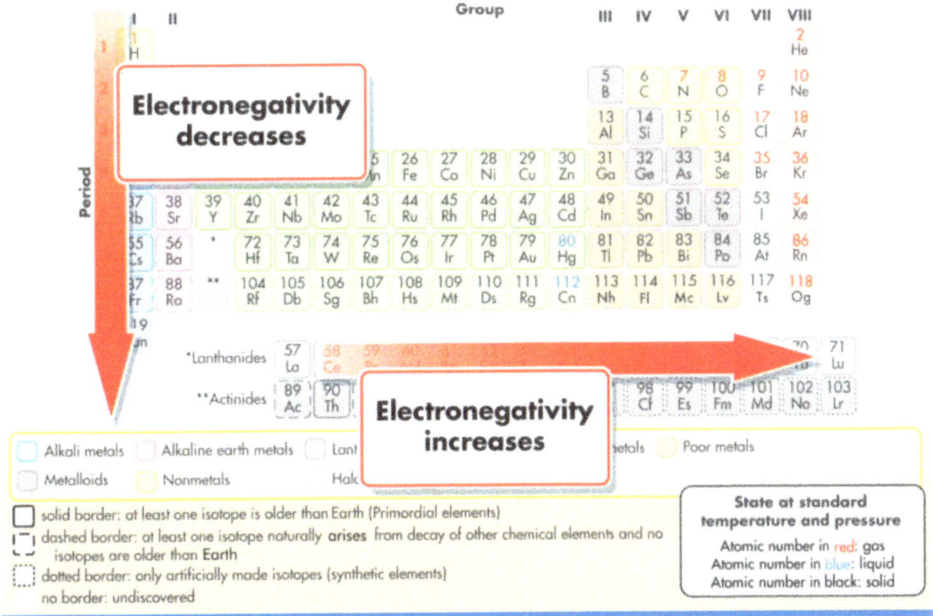


**ELECTRONEGATIVITY CHART**

Electronegativity decreases down a group and increases across a period. How does shielding play a role in electronegativity?



As you move down a group, the distance between the nucleus and the outermost electrons increases. Also, more energy levels filled with electrons shield the outermost electrons from the nucleus. Both factors decrease the effective nuclear charge experienced by the outermost electrons. This decreases the positive pull of the nucleus on electrons, which decreases an element's electronegativity.

Moving from left to right within a period, however, the atoms gain protons and electrons, but no new energy levels form. Therefore, as you move from left to right, the outermost electrons are added to the same energy level. The added protons in the nucleus cause a greater positive charge, which pulls on the electrons. This increasing positive pull makes the elements more electronegative.

### Consider the Explain Question

**How do trends in the periodic table help predict the properties of an element?**

dlc.com/ca9051s

Go online to complete the scientific explanation.

### Check Your Understanding

**How can you use the periodic table to explain the chemical properties of different families of elements?**

dlc.com/ca9052s

## in Action

## Applying Periodic Trends

The different reactivities of atoms allow us to develop materials with varying properties. Inert elements, or those with little to no reactivity with other elements, are used to make products that will resist change and stand up to harsh conditions. More reactive elements are used to make products that change readily, even explosively.

Steel is an alloy of iron, small amounts of carbon, and possibly other elements. In the 1850s, an inexpensive mass-production method, the Bessemer process, was developed to produce steel. By adding different metals to the steel alloy mixture, modern steelmakers have produced steel with very different properties. Adding chromium produces corrosion-resistant stainless steel, and tungsten or cobalt additives increase hardness. Steel is malleable, has high tensile strength, is flexible, and can be adapted to fit the needs of the project. These properties of steel have made it possible to build extremely tall buildings and long bridges that are not only strong, but also resistant to change over long periods of time.

**STEEL FACTORY**

Steel is used in the construction of many different products. Large, steel tubes produced in this factory are used in construction projects. What protects steel from corrosive substances in the environment?

Another example of using an element's reactivity for the advancement of technology-driven resources includes the element hydrogen. Hydrogen is an element that scientists have developed to use as a fuel in cars, buses, and rockets. Hydrogen is abundant in nature and reacts readily with oxygen to produce water. Hydrogen is an ideal element for the production of fuel. An added benefit regarding environmental health to the public is the minimal level of toxicity emitted from the byproducts of burning hydrogen fuel, when compared to other fuels.

## STEM and Periodic Trends

Engineers use the different reactivities of elements to power and protect spacecraft so that humans can safely travel in space. They consider the reactive properties of elements in designing a fuel that will provide enough thrust to overcome the pull of gravity when a rocket blasts off. For decades, NASA has relied on hydrogen gas and fuel cells to deliver crew and cargo to space. Recently, one private company, SpaceX, successfully launched rockets using a kerosene-based fuel which reacts with liquid oxygen, LOX. Blue Horizon, another private company, has launched rockets using peroxide or a mixture of kerosene and peroxide as fuel.

Engineers must also consider the reactivity and other properties of elements when designing the materials to be used for the outer covering of the spacecraft. NASA engineers have developed a silver, metallic-based thermal coating to bond to the thermal protection tiles on the outside of its Orion rocket. This will reduce heat loss as well as limit high temperatures inside the rocket. A titanium skeleton with a carbon fiber skin will reduce the mass of the heat shield's structure. Periodic trends such as melting point and density are some of the important factors that engineers use in designing effective and safe space vehicles.

**HOUSTON, WE HAVE LIFT-OFF**

Engineers design the materials used for a variety of products such as spacecrafts. How do engineers incorporate periodic trends into their development process of spacecraft materials?

CONCEPT 3.1

# Ion Formation

## LESSON OVERVIEW

### Lesson Questions

- How do monatomic ions form?
- How do cations and anions differ?
- How does the formation of ions relate to the octet rule?

### Lesson Objectives

By the end of the lesson, you should be able to:

- Describe how monatomic ions form.
- Distinguish between cations and anions.
- Explain ion formation and its relationship to the octet rule.

### Key Vocabulary

Which terms do you already know?

- [ ] anion
- [ ] cation
- [ ] ion
- [ ] ionic bond
- [ ] monatomic ion
- [ ] valence electron

dlc.com/ca9053s

## Introducing Ion Formation

dlc.com/ca9054s

**Have you ever been to a climbing gym? How does your brain tell your hand to move, once you have decided on your next handhold? How does your foot tell your brain that it is secure, and it is safe to push up on the foothold?**

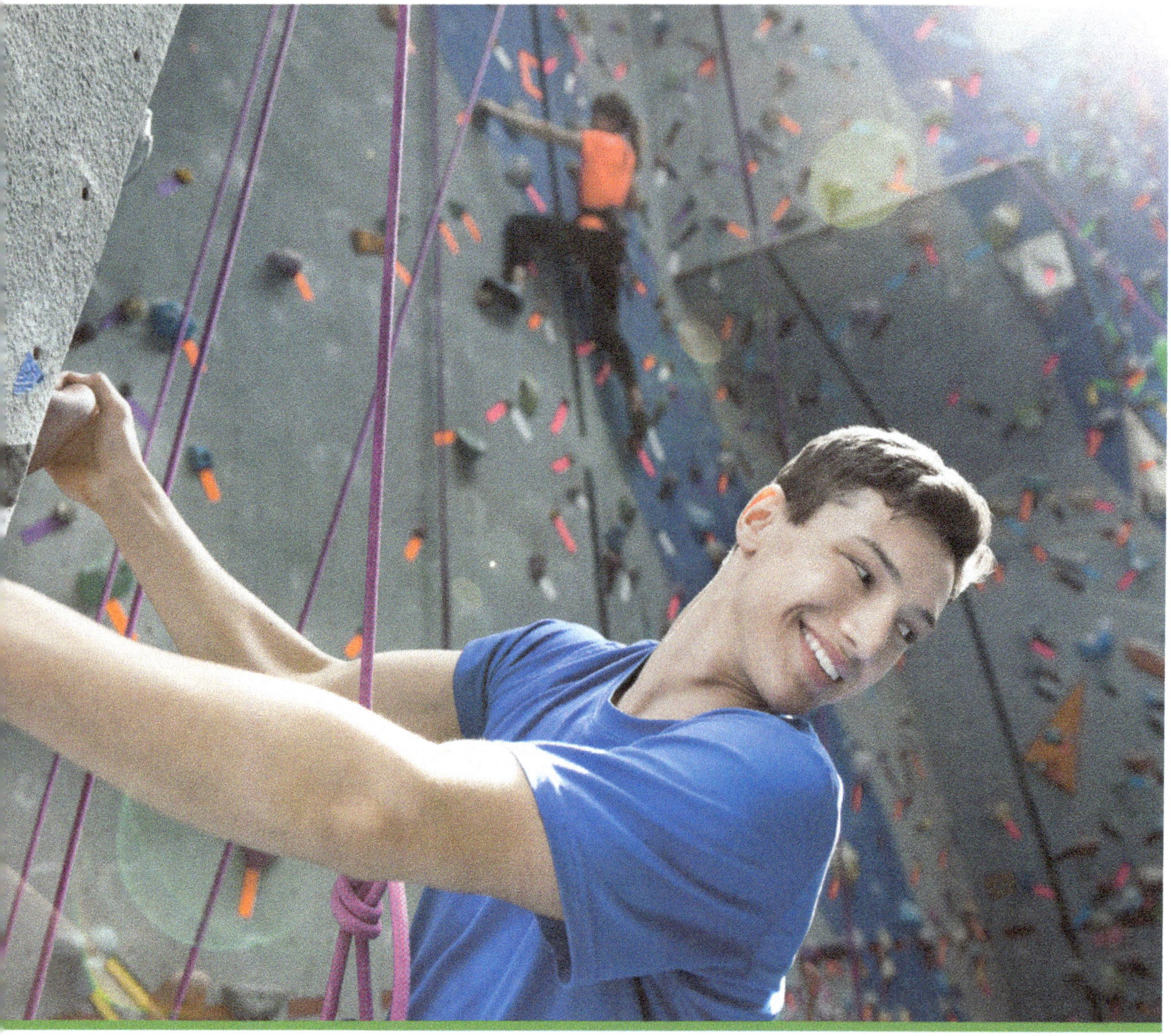

**IONS IN THE HUMAN BODY**

A climber scales a wall in a gym. How do ions help him navigate from handhold to handhold?

**EXPLAIN QUESTION**

**How do ions form, and how does the octet rule determine if an atom becomes a cation or an anion?**

# How Do Monatomic Ions Form?

## Monatomic Ions

Atoms are the smallest units of an element. A neutral atom does not possess a net charge because it contains an equal number of positive protons and negative electrons. For example, an atom of sodium (Na) contains 11 protons in its nucleus and 11 electrons outside its nucleus. The charges on a proton and an electron are equal in magnitude. So, the negative charges of the 11 electrons outside the nucleus balance the positive charges of the 11 protons inside the nucleus. This balance makes the atom neutral.

A neutral atom acquires a charge if its positive and negative charges become unbalanced. This happens when an atom gains or loses one or more electrons. The atom then becomes a charged particle called an **ion**. For example, when an atom of sodium loses one electron, it still contains 11 protons, but only 10 electrons. Since the sodium atom now has one more positive charge than negative charges, it has a net charge of +1. The symbol for this ion is written as $Na^{1+}$, or $Na^{+}$ in shorthand notation.

Many ions consist of a single charged atom. For example, an ion of sodium is one charged sodium atom. An ion that consists of only one charged atom is called a **monatomic ion**.

**SPLIT-SECOND TIMING**

Ions are charged particles. How might charged particles affect the human nervous system?

The chart shows some examples of monatomic ions. Each ion's charge is either positive (+) or negative (–) and has a value between 1 and 4. A value of 1 is represented by a superscript plus or minus sign. Values greater than 1 are shown by placing a number in front of the + or – sign. For example, the calcium ion has a charge of +2, written as $Ca^{2+}$.

| | |
|---|---|
| sodium, $Na^+$ | chromium (III), $Cr^{3+}$ |
| hydrogen, $H^+$ | lead (IV), $Pb^{4+}$ |
| potassium, $K^+$ | bromide, $Br^-$ |
| silver, $Ag^+$ | chloride, $Cl^-$ |
| barium, $Ba^{2+}$ | flouride, $F^-$ |
| calcium, $Ca^{2+}$ | hydride, $H^-$ |
| copper (II), $Cu^{2+}$ | iodide, $I^-$ |
| iron (II), $Fe^{2+}$ | oxide, $O^{2-}$ |
| aluminum, $Al^{3+}$ | sulfide, $S^{2-}$ |

**COMMON MONATOMIC IONS**

A list of ions shows metallic ions (left) and non-metallic ions (right). Can you determine any pattern in the charges of the ions?

In contrast to monatomic ions, polyatomic ions form when a group of atoms are joined together and have an overall positive or negative charge. For example, the hydroxide ion, $OH^-$, contains an atom of oxygen and an atom of hydrogen. There is one more electron than proton in the group, so this polyatomic ion has a charge of –1. Polyatomic ions will be discussed in more detail in another concept.

## Transfer of Electrons

Monatomic ions form when single atoms gain or lose electrons. In general, this happens when one or more electrons are transferred from an atom of one element to an atom of another element. The atom that loses one or more electrons becomes positively charged. The atom that gains electrons becomes negatively charged.

For example, common table salt is a solid compound made up of one atom of sodium (Na) and one atom of chlorine (Cl). The formula NaCl represents this compound. However, to form this compound, each sodium atom transfers one electron to each chlorine atom, so a more accurate representation of table salt would be $Na^+Cl^-$.

# How Do Cations and Anions Differ?

## Cations and Anions

An early model of the atom was developed in 1913 by the Danish scientist Niels Bohr. This model shows the atom as a central nucleus containing protons and neutrons, with the electrons in circular electron shells orbiting at specific distances from the nucleus, like planets orbiting a star. Each electron shell has a different energy level. Those shells closest to the nucleus have a lower energy than those farther from the nucleus. The outermost shell is called the valence shell and electrons present in this shell are called valence electrons.

Positively charged ions are called cations. A **cation** forms when an atom loses one or more valence electrons. Cations usually form from the atoms of metals, such as sodium, potassium, calcium, copper, and iron.

Negatively charged ions are called anions. An **anion** forms when an atom gains one or more electrons in its valence shell. Anions usually form from the atoms of nonmetals, such as fluorine, chlorine, oxygen, and sulfur.

A useful mnemonic to recall the difference is that A Negative **ion** is an ANion.



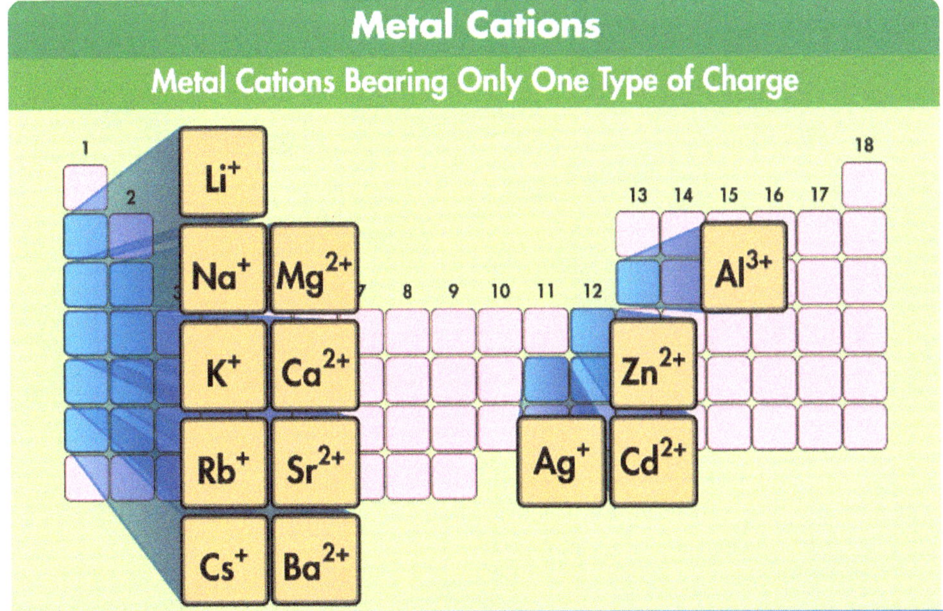


**CATIONS**

Cations, such as the ions shown, are usually formed from the atoms of metals. What explains why these ions have positive charges?

## Predicting the Charge on a Cation or Anion

The charge on a cation or anion can be predicted from the number of valence electrons in an atom. The number of valence electrons can be determined by noting the group in which the atom's element appears in the periodic table. Groups toward the left of the periodic table tend to lose electrons and therefore form cations. With the exception of Group 18, groups toward the right of the periodic table tend to gain electrons and therefore form anions.

For example, the atoms of Group 1 elements (Li, Na, K, Rb, Cs, and Fr) have only one **valence electron**. These atoms readily lose their valence electron. Loss of an electron from an atom of any of these elements will thus produce a cation with a charge of +1. Group 2 elements (Be, Mg, Ca, Sr, Ba, and Ra) have two valence electrons that their atoms tend to lose. In these elements, losing the two valence electrons from their atoms produces cations with a charge of +2.

The charge on an anion is predicted the same way. For example, the atoms of Group 17 elements (F, Cl, Br, I, and At) have seven valence electrons. These atoms tend to gain a single electron. The gain results in the ions of these elements having a charge of –1. The atoms of Group 16 elements (O, S, Se, Te, and Po) have six valence electrons. These atoms tend to gain two electrons, so the ions of these elements have a charge of –2.

**EXPLOSIVE COMBINATION**

Sodium metal reacts violently with water. How does this reaction produce sodium ions?

# How Does the Formation of Ions Relate to the Octet Rule?

## The Octet Rule

The elements in Group 18 of the periodic table (He, Ne, Ar, Kr, Xe, and Rn) are the noble gases. Due to their stable outer electron shells, noble gases do not readily react with other elements.

The outer valence shell of each of the noble gas atoms holds a full complement of electrons. Excluding helium, each of the noble gas atoms has an outer valence shell consisting of two *s* orbitals and three *p* orbitals. Each of these four orbitals contains two electrons. Because their valence shells are complete, the atoms are stable. Chemical reactions tend to occur so that the atoms of non-noble elements have a stable outer shell of eight electrons as well. Since a set of eight is an octet, this principle is called the octet rule.

For example, an atom of sodium holds one electron in its outermost valence shell. The next inner shell holds eight electrons. When a sodium atom loses the electron in its outermost shell, a complete valence shell remains. Since the sodium atom has lost a negative charge, its **ion** has a net charge of +1 and is written as $Na^{+}$. Metals such as sodium typically lose electrons to form a complete valence shell. Therefore, metals usually form positively charged ions.

How does the octet rule affect elements at the other end of the periodic table? This is where nonmetals reside. An atom of chlorine, for example, holds seven electrons in its valence shell. When it gains an electron, its valence shell will have the stable configuration of eight electrons. The octet rule is again followed.

Nonmetals like chlorine typically gain electrons to form a complete valence shell. Therefore, nonmetals form negatively charged ions.

## Exceptions to the Octet Rule

There are many exceptions to the octet rule. Transition metals, for example, may not follow the octet rule because they form bonds that involve *d* or *f* orbitals. A complete outer shell for these elements would therefore include more than eight electrons.

Another exception to the octet rule involves the element helium (He). Helium is a very small atom. It has only two protons in its nucleus, so helium needs only two electrons to make a neutral atom. These two electrons make up helium's valence shell and give helium a complete, stable electron configuration. To achieve this configuration (and become ions) the atoms of metallic elements close to helium in the periodic table lose electrons so that their valence shells contain two electrons.

For example, lithium (Li) has a total of three electrons. Two of these lie in an inner shell. One lies in lithium's valence shell. So, to achieve a complete, stable outer shell of two electrons like that of helium, the lithium atom loses its single outermost electron. The lithium ion then has a charge of +1 ($Li^+$).

An atom of beryllium (Be) has a total of four electrons. Two are in the inner shell, and two are in the outer shell. To achieve a stable valence shell, the beryllium atom can lose its two outermost electrons. The beryllium ion then has a charge of 2 ($Be^{2+}$).



**EXCEPTIONS TO THE OCTET RULE**

Helium holds only two electrons ($e^-$) in its valence shell. To achieve stability, the atoms of metals close to helium in the periodic table lose electrons, resulting in a valence shell identical to that of helium. Beryllium (Be) loses two electrons to become a positive ion with a charge of +2. How does lithium change to achieve stability?

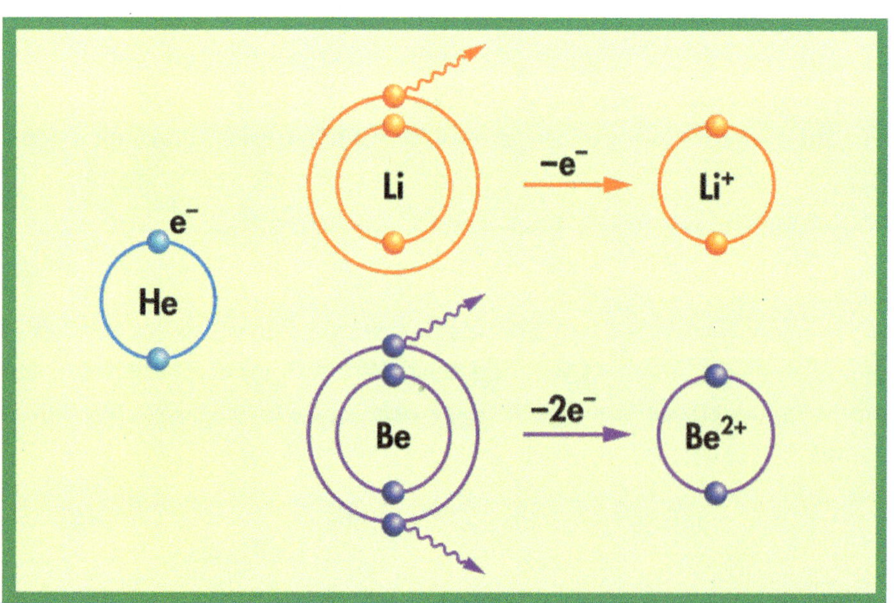


**Consider the Explain Question**

**How do ions form, and how does the octet rule determine if an atom becomes a cation or an anion?**

Go online to complete the scientific explanation.

dlc.com/ca9055s

**Check Your Understanding**

**How is a neutral atom similar to or different from ions of the same element?**

dlc.com/ca9056s

## Applying Ion Formation

To run, kick a soccer ball, or shoot a basket, muscles in your legs and arms must be activated. The muscles must rapidly switch from a resting state to an active state. When this happens, your muscles contract and you move. The process of activating muscle is a highly complex synchronized series of related events.

**MUSCLES AND ION ACTIVITY**

Playing soccer takes energy and coordination. How do ions help muscles contract?



In the resting state, a protein called tropomyosin blocks the activation of muscle fibers. Think of this protein as analogous to insulating material on a wire. While this insulating material is intact, it blocks messages from nerve cells to muscle cells. The muscle will not contract.

However, as a nerve impulse leaps from a nerve cell to a muscle cell, it unblocks messages from the nerve cells to the muscle fiber. As the nerve impulse reaches the muscles, it triggers release of calcium ions ($Ca^{2+}$) from a part of the muscle. The calcium ions attach to the tropomyosin molecules. While calcium ions are attached to tropomyosin, the molecule cannot stop the muscle fiber from contracting. The muscle then contracts and you run, kick, or shoot a basket. All of this, of course, happens in a fraction of a second.

## STEM and Ion Formation

To conserve petroleum and reduce air pollution, scientists and engineers have designed cars that run partly or entirely on electricity. Cars that run only on electricity have no starter under the hood, nor do they have spark plugs or any of the other parts of a conventional internal combustion engine that runs on gasoline. Instead, an electric car is powered by a motor that converts electrical energy into mechanical energy. Electrical cars cost less to operate per mile than gasoline-powered cars.

The electricity that powers the car is produced by a rechargeable lithium-ion battery pack, which is comprised of a number of individual lithium-ion batteries connected together. In each battery, lithium (Li) is converted into lithium ions ($Li^+$) with the release of electrons ($e^-$). The chemical equation for this reaction is:

$$Li \rightarrow Li^+ + e^-$$

This reaction produces a flow of charged particles, or an electric current, that runs the motor. In turn, the motor makes the car move.

When the battery is depleted, it can be recharged using a public charging station, or it can be recharged at home with electricity from the grid. Solar-powered charging stations are being built that are virtually emission free.

**GREEN REVOLUTION: ELECTRIC VEHICLES**

Electric cars seem to be the future of transportation. However, as with any new technology, there are significant challenges. What are the challenges faced by the 'car of the future'?

Electric vehicles represent a growing field that requires workers with knowledge not only of electricity, but also chemistry, mechanics, industrial design, materials, and computer software and hardware. Workers who design both the vehicles and the stations required to recharge them generally have an advanced degree in a field of engineering. Technicians and skilled tradespeople typically require a two- to four-year degree, and are valued partners in design, prototype testing, and modification of the vehicles.

Electrolytes are ions that mammalian bodies require in order to regulate nerve and muscle function. Horses that are working hard, as in endurance racing, lose high levels of electrolytes through their sweat, which can then lead to muscle cramping in the horse. A research veterinarian is collecting data on the amounts of specific electrolytes lost from a group of competing horses on a race day, as well as data about the frequency of muscle cramps in the horses throughout the race.

**ENDURANCE RACE**
Endurance racing for horses includes long distance events of 50 miles or more over natural terrain—no race tracks. Horses racing long distances generate a lot of sweat. How are ions related to the sweating of horses?

CONCEPT 3.2

# Ionic Bonding

## LESSON OVERVIEW

dlc.com/ca9057s

### Lesson Questions

- How do ionic bonds form in binary compounds?
- How is electronegativity used to determine which atoms form ionic bonds?

### Lesson Objectives

By the end of the lesson, you should be able to:

- Illustrate how ionic bonds are formed in binary compounds.
- Use electronegativity to determine which atoms form ionic bonds.

### Key Vocabulary

Which terms do you already know?

- [ ] anion
- [ ] binary compound
- [ ] cation
- [ ] covalent bonding
- [ ] crystal lattice
- [ ] double bond
- [ ] electronegativity
- [ ] evaporation
- [ ] formula unit
- [ ] ion
- [ ] ionic bond
- [ ] molecular bonding
- [ ] polyatomic ion
- [ ] resonance structure
- [ ] single bond
- [ ] triple bond
- [ ] valence electron
- [ ] valence shell

## Deconstructing Salts

**You are probably very familiar with salt in two different forms: the salt you shake onto your popcorn, and the salt in the ocean. But, have you ever thought about whether those two types of salt are the same?**

dlc.com/ca9058s

**EXPLAIN QUESTION**

**How do atoms form ions that then combine to form crystals?**

**SALT WATER**

How can salt in seawater be the same as the salt at home?

# How Do Ionic Bonds Form in Binary Compounds?

## Binary Ionic Compounds

An ionic compound is formed when two electrically charged atoms, or ions, are held together by their opposite charges. This bond is called an **ionic bond**.

Valence electrons are those that inhabit the outermost electron shell of an atom. An ionic bond is formed when one or more valence electrons move from a metal atom to a nonmetal atom. In the process of ionic bond formation, the metal atom becomes positively charged because it contains more positively charged protons in its nucleus than negatively charged electrons outside of its nucleus. An atom that has acquired a positive charge is called a **cation**.

In contrast, the nonmetal atom becomes negatively charged because it contains fewer positively charged protons in its nucleus than negatively charged electrons outside of its nucleus. An atom that has acquired a negative charge is called an anion. When a cation and an **anion** are chemically bonded through the attraction of their opposite electrical charges, an ionic compound is formed.

The simplest ionic compounds consist of the atoms of only two elements. Such compounds are called binary ionic compounds. The term *binary* refers to the quantity two.

Sodium chloride (NaCl), which is table salt, is a common binary ionic compound. The following chemical equation shows how this ionic compound forms.

$$2Na + Cl_2 \rightarrow 2Na^+ + 2Cl^-$$

**SALT CRYSTALS**

Various ions combine to form salts. Is it possible to predict whether two ions will form a salt?

An electron from each of two sodium atoms is transferred to each of two chlorine atoms.

Magnesium chloride ($MgCl_2$) is another example of a binary ionic compound. In this case, a single magnesium atom transfers two electrons, one to each of two chlorine atoms.

$$Mg + Cl_2 \rightarrow Mg^{2+} + 2Cl^-$$

Notice that in both equations, the positive and negative charges are balanced. In other words, overall, no electrons are gained or lost.

aluminum flouride, $AlF_3$
copper oxide, $Cu_2O$
iron oxide, $Fe_2O_3$
lead flouride, $PbF_2$
lithium flouride, LiF
magnesium chloride, $MgCl_2$
potassium flouride, KF
sodium chloride, NaCl
sodium sulfide, $Na_2S$
sodium nitride, $Na_3N$
zinc sulfide, ZnS

**IONIC BINARY COMPOUNDS**

Electron charges are balanced in ionic binary compounds, to produce a total charge of zero. What were the charges of copper and oxygen before they formed copper oxide?



## Ionic Compounds with Polyatomic Ions

Not all ionic compounds are binary compounds. Some ionic compounds are formed by using both ionic bonding and covalent bonding. In order for this to occur, polyatomic ions are used. Polyatomic ions are more complex ions that form when two or more atoms are covalently bonded together. The resulting ion contains a charge because the total number of electrons from the bonded set of atoms does not produce a stable structure. In short, the resulting ion no longer has an equal number of protons and electrons. Even though more than two types of elements are being used to form the ionic bond, the process of ionic bonding, the transfer of valence electrons, is still applied. Chemists use parentheses to indicate that the atoms in the polyatomic ion are bonded together, as seen in the following example.

An example of an ionic compound using a positively charged polyatomic ion is ammonium chloride: $NH_4Cl$. In this case, two polyatomic ions of ammonium are needed to transfer their available **valence electron** to the two atoms of chlorine. When this occurs, each atom of chlorine gains an electron and becomes the anion, chloride.

$$2(NH_4)^+ + Cl_2 \rightarrow 2(NH_4)Cl$$

**COMMON POLYATOMIC IONS**

This table shows some common polyatomic ions. Can you explain the differences between nitrate and nitrite and also between sulfate and sulfide?

| Some Common Polyatomic Ions | | |
|---|---|---|
| **Name** | **Formula** | **Charge** |
| Acetate | $C_2H_3O_2^-$ | 1– |
| Ammonium | $NH_4^+$ | 1+ |
| Carbonate | $CO_3^{2-}$ | 2– |
| Cyanide | $CN^-$ | 1– |
| Bicarbonate | $HCO_3^-$ | 1– |
| Hydroxide | $OH^-$ | 1– |
| Nitrate | $NO_3^-$ | 1– |
| Nitrite | $NO_2^-$ | 1– |
| Peroxide | $O_2^{2-}$ | 2– |
| Phosphate | $PO_4^{3-}$ | 3– |
| Sulfate | $SO_4^{2-}$ | 2– |
| Sulfide | $S^{2-}$ | 2– |
| Tartrate | $C_4H_4O_6^{2-}$ | 2– |

## Representing Ions and Ionic Compounds as Lewis Structures

Lewis structure diagrams can be used to represent ions and ionic compounds. These diagrams are a useful tool for visualizing the outer electron configuration of ions. If an atom gains electrons, its ion will have a negative charge. For example, if a chlorine atom gains one electron to become a chloride ion, it will have a charge of -1. If an atom loses electrons, its ion will have a positive charge. For example, when a calcium atom forms an ion, it loses two electrons and therefore has a charge of +2. When drawing a Lewis structure of an ion, the structure is usually placed inside a bracket, and the resulting charge on the ion is placed outside the bracket. Since all the valence electrons are transferred to the cation, the outer shell electrons are only shown on negative ions.

$[:\ddot{\mathrm{Cl}}:]^{-}$ chloride ion

$[\mathrm{Na}]^{+}$ sodium ion

$[:\ddot{\mathrm{O}}:]^{2-}$ oxide ion

$[\mathrm{Ca}]^{2+}$ calcium ion

**USING LEWIS STRUCTURES TO REPRESENT IONS**

These are the Lewis structures of some common ions. Which structures are anions?

When a chemical reaction occurs between two elements that form a compound with an ionic bond, one element, usually a metal, donates one or more electrons to another (a nonmetal). The metal loses its outer valence electrons. This provides it with a full outer shell. The other element in the compound uses these electrons to fill its outer shell.

Because ionic compounds are held together by electrostatic forces, there is no sharing of electrons. Lewis structures of ionic compounds therefore keep the ions separate; each is usually depicted as being contained within its own set of brackets.

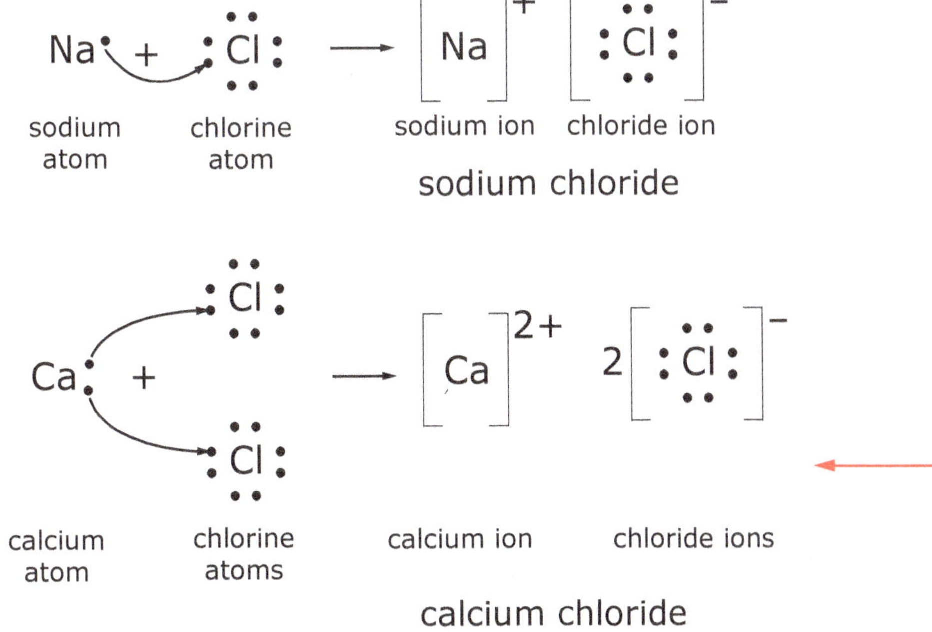


**USING LEWIS STRUCTURES TO REPRESENT IONIC COMPOUNDS**

In an ionic compound, sodium donates its electron to chlorine to provide both elements with a stable configuration and an electrostatic charge that holds the ions together. Why does a calcium ion need two chloride ions to form?

Lewis structures can be used to depict electron configuration in more complex polyatomic ions. In this case, all the atoms that make up the ion are placed within a bracket and the overall charge on the ion is placed outside the bracket.

**LEWIS STRUCTURES OF COMPLEX IONS**

These are the Lewis structures of some common complex ions. Can you think of any common household substances that make use of these ions?

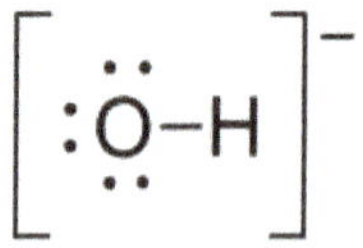

hydroxide ion

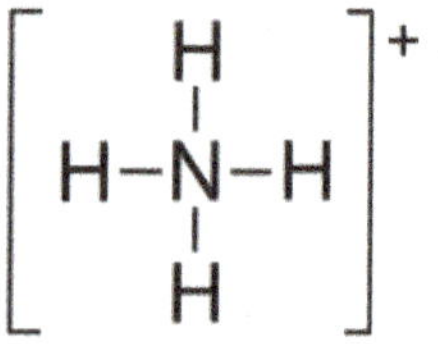

ammonium ion

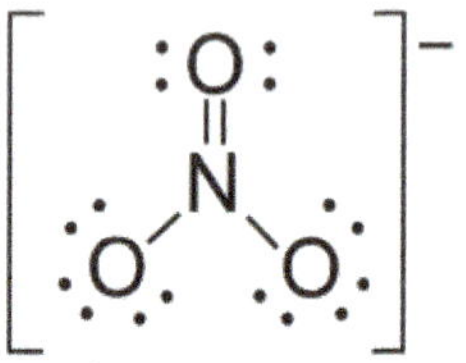

nitrate ion

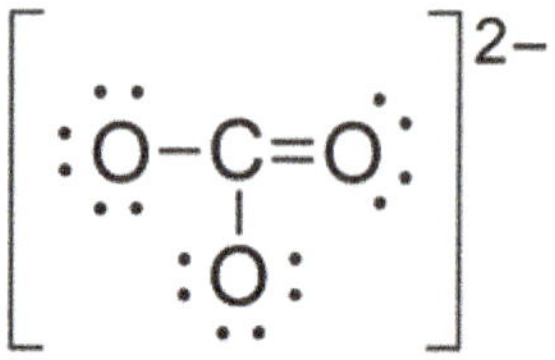

carbonate ion

## How Is Electronegativity Used to Determine Which Atoms Form Ionic Bonds?

### Electronegativity

Which elements are likely to form ionic bonds? The answer depends on the **electronegativity** of each element. Electronegativity is a measure of the attraction of an atom for electrons in a chemical bond.

If the difference in electronegativity between two atoms is 1.7 or greater, the atoms will form an **ionic bond**. For example, fluorine (which has an electronegativity of 4.0) will form an ionic bond with nickel (which has an electronegativity of 1.8). The difference between the two values is 2.2. On the other hand, phosphorous (which has an electronegativity of 2.1), will not form an ionic bond with nickel because the difference between these two values is only 0.3.

In general, elements with the lowest electronegativity are found among the metals in Groups 1 and 2 of the periodic table. Elements with the highest degree of electronegativity are found among the nonmetals, especially those in Groups 16 and 17. As a matter of fact, the element with the highest electronegativity (4.0) is fluorine, which is located at the top of Group 17.

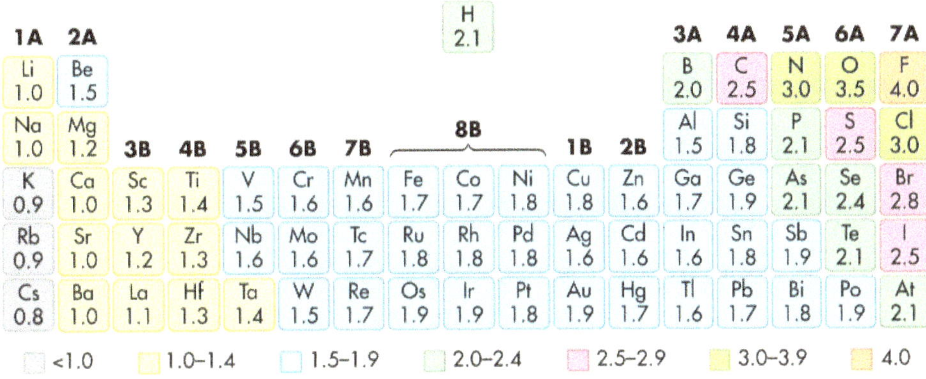

H 2.1
1A 2A
Li 1.0 Be 1.5
Na 1.0 Mg 1.2
3B 4B 5B 6B 7B 8B 1B 2B
3A 4A 5A 6A 7A
B 2.0 C 2.5 N 3.0 O 3.5 F 4.0
Al 1.5 Si 1.8 P 2.1 S 2.5 Cl 3.0
K 0.9 Ca 1.0 Sc 1.3 Ti 1.4 V 1.5 Cr 1.6 Mn 1.6 Fe 1.7 Co 1.7 Ni 1.8 Cu 1.8 Zn 1.6 Ga 1.7 Ge 1.9 As 2.1 Se 2.4 Br 2.8
Rb 0.9 Sr 1.0 Y 1.2 Zr 1.3 Nb 1.6 Mo 1.6 Tc 1.7 Ru 1.8 Rh 1.8 Pd 1.8 Ag 1.6 Cd 1.6 In 1.6 Sn 1.8 Sb 1.9 Te 2.1 I 2.5
Cs 0.8 Ba 1.0 La 1.1 Hf 1.3 Ta 1.4 W 1.5 Re 1.7 Os 1.9 Ir 1.9 Pt 1.8 Au 1.9 Hg 1.7 Tl 1.6 Pb 1.7 Bi 1.8 Po 1.9 At 2.1
<1.0
1.0–1.4
1.5–1.9
2.0–2.4
2.5–2.9
3.0–3.9
4.0


**ELECTRONEGATIVITY OF THE ELEMENTS**

Metals have relatively low values for electronegativity, and nonmetals have relatively high values for electronegativity. Which elements have the highest and lowest electronegativities?

## Crystal Lattices

Compounds held together by ionic bonds form three-dimensional geometric solid shapes called crystals.

The atoms of an ionic compound are linked in a particular structural pattern. This structural pattern of atoms is called a **crystal lattice**. In sodium chloride, atoms of sodium and chlorine form a lattice that produces a cubic crystal. A variety of crystal shapes are produced by other ionic substances. The shape of the crystals is determined partly by the ratio and sizes of the positive and negative ions in the compound.

**Ionic Compounds**

*Structure of Solid NaCl*

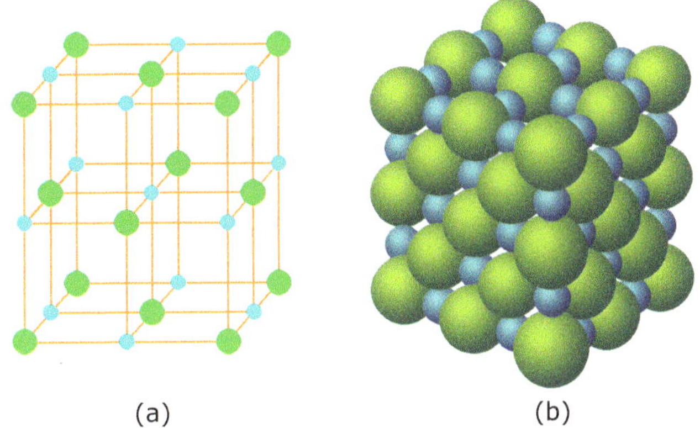

(a) (b)

**SOLID SODIUM CHLORIDE STRUCTURE**

In solid form, ionic compounds such as common salt (NaCl) form arrangements of ions called crystals. Why are there no sodium chloride molecules?

Because multiple ionically bonded atoms are attracted to each other in a crystal lattice, they do not form individual molecules. The smallest unit of an ionic compound is called a **formula unit**. This is the smallest whole number ratio that exists between the cations and anions in an ionic compound.

Crystals tend to be brittle and have high melting and boiling points. For example, the melting and boiling points of sodium chloride ($NaCl$) are 801°C and 1,413°C. Those of lithium fluoride ($LiF$) are 870°C and 1,676°C. In general, the stronger the ionic bonds between atoms in a compound, the higher their melting and boiling points. Based on this principle, scientists have determined that the ionic bonds holding lithium and fluoride ions together are stronger than the ionic bonds holding sodium and chloride ions together.

Another characteristic of ionic compounds is their ability to conduct electric current when melted. The cations flow to one electrode and the anions to another. This produces a flow of electricity. Ionic compounds also conduct electricity if they are dissolved in water because the ions are free to move in the solution.

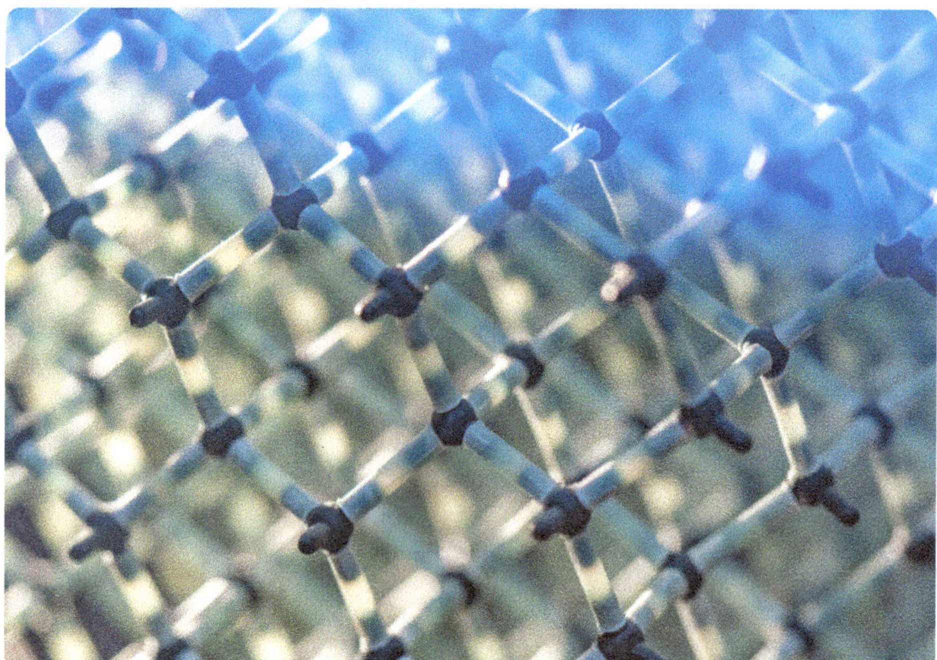

**CRYSTAL LATTICE MODEL**

An ionic crystal is formed by ions bound together by electrostatic forces. The regular arrangement of ions forms a structure called a crystal lattice. What are the properties of compounds with a crystal lattice structure?



### Consider the Explain Question

**How do atoms form ions that then combine to form crystals?**

Go online to complete the scientific explanation.

dlc.com/ca9059s

### Check Your Understanding

**How does the electronegativity of an element determine the type of bond formed with atoms of another element?**

dlc.com/ca9060s

## in Action

### Applying Ionic Bonding

Plants require 16 elements in order to grow and reproduce. Carbon, hydrogen, and oxygen are supplied by the atmosphere and water. The remaining nutrients are found in the soil. Those nutrients required in large amounts are called macronutrients and include nitrogen, phosphorus, potassium, magnesium, calcium, and sulfur. The other seven elements are needed in small amounts; they are called micronutrients. Many micronutrients, such as iron, copper, and zinc, are metals.



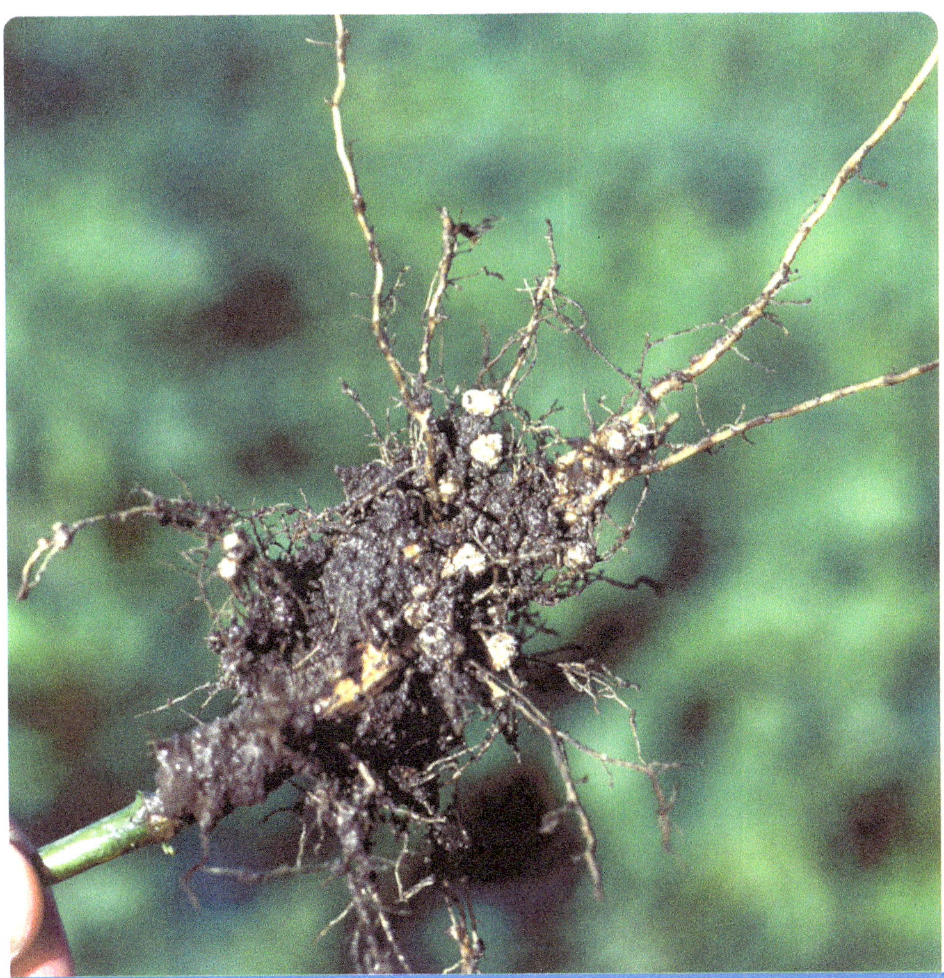

**IONIC BONDS IN NATURE**

How does the nature of ionic bonds influence the availability of nutrients for plants?

The soil nutrients are dissolved in water and then absorbed by the roots of the plant. Most nutrients are absorbed as ions, making the balance of ions in the soil very important. In addition, the pH of the soil plays a role in the availability of ions. In alkaline soil, phosphorus and calcium ions bond and become unavailable for the plants. Similarly, in acidic conditions, phosphorus, potassium, sulfur, and magnesium ions can be leached out of the soil. Many of the micronutrients are antagonistic to other nutrients; high levels prevent the uptake of other nutrients by plant roots. When gardeners apply fertilizers to plants, in many cases they are applying salts comprised of required nutrients. Recall that salts are ionic compounds that can disassociate into ions in solution.

Clay in soil helps to maintain a balance of nutrient ions available for plants. Clay particles possess two characteristics that help them support plant life. First, clay particles have an extremely large surface area for their mass. According to one source, if clay particles weighing 450 g (1 lb) were arranged like tiny tiles on a floor, they would cover an area of about 24,000 $m^2$—approximately the area of 4.5 football fields.

Second, the charge on the surface of clay particles is negative. Therefore, cations are attracted to the surface of the clay particles and stick there weakly. In the presence of water in the soil, the roots of plants can then absorb the cations from the clay and use them to support their life processes.

## STEM and Ionic Bonds

Crystallography is the study of the arrangement of particles and bonding of atoms in crystalline solids. Compounds formed by ionic bonds form crystal lattices in which the arrangement of the ions is repeated in a regular pattern. The ionic compounds do not contain individual molecules.

Crystals have characteristic geometrical shapes that are related to the arrangements of ions. In early studies, crystallographers learned about the shapes of these compounds by measuring the angles between the faces of crystals. They used the optic properties of crystals, measuring the diffraction angles and frequencies of light that passed through them. Modern X-ray diffraction crystallography, however, changed how scientists study crystals.

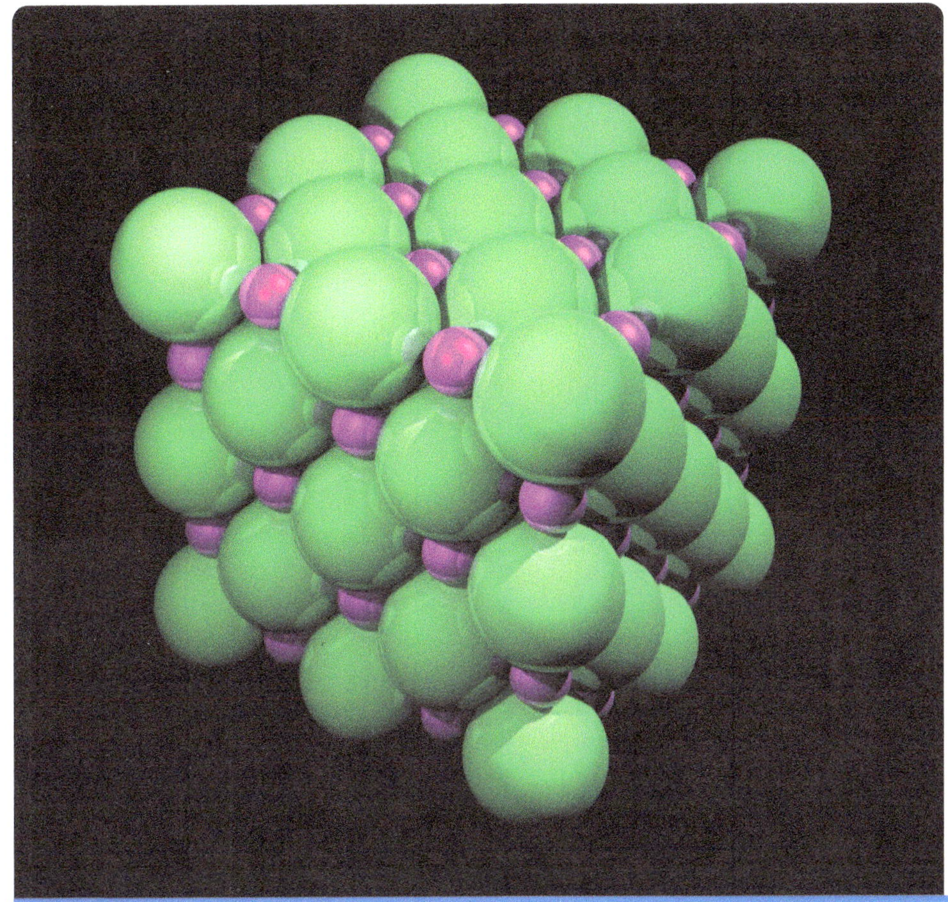

**A CAREER IN CRYSTALS**

The shape of a sodium chloride crystal is related to its crystal lattice structure. How could you determine which spheres are the sodium atoms?



When an X-ray beam is directed at a crystal, a diffraction pattern is produced. The diffraction pattern is related to the arrangement of the particles in the crystal. Using the diffraction pattern to predict the arrangement of a compound's ions can be tricky. The task has been compared to building a piano based only on the sounds it makes while it is falling down a flight of stairs. It is a complex procedure involving mathematical skills. This once tedious task, however, has been made much easier by advances in computer technology.

Although ionic crystals seem like the obvious subject of study in this field, the technique can even be applied to studies of large, complex organic molecules. In fact, the double helix structure of DNA was deduced, in part, from crystallographic data. Today, crystallographers work in many fields, including chemistry, physics, geology, biology, materials science, medicine, forensics, and metallurgy.

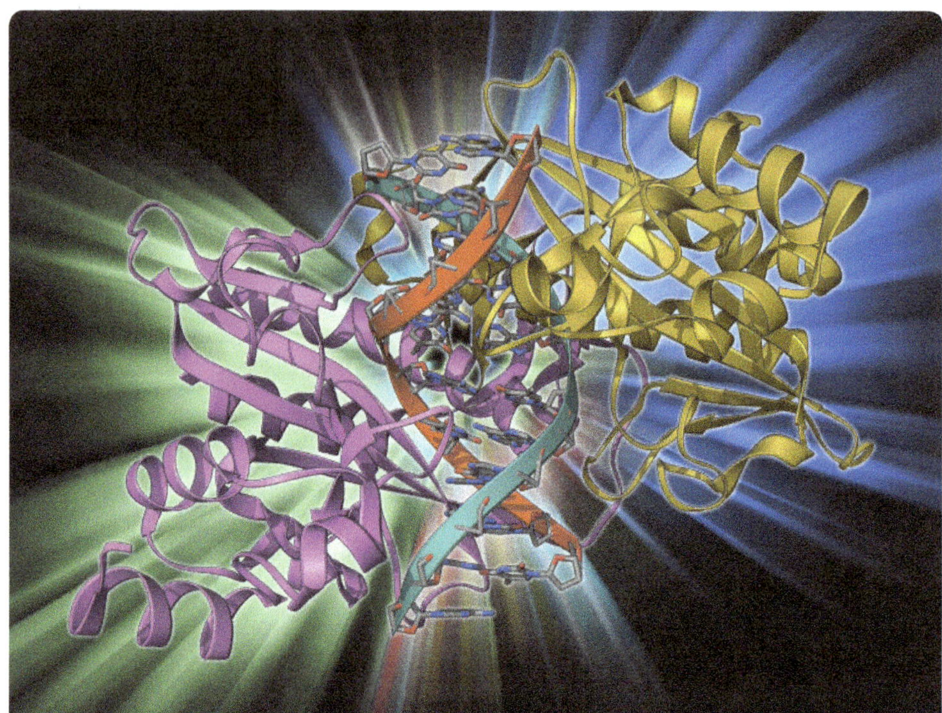

**CRYSTALLOGRAPHY OF COMPLEX MOLECULES**

This model of an enzyme (pink and yellow) is based on data from X-ray crystallography. Do you recognize the blue and red structure?



A person choosing a career as a crystallographer can expect to work in a laboratory setting. The work involves sophisticated technical instruments and computers. Chemistry, physics, and biology are important high school courses for a person interested in a career in crystallography. Mathematics, including calculus, is also necessary. Undergraduate college preparation includes chemistry, physics, mathematics, and advanced computer science and programming. Laboratory experience will be very valuable. Many crystallographers acquire an advanced degree, especially those who plan to work as research scientists. Some employment opportunities include research, scientific journal editing, materials engineering, or sales and marketing.

CONCEPT 3.3

# Covalent Bonding



## LESSON OVERVIEW

### Lesson Question

- How does the formation of covalent bonds relate to the octet rule?

### Lesson Objective

By the end of the lesson, you should be able to:

- Explain covalent bond formation and its relationship to the octet rule.

### Key Vocabulary

Which terms do you already know?

- [ ] carbon compound
- [ ] coordinate covalent bond
- [ ] covalent bond
- [ ] electronegativity
- [ ] hybrid bonding
- [ ] Lewis (electron) dot structure
- [ ] molecule
- [ ] octet rule
- [ ] resonance structure
- [ ] single bond
- [ ] $sp^2$ orbital
- [ ] $sp^3$ orbital
- [ ] sp orbital
- [ ] triple bond
- [ ] valence electron

dlc.com/ca9061s

## Thinking about Covalent Bonding

dlc.com/ca9062s

**Is your hair curly, or is it straight? What makes curly hair different from straight hair? There are two factors involved in hair curliness: how it emerges from the skin, and its chemistry.**

**CHANGING HAIR**

How can straight hair be made curly?

**EXPLAIN QUESTION**

**How does the structure of a water molecule demonstrate the octet rule?**

# How Does the Formation of Covalent Bonds Relate to the Octet Rule?

## Covalent Bonds and Electronegativity Difference

When two atoms form a chemical bond, the difference in their electronegativities can be used to classify the type of bond that forms. **Electronegativity** measures the ability of atoms to attract electrons. An electronegativity difference of 1.7 or greater indicates a transfer of one or more electrons and the formation of an ionic bond. Ionic bonds often form when metals bond with nonmetals.

An electronegativity difference below 1.7 means that neither atom attracts electrons strongly enough to allow the transfer of an electron. Instead, the atoms share electrons and form a **covalent bond**. Covalent bonds form between two nonmetal atoms because the atoms have generally similar electronegativities. The covalent bonds can be classified as nonpolar or polar bonds based on the equal or unequal sharing of the electrons.

Nonpolar covalent bonds form when the difference in electronegativity between bonding atoms is between 0.0 and 0.4. The electrons are shared equally between the atoms.

Polar covalent bonds form when the electronegativity difference is between 0.4 and 1.7. This results in unequal sharing of electrons. One atom with a higher electronegativity draws the bonding electrons toward itself, pulling those electrons away from the atom with the lower electronegativity value. The atom with the higher electronegativity becomes partially negatively charged and the atom with the lower electronegativity becomes partially positively charged. This makes the bond polar. Examples of atoms with a polar bond are water ($H_2O$) and ammonia ($NH_3$).

**AMMONIA**

In an ammonia molecule, electrons in the hydrogen atoms are pulled towards the nitrogen atom. Which type of atom in this molecule is partially positively charged?

## The Octet Rule and Lewis Structures

The elements of Group 18 are the noble gases. They are nonreactive nonmetals. These elements exist naturally as individual atoms. What makes these atoms unlikely to form chemical bonds?

Think about the number of valence electrons that are in the outermost energy level of an atom. Each noble gas atom has eight valence electrons, except for the helium atom which has only two valence electrons.

A convenient model to show the valence electrons in an atom is a Lewis structure. In this model, an atom's nucleus and inner electrons are represented by its chemical symbol. A dot is used for each **valence electron**. For example, the Lewis structure for a neon atom begins with the chemical symbol, Ne. This symbol is then surrounded by two dots above, below, to the left, and to the right. The dots represent neon's eight valence electrons.

**NEON**

Neon is a noble gas and therefore has a full valence shell of eight electrons. What does this mean for neon's ability to create covalent bonds?

Atoms that have eight valence electrons (and helium with two) do not tend to react with other atoms. Many other atoms, however, tend to form chemical bonds that put eight electrons in their outermost energy levels. The **octet rule** describes the tendency of atoms to gain, lose, or share electrons until they have eight valence electrons.

## Single, Double, and Triple Bonds

The Lewis structure of a fluorine atom shows that each fluorine atom has seven valence electrons.

**FLUORINE**

Fluorine has an almost full valence shell of seven electrons. How does this observation explain fluorine's reactivity?

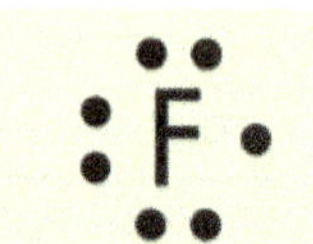

A fluorine atom needs only one more electron to complete its octet. If a fluorine atom shares one of its electrons with another fluorine atom, the valence shells of both atoms become filled. The two shared electrons form a **single bond** between the atoms.

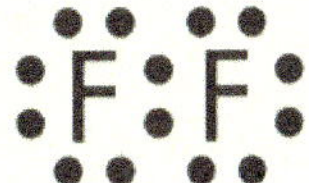

**FLUORINE GAS**

Fluorine gas consists of two fluorine atoms with a bond allowing them to share a single pair of electrons. How does this bond compare to unbounded atoms?

A single covalent bond is composed of two shared electrons. In the Lewis structure for the fluorine **molecule** formed by this bond, a line can be used to represent the shared electron pair.

**SINGLE BOND**

A line can be used to represent the shared electron pair. The bonding between which other elements could be represented this way?

The Lewis structure for oxygen shows that each atom has six valence electrons. An oxygen atom needs two more electrons to complete its octet.

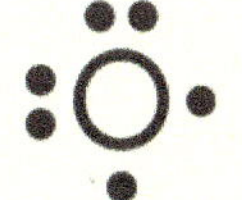

**OXYGEN**

Oxygen has an almost full valence shell of six electrons. What does this arrangement mean for oxygen's reactivity?

A single bond with another oxygen atom would not fill the octet of either atom. Each atom would only have seven electrons in its valence shell. Instead, oxygen atoms form a double bond. They share two pairs of electrons between them. In this way, each oxygen atom completes its octet.

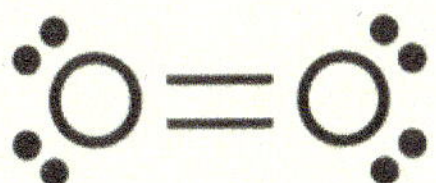

**DOUBLE BOND**

Oxygen gas consists of two oxygen atoms with a double bond connecting them. How does a double bond compare to a single bond?

A **triple bond** consists of three pairs of electrons shared by two atoms. What kinds of atoms can form a triple bond? An atom must have room for at least three electrons in its valence shell if it is to form a triple covalent bond. A nitrogen atom has five valence electrons, so it needs three more electrons to complete its octet.

**NITROGEN**

Nitrogen's Lewis structure shows that it only has five electrons in its valence shell. How does this arrangement affect its reactivity?

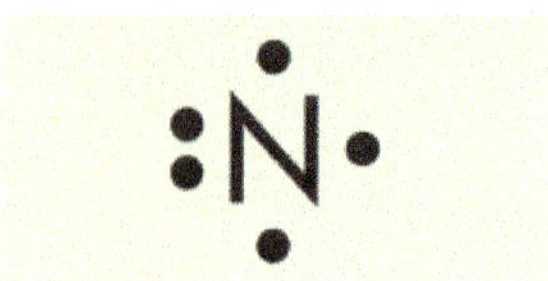


Two nitrogen atoms can form a nitrogen molecule by sharing three pairs of electrons in a triple covalent bond. The Lewis structure for the nitrogen molecule shows the triple bond as three lines between the nitrogen symbols.

**TRIPLE BOND**

Nitrogen gas consists of two nitrogen atoms with a triple bond connecting them. How does a triple bond compare to a single bond or a double bond?

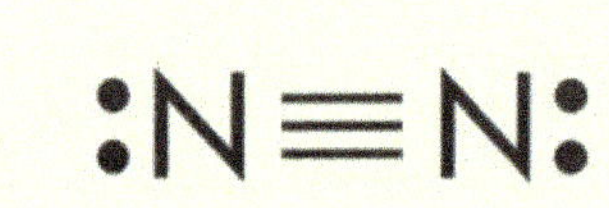


## Coordinate Covalent Bonds

Covalent bonds form when an atom shares valence electrons with another atom. For example, in a molecule of methane, $CH_4$, a carbon atom forms a single bond with each of four hydrogen atoms. Each single bond is made up of one electron from the carbon atom and one electron from a hydrogen atom.

**METHANE**

Methane consists of a carbon atom forming covalent bonds with four hydrogen atoms. How many electrons are originally in carbon's valence shell?

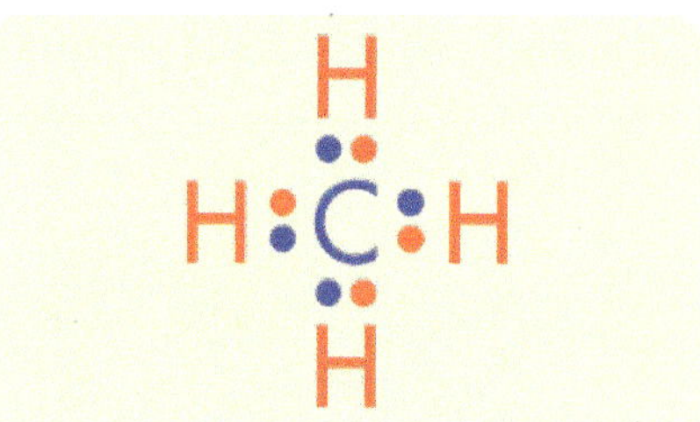

Likewise, each single bond in ammonia, $NH_3$, is composed of one electron from the nitrogen atom and one electron from a hydrogen atom.

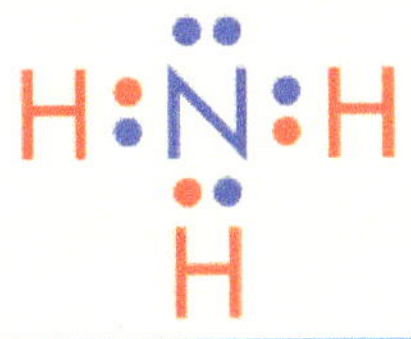


**AMMONIA**

Ammonia consists of a nitrogen atom forming covalent bonds with three hydrogen atoms. How many electrons are originally in nitrogen's valence shell?

The nitrogen atom in $NH_3$ has a complete octet. Can the nitrogen atom form a fourth bond to make the polyatomic ion ammonium ($NH_4^+$)? Yes. The nitrogen atom in ammonia has a nonbonding pair of electrons. The nitrogen atom uses this pair of its own electrons to form a single bond with a hydrogen ion, which has no electrons. A covalent bond in which both electrons come from one atom is a **coordinate covalent bond**.

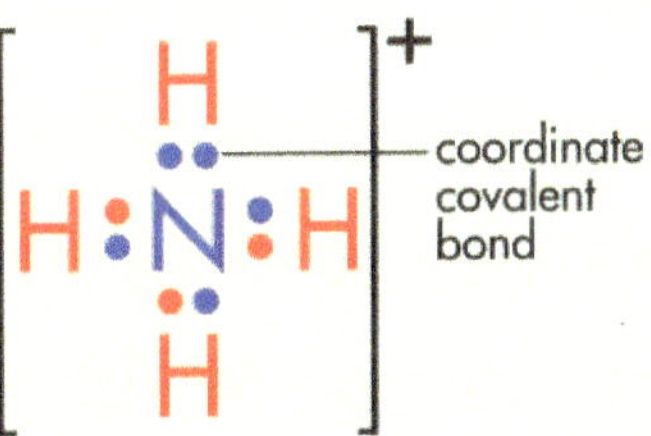


**COORDINATE COVALENT BOND**

Coordinate covalent bonds consist of bonds in which both electrons come from one atom. How does polyatomic ion ammonium compare with ammonia?

## Resonance Structures

Lewis structures often show the correct bonding in a molecule. But as with any model, Lewis structures have their limitations. In some molecules that have double or triple bonds, the locations of the bonds may vary. We say that these molecules exhibit resonance. They cannot be represented by only one Lewis structure. Resonance structures represent all of the possible positions of electrons in a molecule's chemical bonds.

For example, ozone, $O_3$, is an important molecule for life on Earth. Ozone helps to shield Earth's surface from harmful ultraviolet light. The Lewis structure of an ozone molecule has a central oxygen atom joined to one oxygen atom by a single bond and to another oxygen atom by a double bond.

**OZONE**

Ozone consists of three oxygen atoms. How are these oxygen atoms bonded to each other?

**OZONE SECOND STRUCTURE**

Ozone consists of three oxygen atoms, one pair bonded by a single bond and the other pair bonded by a double bond. How does this ozone structure compare to the prior arrangement?

The single bond is a coordinate covalent bond because both electrons come from the central oxygen atom. But this is not the only possible configuration for this molecule. A second Lewis structure can be drawn in which the single and double bonds are switched.

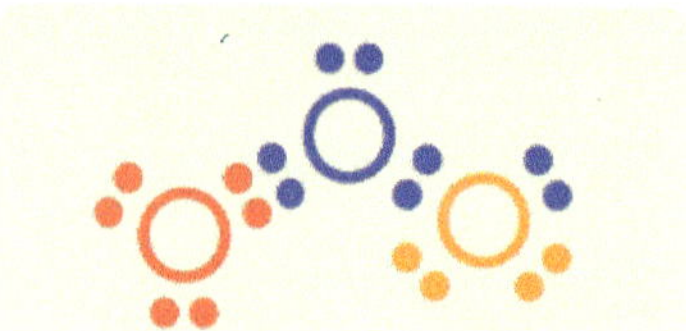

Thus, ozone has two resonance structures. Studies of ozone show that both bonds are identical. Neither bond is a single bond or a double bond. Instead, an ozone molecule is a mixture of the two resonance structures. The bonds have characteristics that fall somewhere between those of single bonds and those of double bonds.

## Covalent Bonding and Properties of Substances

Many covalently bonded compounds, such as methane and carbon dioxide, are gases at room temperature and standard pressure. Others, such as water, are liquids at room temperature and standard pressure. When atoms bond covalently, they can form molecules of simple compounds that have low melting points and low boiling points.

However, some atoms that bond covalently do not form individual molecules. Instead, they form a large network of bonded atoms. These substances are referred to as network solids or network compounds. For example, silicon dioxide or silica, $SiO_2$, does not exist as individual molecules. Each silicon bonds with four oxygen atoms, and each oxygen atom bonds with two silicon atoms. The result is a network compound that is very hard and has a high melting point, which is why glass is made from silica.

**QUARTZ AND SILICON DIOXIDE**

Quartz is a network solid composed of almost pure silica. The silicon and oxygen atoms in silicon dioxide form a strong network of covalent bonds. What would you predict are some properties of silicon dioxide?

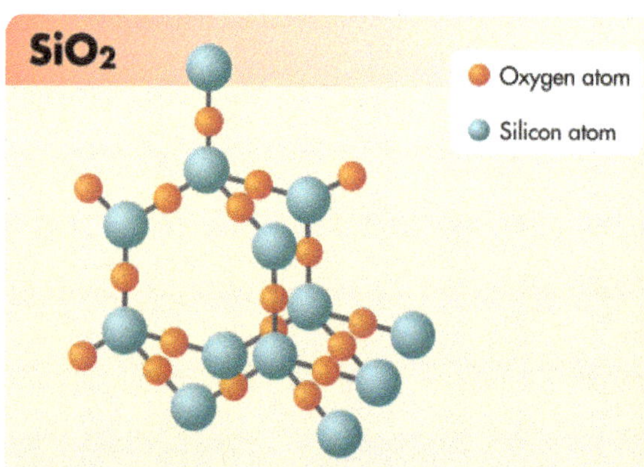

## Hybridization in Carbon Atoms

Each carbon atom can form up to four bonds with other atoms. The Lewis structure of carbon shows us how this works. Carbon has four single electrons that can be used to form chemical bonds.

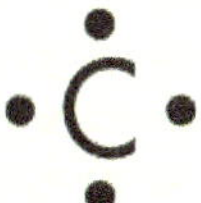


**CARBON**

The carbon atom has four electrons in its valence shell. How does this affect carbon's ability to form bonds?

A carbon atom can complete its octet in several ways:

- four single bonds
- two double bonds
- a double bond and two single bonds
- a triple bond and one single bond

However, the orbital diagram for a carbon atom tells a different story. The valence electrons are not arranged as four single electrons. Two electrons are paired in the 2*s* orbital. The other two electrons are unpaired in two 2*p* orbitals.

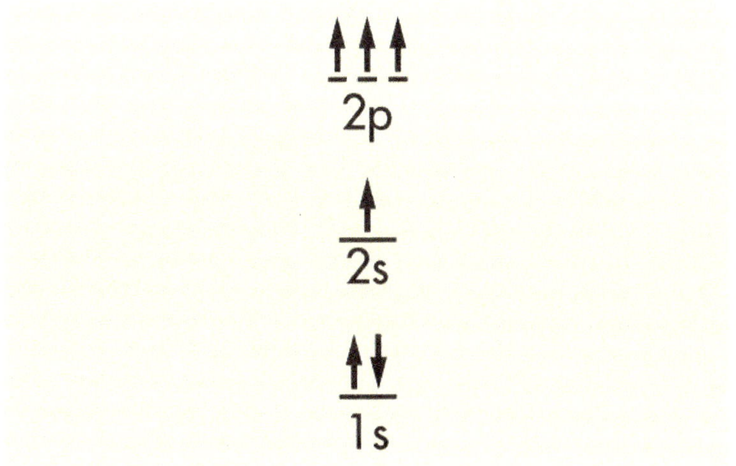


**METHANE ORBITAL DIAGRAM**

The orbital diagram conveys information about the number of electrons in the valence shell of an atom or molecule. How does this compare to the orbital diagram of carbon?

So how does the carbon atom in a methane molecule form four identical single bonds?

First, an electron from the 2*s* orbital enters the empty 2*p* orbital.

**CARBON ORBITAL DIAGRAM**

The orbital diagram conveys information about the number of electrons in the valence shell of an atom or molecule. How does this diagram compare to the Lewis dot diagram of carbon?

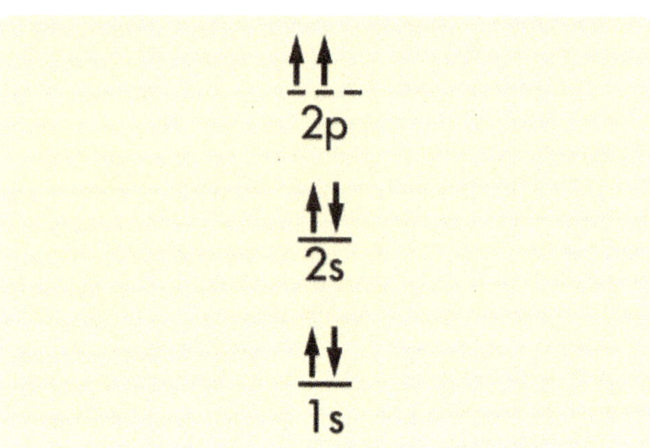


This gives the carbon atom four unpaired electrons. But these electrons are in different types of orbitals, so they cannot form four identical bonds. However, the orbitals are combined in a process called hybridization. The result is four identical orbitals.

**HYBRIDIZATION**

The orbitals here are combined in a process called hybridization. How does this relate to the orbital diagram of methane?

$sp^3$

1s

The orbitals are equal in energy. Because they formed from one *s* and three *p* orbitals on the same energy level, these four orbitals are called $sp^3$ hybrid orbitals. They allow the carbon atom to form four identical single bonds.

To form a double bond and two single bonds, a carbon atom needs to keep one electron in a *p* orbital. The *s* and two of the *p* orbitals hybridize. So, the carbon atom in methanal, $H_2CO$, uses $sp^2$ hybrid orbitals.

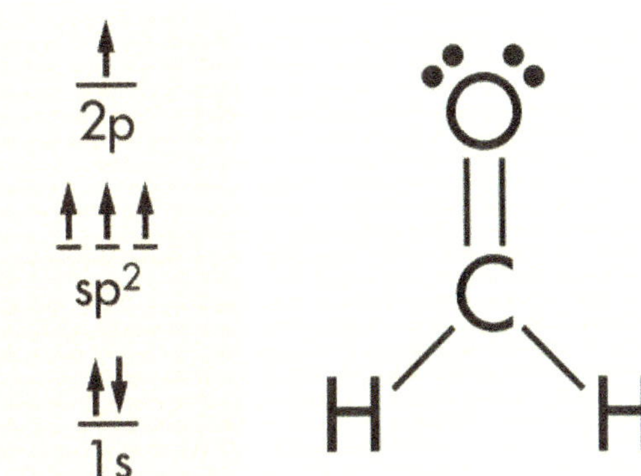


***sp* HYBRID ORBITALS**

Methanal forms a double bond and two single bonds. What needs to happen to form these bond types?

To form a triple bond or two double bonds, a carbon atom needs to keep two electrons in *p* orbitals. The *s* and one *p* orbital hybridize. In this case, the carbon atoms in hydrogen cyanide, HCN, and in carbon dioxide, $CO_2$, use *sp* hybrid orbitals.

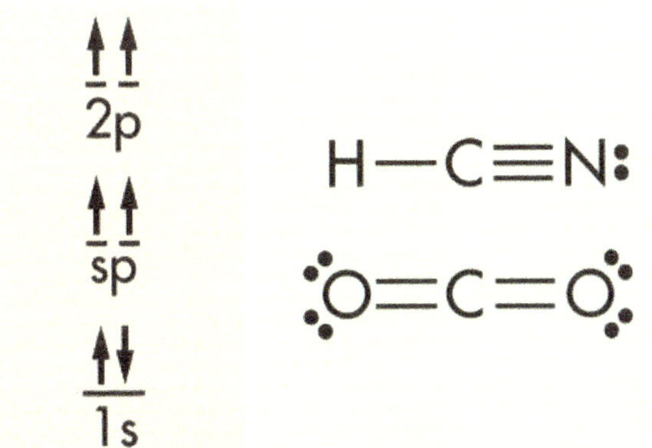


**$sp^2$ HYBRID ORBITALS**

Hydrogen cyanide, HCN, and carbon dioxide use *sp* hybrid orbitals. How would you describe the properties of these bonds?



## Consider the Explain Question

**How does the structure of a water molecule demonstrate the octet rule?**

Go online to complete the scientific explanation.

dlc.com/ca9063s

## Check Your Understanding

**Can you model what happens to electrons during the formation of covalent bonds?**

dlc.com/ca9064s

# STEM in Action

## Applying Covalent Bonding

Have you ever had to breathe oxygen from an oxygen tank or used nitrous oxide—laughing gas—as an anesthetic in a dentist's office? Have you ever been scuba diving and breathed air from a tank, or cooked a hamburger on a gas grill that used a propane tank? These events all involve gases that have been compressed and stored. Because covalent compounds have lower intermolecular forces than ionic compounds of the same size, covalent molecules such as nitrous oxide, oxygen, and propane exist as gases at room temperature. Many of the uses for covalent substances are based on the properties of covalent molecules. Covalent substances generally exist as individual molecules and are more likely to be soft solids, liquids, or gases. Ionic compounds exist as a lattice of ions in which many ions of opposite charge bond together. Thus, ionic compounds are, in general, hard and brittle solids.

**PROPANE GAS**

This grill uses propane gas from a small storage tank. Why is propane a gas at room temperature?

## Lesson Objectives

By the end of the lesson, you should be able to:

- Predict the type of bond between atoms of a molecule from the electronegativities of those atoms.
- Explain the intermolecular forces between polar molecules and their relative strengths.
- Use Lewis structures to represent covalent molecules and polyatomic ions.
- Use Lewis structures and VSEPR to predict the polarities, geometries, and bond angles of covalent molecules.
- Use Lewis structures and VSEPR to explain the unique properties of water.

## Key Vocabulary continued

- ☐ nonpolar covalent bond
- ☐ octet rule
- ☐ polar covalent bond
- ☐ polarity
- ☐ polyatomic ion
- ☐ surface tension
- ☐ tetrahedron
- ☐ valence shell
- ☐ valence electron
- ☐ VSEPR theory
- ☐ water



The oxygen gas in an oxygen tank is just one element, but oxygen, $O_2$, exists in nature as a **molecule** of two covalently bonded oxygen atoms. Nitrous oxide molecules, $NO_2$, consist of two nitrogen atoms and one oxygen atom that are covalently bonded. The gas in a scuba tank is usually compressed air, which is a mixture of several different gases. Most of the compressed air is nitrogen, $N_2$, which also exists in nature as a molecule of two covalently bonded nitrogen atoms. Compressed air also contains other gases found in air—oxygen, argon, and carbon dioxide ($CO_2$). Argon exists as a single atom, but carbon dioxide is a covalently bonded molecule. Propane, $C_3H_8$, is a molecule of three covalently bonded carbon atoms covalently bonded to eight hydrogen atoms.

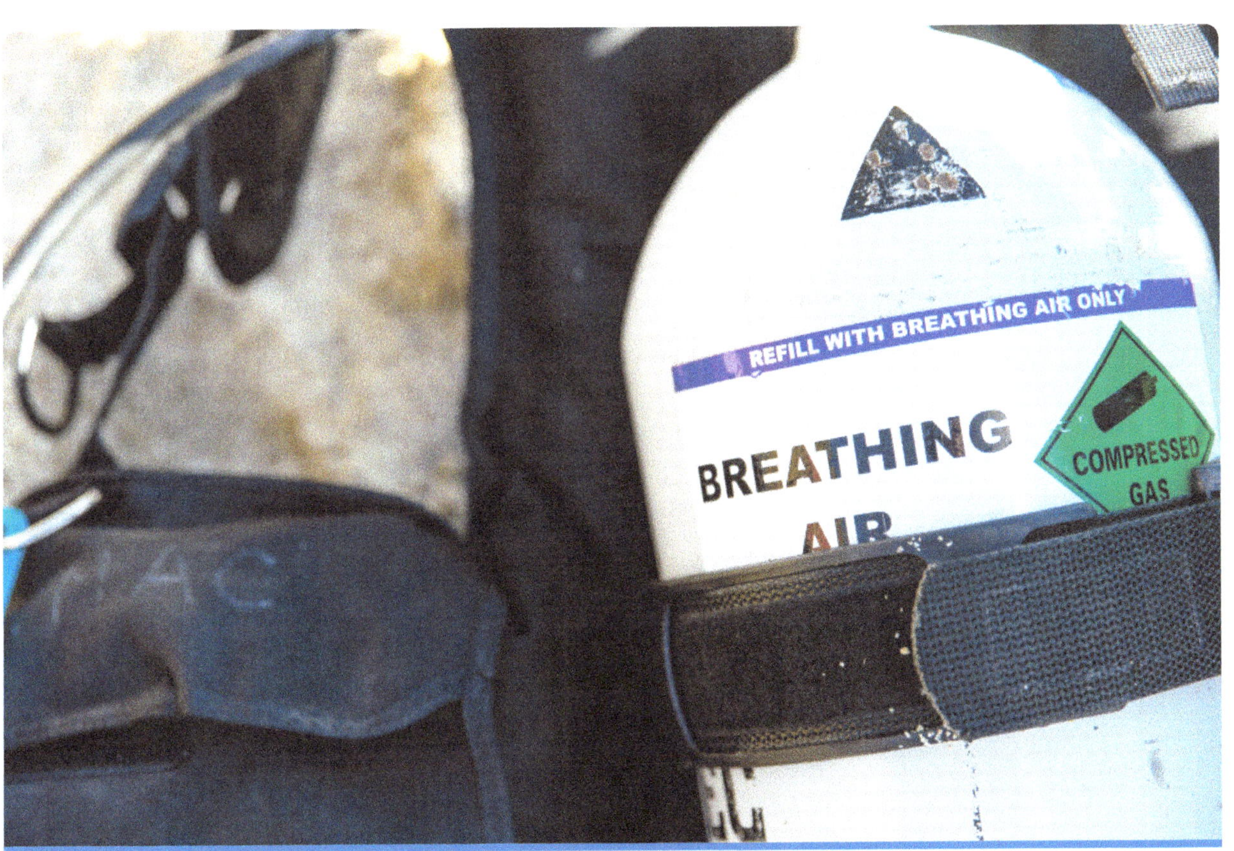


**BREATHING AIR**

Why do you think air has to be pressurized for a diver to use it?

### STEM and Covalent Bonding

Cyanoacrylate glues are found in many homes, hospitals, and even at crime scenes. Once applied, cyanoacrylates react with the thin film of moisture that usually exists on a surface. Double bonds between carbon atoms in the glue are broken during the reaction. Then, one molecule forms a new **single bond** to another molecule, which forms a new single bond to another molecule, and so on. Long chains, or polymers, of cyanoacrylate molecules join the glued pieces together.

ELABORATE

The ability of cyanoacrylate glues to bond to skin led to their use in medicine. These compounds are used to close wounds without stitches. They have become very important in battlefield medicine.

Forensic scientists use chemistry in a number of ways to help solve crimes. A forensic scientist might use cyanoacrylate fuming to develop latent fingerprints and make them visible. A latent fingerprint comes from materials such as sweat and oils that are left behind when a finger touches a surface. Fresh fingerprints can adequately be developed by fingerprint powders because these powders stick to sweat. However, a forensic scientist knows that cyanoacrylate adheres to skin components that do not evaporate, such as oils, proteins, and fatty acids. This glue enables the scientist to lift fingerprints that are old or are on uneven, multicolored, or otherwise difficult surfaces. Covalent bonding between atoms helps cyanoacrylate glues to be truly super.

Fire scene investigations often rely on forensic chemists to determine if certain compounds called accelerants were used by arsonists to accelerate the development of a fire. Commonly used accelerants are covalently bonded compounds such as gasoline or kerosene. When a forensic chemist takes samples at a crime scene, he or she does not know the nature of the sample until it is analyzed. If chemical tests reveal evidence of accelerants, investigators use this information to narrow the suspects by looking at who had access to the accelerant found at the scene. Sometimes forensic scientists are asked to testify in court trials and explain their findings.

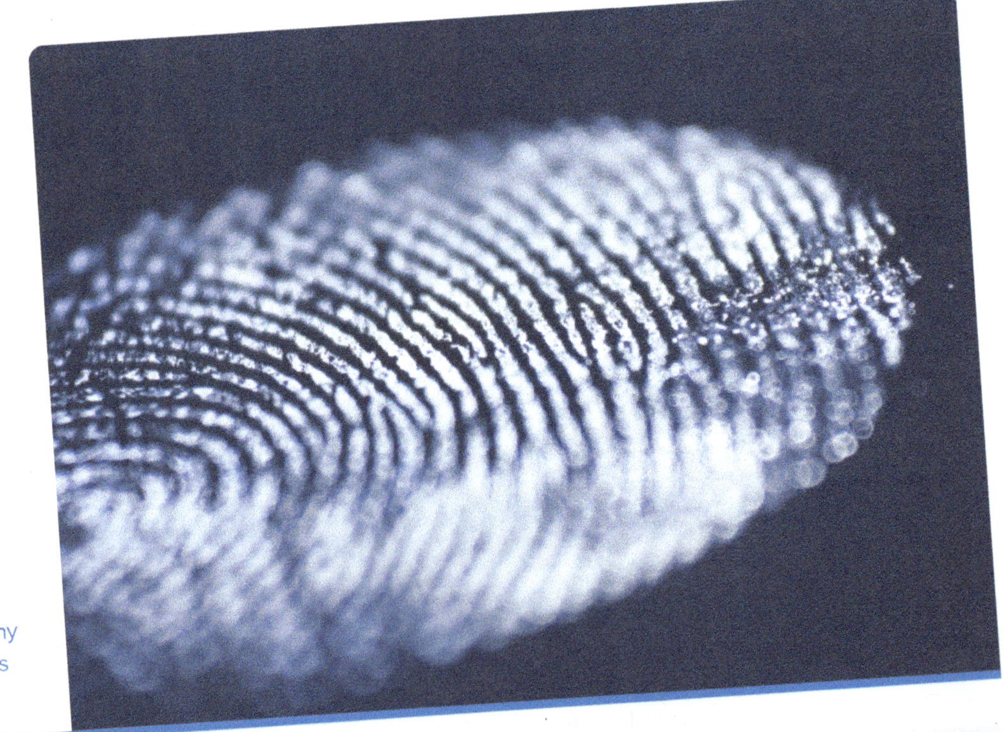

**FINDING FINGERPRINTS** Fingerprints can help investigators to place a suspect at a crime scene. How does cyanoacrylate glue help to reveal latent fingerprints, even many days after a crime was committed?





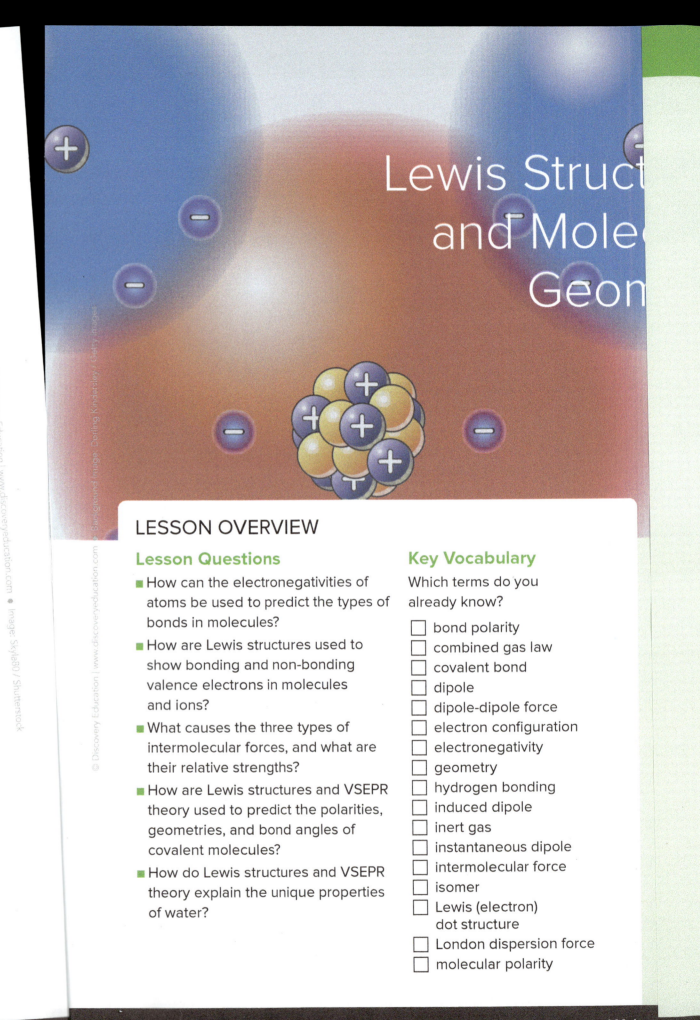

# Lewis Struct[ures] and Mole[cular] Geom[etry]



## LESSON OVERVIEW

### Lesson Questions

- How can the electronegativities of atoms be used to predict the types of bonds in molecules?
- How are Lewis structures used to show bonding and non-bonding valence electrons in molecules and ions?
- What causes the three types of intermolecular forces, and what are their relative strengths?
- How are Lewis structures and VSEPR theory used to predict the polarities, geometries, and bond angles of covalent molecules?
- How do Lewis structures and VSEPR theory explain the unique properties of water?

### Key Vocabulary

Which terms do you already know?

- ☐ bond polarity
- ☐ combined gas law
- ☐ covalent bond
- ☐ dipole
- ☐ dipole-dipole force
- ☐ electron configuration
- ☐ electronegativity
- ☐ geometry
- ☐ hydrogen bonding
- ☐ induced dipole
- ☐ inert gas
- ☐ instantaneous dipole
- ☐ intermolecular force
- ☐ isomer
- ☐ Lewis (electron) dot structure
- ☐ London dispersion force
- ☐ molecular polarity

## Constructing a Chemical Compound

**Have you ever assembled something that came with a set of instructions? If so, did the instructions list the exact number of pieces along with an explanation for how those pieces fit together? Similarly, molecules are made up of specific components arranged in a specific way.**

dlc.com/ca9066s

### EXPLAIN QUESTION

**What does the molecular geometry of a chemical compound tell you about its properties?**

**WATER MOLECULES**

How many atoms and bonds are present in water's structure? What is the shape of the molecule?

# How Can the Electronegativities of Atoms Be Used to Predict the Types of Bonds in Molecules?

## Electronegativity

**Electronegativity** is the measure of an atom's ability to attract electrons. The unit of measurement for electronegativity is called the Pauling, named after an important American chemist.

Nonmetals have a greater attraction for electrons because their **valence shell** electrons are closer to a complete octet. In the periodic table, electronegativity increases from left to right and from bottom to top. The first three periods of the periodic table show these trends. The inert gases are not included because they rarely form bonds.

The difference in the electronegativity of two atoms is a good predictor of the type of bond that will form between them. If the difference is large, the bond will be ionic because the bonding electrons will be transferred completely to the more electronegative atom. If the difference is small, the bond will be nonpolar covalent, and the bonding electrons will be shared equally. If the difference is moderate, the bond will be polar covalent, and the bonding electrons will be shared; however, the electrons in the bond will be pulled toward the more electronegative atom. In a molecule with more than two atoms, the electronegativity difference is calculated separately for each bond between two atoms. The table gives the approximate ranges for these three types of bonds.

| Electronegativity Difference | Bond Type |
|---|---|
| Less than 0.4 | Nonpolar covalent |
| 0.4 to 1.7 | Polar covalent |
| Greater than 2.0 | Ionic |

**ELECTRONEGATIVITY AND BOND TYPES**

Electronegativity determines bond types. What type of bonding would you expect to find in an oxygen molecule?

The dividing lines between the three types of bonds are not definite. Sometimes classification depends on the atoms involved.

Increasing →

Increasing ↑

| 1<br>H<br>2.1 | | | | | | |
|---|---|---|---|---|---|---|
| 3<br>Li<br>1.0 | 4<br>Be<br>1.5 | 5<br>B<br>2.0 | 6<br>C<br>2.5 | 7<br>N<br>3.0 | 8<br>O<br>3.5 | 9<br>F<br>4.0 |
| 11<br>Na<br>0.9 | 12<br>Mg<br>1.2 | 13<br>Al<br>1.5 | 14<br>Si<br>1.8 | 15<br>P<br>2.1 | 16<br>S<br>2.5 | 17<br>Cl<br>3.0 |

**PERIODIC TREND: ELECTRONEGATIVITY**

Some versions of the periodic table show the electronegativities of the elements. What is the most electronegative element?

### Electronegativity: Sample Problem

Use the electronegativity values listed in the periodic table to predict the type of bonding in each of these molecules.

1. MgO 2. $Cl_2$ 3. $SO_2$

**Solution:**

MgO: Electronegativity difference = 3.5 − 1.2 = 2.3 Paulings. The bond is ionic.

$Cl_2$: Electronegativity difference = 3.0 − 3.0 = 0.0 Paulings. The bond is nonpolar covalent.

$SO_2$: Electronegativity difference = 3.5 − 2.5 = 1.0 Paulings. The bond is polar covalent.

## How Are Lewis Structures Used to Show Bonding and Non-Bonding Valence Electrons in Molecules and Ions?

### Lewis Structures

A Lewis structure is a kind of shorthand for showing the details of various types of bonds. This method is especially useful for showing covalent bonding between nonmetal atoms, as is found in most organic molecules.

Remember that atoms are most stable when they lose, gain, or share electrons to complete the outer (valence) shell. Noble gases are stable and unreactive because their outer shell of electrons is already full. For the second and third rows of the periodic table, a complete valence shell has eight electrons, and for the first row, a complete **valence shell** has only two electrons.

**HYDROGEN AND NITROGEN DOT STRUCTURES**

Hydrogen has only one electron. Nitrogen has five electrons. How many electrons are needed to complete each of their outer shells?

To use Lewis structures to predict bonding in a molecule, begin with the structures of the neutral atoms. For example, consider the structures for nitrogen and hydrogen, which combine to form ammonia.

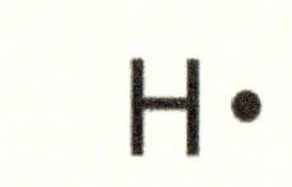

**AMMONIA MOLECULE $NH_3$**

In this electron dot structure, all of the available electrons are arranged around nitrogen. Why is N the best choice for the central atom?

Nitrogen is three electrons short of a complete shell, and hydrogen is one short. The following combination will give them both complete, stable valence shells:

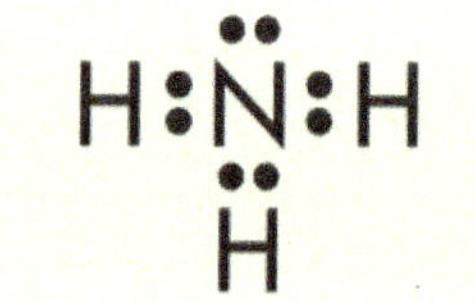

**AMMONIA MOLECULE $NH_3$**

The electrons for the ammonia molecule can be depicted in two different ways. Why are the two electrons above the symbol, N, not replaced with a line?

The pairs of electrons between atoms represent bonds, and the other pair (shown above the N) is a non-bonding pair. Showing bonds as a line between atoms is equivalent to showing two dots.

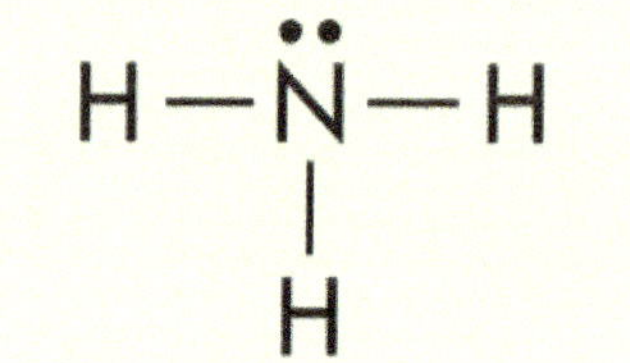

Bonds can also have more than two electrons. Double bonds have four electrons and are represented by two lines. Triple bonds have six electrons and are represented by three lines.

Double and triple bonds are less flexible than single bonds. Double bonds in general have higher bond energies and are shorter than single bonds. Triple bonds have even more energy and are shorter than double bonds.

These are the Lewis structures for carbon dioxide and hydrogen cyanide.

$$H-C\equiv N:$$

**HYDROGEN CYANIDE**

Hydrogen cyanide was first isolated from the pigment Prussian blue. What is the Lewis structure for the cyanide ion?

$$:\ddot{O}=C=\ddot{O}:$$

**CARBON DIOXIDE**

The Lewis structure for carbon dioxide contains double bonds. How do the length and strength of these bonds compare to the single bonds in methane, $CH_4$?

Ions can also be represented by Lewis structures. For example, the non-bonding electron pair in $NH_3$ can attract a positively charged hydrogen ion to produce the stable ammonium ion, $NH_4^+$.

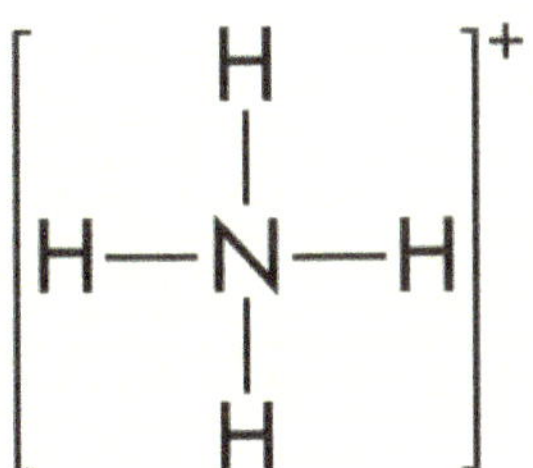


**AMMONIUM ION LEWIS STRUCTURE**

The ammonium ion often forms when ammonia, $NH_3$, reacts with an acid. What species is contributed by the acid?



## Lewis Structures: Sample Problem 1

Draw the Lewis structure for $CH_2Br_2$.

**Solution:**

1. **Determine the type and number of atoms in the molecule.**

   The formula shows one carbon atom, two hydrogen atoms, and two bromine atoms.

2. **Write electron dot notations for each element in the compound.**

   We can use the periodic table to determine the number of electrons in the valence shell of each element. Carbon has 4, hydrogen has 1, and bromine has 7.

3. **Find the total number of available valence electrons in the atoms that will combine.**

   - C: $1 \times 4e^- = 4e^-$
   - H: $2 \times 1e^- = 2e^-$
   - Br: $2 \times 7e^- = 14e^-$
   - Electrons available for molecule: $20e^-$

4. **Arrange the atoms to form the molecule**

   Carbon will be the central atom. If carbon is not present, the atom with the lowest electronegativity will be the central atom. Hydrogen is never the central atom. Use electron pairs to form bonds.

   Br
   ••
   H : C : H
   ••
   Br

5. **Add the remaining electrons around the atoms in the molecule so that each hydrogen shares one pair of electrons and other atoms have an octet.**

   Be sure to use only available electrons.

6. Arranging the elements so that they share valence electrons in a manner that fills each valence shell, we can draw the following:

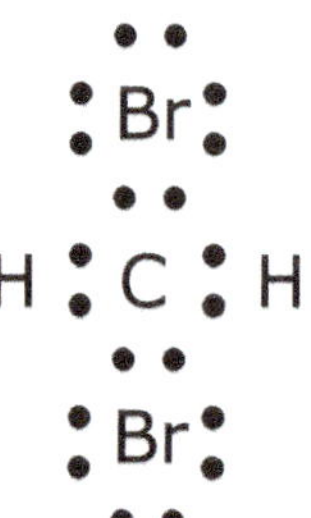


Double check to be sure that the number of valence electrons equals the number of available electrons. Be sure all atoms except hydrogen have a full octet.

## What Causes the Three Types of Intermolecular Forces, and What Are Their Relative Strengths?

### Dipoles

A **polar covalent bond** causes the more electronegative atom to have a partial negative charge and the other atom to have a partial positive charge. These partial charges are less than the charge on an electron or proton. This charge separation is called a **dipole**.

If each polar molecule has only two atoms, there will be attractive forces between the molecules. These dipole-dipole forces are of moderate strength. As we will see later, a molecule with more than two atoms may or may not have an overall dipole, depending on its **geometry**.

## Intermolecular Forces

Intermolecular forces are forces of attraction between separate molecules. There are three main types of intermolecular forces:

- The partial negative charge on an atom in a polar covalent molecule causes it to be attracted toward atoms with a partial positive charge on other molecules. These are dipole-dipole forces.
- Hydrogen atoms with a partial positive charge are attracted to electronegative atoms, such as oxygen and nitrogen, on other molecules. This is the strongest of the intermolecular forces and is called **hydrogen bonding**.
- Because electrons are always on the move, at any particular instant they may be unevenly distributed. This causes instantaneous dipoles, which create weak attractions between molecules. These London dispersion forces are the weakest intermolecular forces. These forces can arise even in completely nonpolar molecules.



**London Dispersion Forces**
- instantaneous dipoles
- weakest intermolecular force

**Dipole-Dipole Forces**
- negative charge on one atom attracts to the positive charge on another
- moderate intermolecular force

**Hydrogen Bonds**
- hydrogen has positive charge
- bonds with electronegative atoms
- strongest intermolecular force

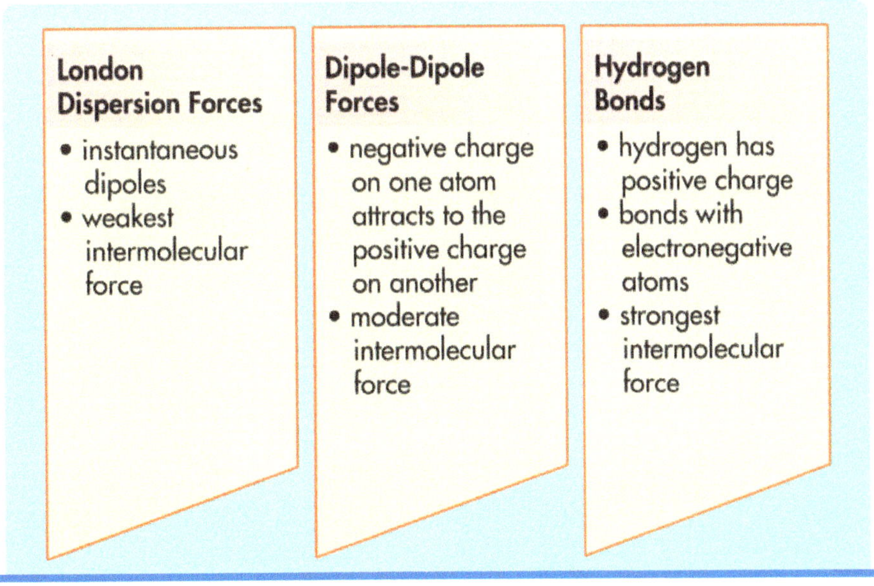

**INTERMOLECULAR FORCES**
There are three types of intermolecular forces. Which of these forces is the strongest?

The electron configurations of elements and their positions on the periodic table can be used to predict the shapes of the molecules they form. In turn, molecular shape can be used to predict the strengths of intermolecular forces. The same information can be used to predict some of the physical properties of these compounds. Those molecules that have shapes conducive to the formation of strong intermolecular forces will have higher melting and boiling points than those that do not. This is because more energy is required to overcome the intermolecular forces that hold the molecules together. A general rule for compounds with similar molecular masses is therefore that the stronger the intermolecular forces, the higher the melting and boiling points.

**Water** exemplifies the effect of intermolecular forces on melting and boiling points. The hydrogen bonding between polar water molecules accounts for water's relatively high melting and boiling points. Other compounds of similar size are usually gases at room temperature.

Intermolecular forces also impact the solubilities of substances in water. This is because water has polar molecules that readily attract the molecules of other polar molecules.

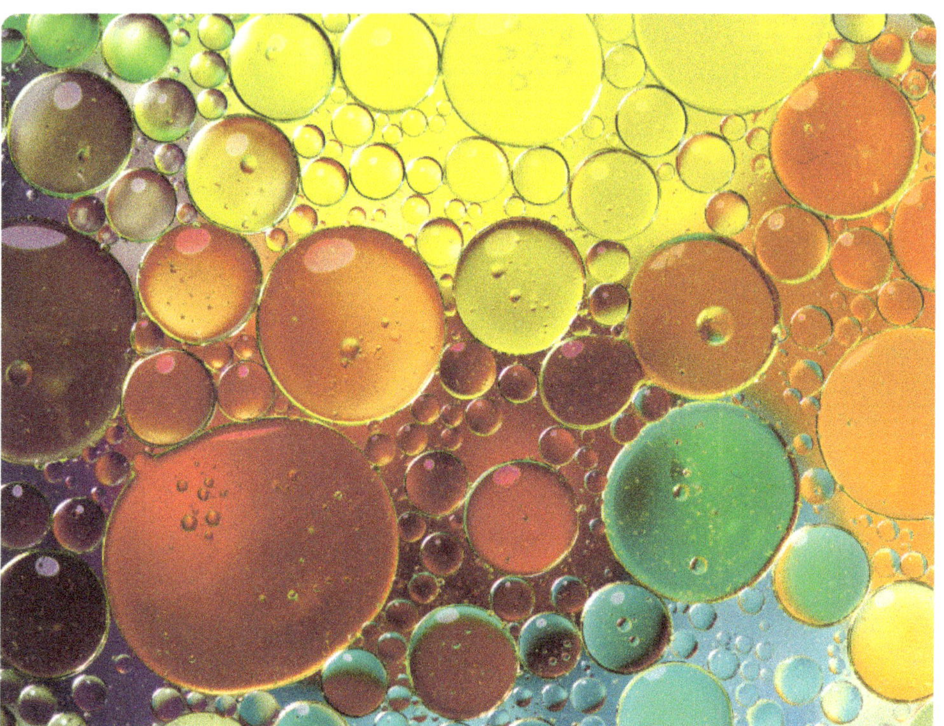

**OIL AND WATER**

When combined, oil and water form a mixture. Why does the oil form spherical structures?



## How Are Lewis Structures and VSEPR Theory Used to Predict the Polarities, Geometries, and Bond Angles of Covalent Molecules?

### The VSEPR Model

VSEPR stands for **Valence Shell** Electron Pair Repulsion. Because electrons all have negative charges, they repel each other. That is why the electrons and bonds in Lewis structures are spread evenly around the symbols. That is also why electrons don't pair until the valence shell is more than half full.

| Electron Regions, shape, & hybridization | | Bonding Regions | Lone Pairs | Electron Region Geometry | Molecular Geometry | Examples |
|---|---|---|---|---|---|---|
| 2 | —X— sp | 2 | 0 | linear | linear | $BeF_2$, $CO_2$ |
| | | 1 | 1 | | linear | CO, $N_2$ |
| 3  | $sp^2$ | 3 | 0 | trigonal planar | trigonal planar | $BF_3$, $CO_3^{2-}$ |
| | | 2 | 1 | | bent | $O_3$, $SO_2$ |
| | | 1 | 2 | | linear | $O_2$ |
| 4  | $sp^3$ | 4 | 0 | tetrahedral | tetrahedral | $CH_4$, $SO_4^{2-}$ |
| | | 3 | 1 | | trigonal pyramidal | $NH_3$, $H_3O^+$ |
| | | 2 | 2 | | bent | $H_2O$, $ICl_2^+$ |
| | | 1 | 3 | | linear | HF, $OH^-$ |
| 5  | $sp^3d$ | 5 | 0 | trigonal bipyramidal | trigonal bipyramidal | $PF_4$ |
| | | 4 | 1 | | seesaw | $SF_4$, $TeCl_4$, $IF_4^+$ |
| | | 3 | 2 | | T shaped | $ClF_3$ |
| | | 2 | 3 | | linear | $I_3^-$, $XeF_2$ |
| 6  | $sp^3d^2$ | 6 | 0 | octahedral | octahedral | $SF_6$, $PF_6^-$, $SiF_6^{2-}$ |
| | | 5 | 1 | | square pyramidal | $BrF_5$, $SbCl_5^{2-}$ |
| | | 4 | 2 | | square planar | $XeF_4$, $ICl_4^-$ |

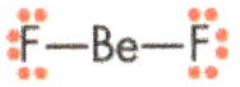

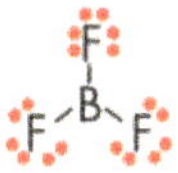

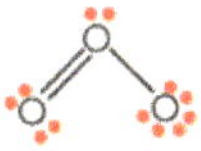

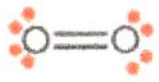

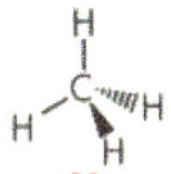

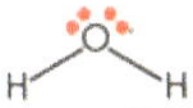

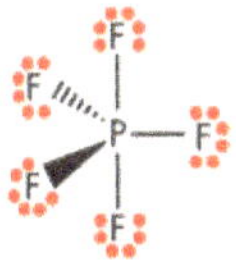

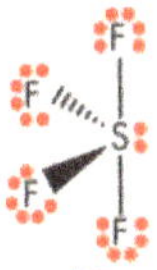

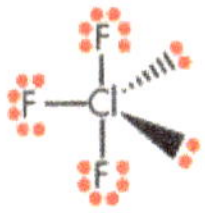

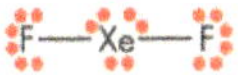

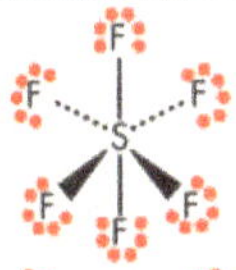

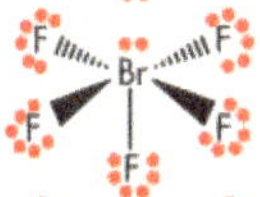

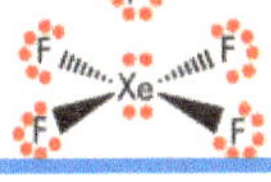

**ELECTRON AND MOLECULAR GEOMETRY**

This table shows the shapes of molecules based on VSEPR theory. Why are $BeF_2$ and $N_2$ the same shape?

Remember that Lewis structures are simpler than real molecules in several ways. For one thing, Lewis structure diagrams are always two-dimensional. The VSEPR model considers the third dimension instead of a flat plane. VSEPR explains that electrons can actually move in all directions, and this allows them to move even farther apart. Look at the Lewis structure for methane, $CH_4$.

**METHANE LEWIS STRUCTURE**

Methane has four single bonds in this two-dimensional model. What might affect the molecule's structure when seen in three dimensions?

H
|
H—C—H
|
H

The angle between the bonds in this drawing is 90°. The true shape of the $CH_4$ molecule, however, is three-dimensional. VSEPR shapes are based on the **geometry** around the central atom of the molecule. The bonds in methane actually form angles of 109.28°. The hydrogen atoms are at the corners of a three-sided pyramid called a **tetrahedron**. The angle between the bonds is called the tetrahedral angle. The shape of a methane molecule is tetrahedral.

**METHANE TETRAHEDRAL STRUCTURE**

In methane, the bonding pairs of electrons push the hydrogen atoms as far away from each other as possible. What would happen to the H—C bond angles if one hydrogen was replaced by a chlorine atom?

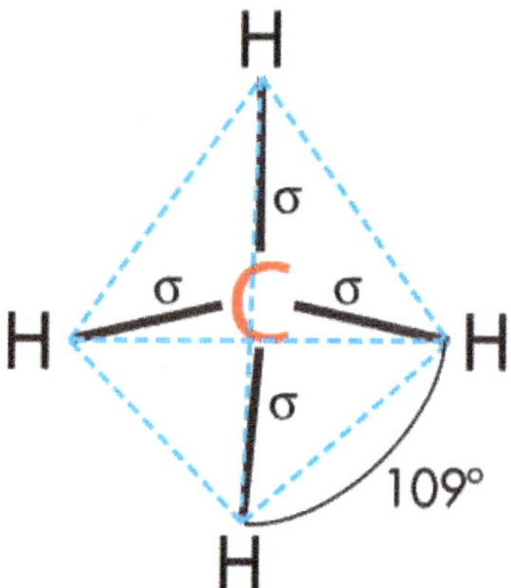


Now consider the Lewis structure of **water**.

**WATER LEWIS STRUCTURE**

The water molecule has two pairs of unbonded electrons. Would you expect this to be a polar molecule?

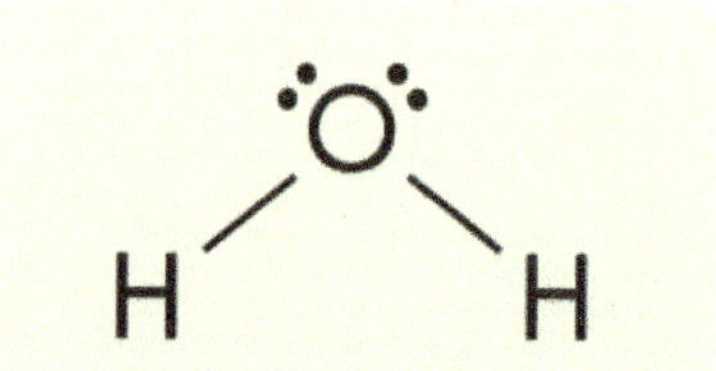

We can expect all the electron pairs to repel each other, whether they are bonding pairs or non-bonding pairs. However, the repulsion between non-bonding pairs and other pairs of electrons is stronger than between two bonding pairs because two nuclei are exerting a pull on the electrons in a bonding pair. The angle between the bonds will increase (in this case, to 104.5°) due to this repulsion. Because there are only three atoms involved, the molecule won't be a tetrahedron. The shape of the water molecule is bent.

To incorporate **VSEPR theory** into your diagrams, first draw the Lewis structure of a molecule. Then mentally picture the bonds and electron pairs moving as far away from each other as possible.

## Bond Angles

Combining Lewis structures with VSEPR theory, we can determine the bond angles and geometry of a molecule. We can also use this information to tell whether it is a polar molecule. In general, there are three situations to consider:

1. A molecule can be nonpolar because it has nonpolar bonds.
2. A molecule can have polar bonds, but be nonpolar because its dipoles are arranged in a way so that they cancel each other.
3. A molecule can be polar because its geometry causes one part of the molecule to be more electronegative than the other.

Chlorine, $Cl_2$, is an example of a nonpolar molecule with nonpolar bonds. Carbon dioxide, $CO_2$, is an example of a nonpolar molecule with polar bonds that cancel each other out.

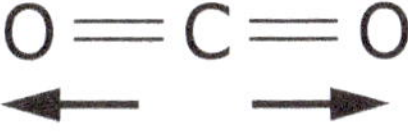


**CARBON DIOXIDE**
Carbon dioxide is a nonpolar molecule, yet it has polar bonds. How does this occur?

The bonds are polar, but the molecule is not because of the 180° angle of the bonds.

Water is an example of a polar molecule. The angle between the bonds thrusts the electronegative oxygen atom to one side and the less electronegative hydrogen atoms toward the other. The result is an electron-rich side of the molecule, which has a slightly negative charge.

## Molecular Geometry: Sample Problem

The chart shows the possible molecular geometries that compounds can show and their resulting Lewis structures. These geometries are governed by VESPR theory and bond angle.

What are the geometries, polarities, and bond angles of ammonia, $NH_3$, and the ammonium ion, $NH_4^+$?

**Solution:**

The Lewis structures, as we saw before, are shown below.

**AMMONIA, $NH_3$**

This shows the Lewis structure of ammonia. How does it differ from the ammonium ion?

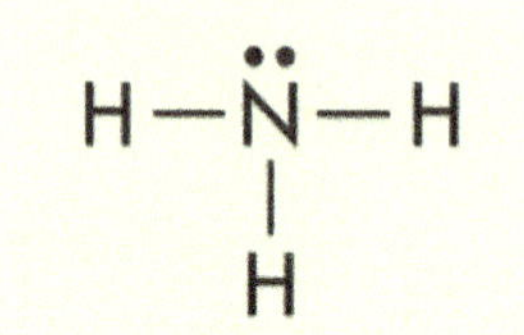


The N—H bonds are all polar because the **electronegativity** difference is greater than 0.4.

Because of electron repulsion, the unbonded electrons and the three bonds of $NH_3$ will form a trigonal pyramidal molecule.

**VSEPR SHAPE OF $NH_3$ MOLECULE — TRIGONAL PYRAMIDAL**

The Lewis structure of ammonia shows the impact of the unbonded electron pair. How might it change if another H were removed?

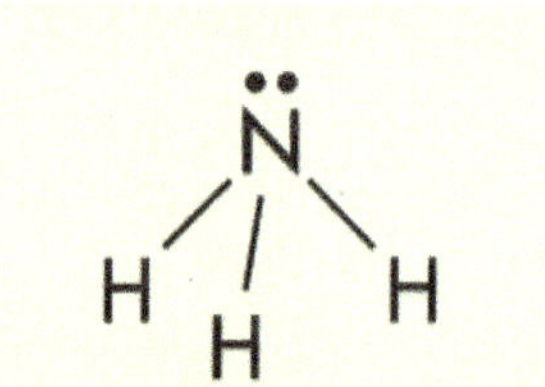


Since the electrons are not part of the molecular structure, the shape will be a flattened, three-sided pyramid. $NH_3$ will be polar because the side with the unbonded electron pair is more electronegative than the other three sides.

**AMMONIUM ION, $NH_4^+$**

The ammonium ion has four identical bonds. How does that affect its shape?

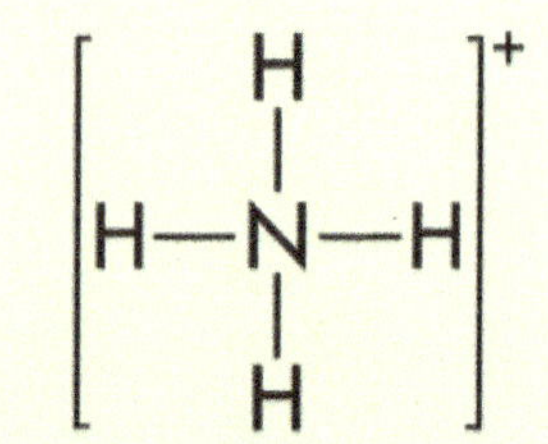

In the ammonium ion, all the bonds are polar, but equivalent. They will be separated symmetrically by tetrahedral angles. Although they are polar, the dipoles will cancel to produce a charged, nonpolar ion with a tetrahedral shape.

## How Do Lewis Structures and VSEPR Theory Explain the Unique Properties of Water?

### The Molecular Structure of Water

The molecular **geometry** of a compound can help explain the properties of the compound. We think of **water** as a very common substance, but its molecular structure makes it unique in some ways. The **polarity** of the water molecule causes strong hydrogen bonds in all directions. Under the surface of a sample of water, the attractive forces act in all directions. On the surface, however, the forces act downward and to the sides, toward other water molecules. This creates **surface tension**, which minimizes the surface area and creates a tight layer of molecules on the surface of the water. It also causes water droplets to form rounded beads.

**THE STRUCTURE OF WATER**

Water's properties make it very important to life and physical processes on Earth. What causes water's unique properties?



In addition to creating surface tension, the **hydrogen bonding** in water also helps to explain the unusually low vapor pressure in water. As was explained earlier in this concept, hydrogen bonding is the strongest **intermolecular force**. This strong force helps to hold water molecules together tightly. This has a large impact on how water behaves on the surface of Earth. Since the vapor pressure water is low, it does not readily evaporate from the surface of Earth.

Another unique property of water is that the density of ice is less than that of liquid water at the freezing point. Therefore, unlike almost all other materials, water bodies such as lakes and streams freeze from the top down. This is an advantage for life in marine and freshwater ecosystems, where water remains in liquid form at the bottom of a lake or ocean even when it freezes at the surface. This is also why ice floats in water. If ice is placed in a glass and then the glass is completely filled with water, the ice will float above the lip of the glass. If the glass of ice water were left untouched and the ice allowed to melt, the water in the glass would not overflow.

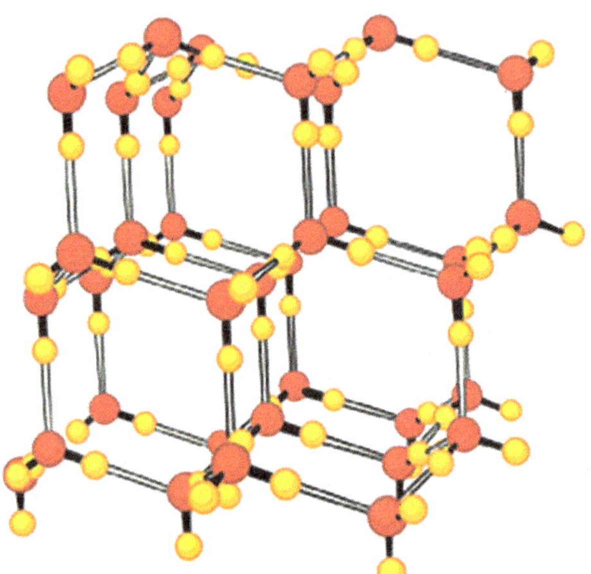

**WATER EXPANDS WHEN FROZEN**

When water freezes, the molecules form a structure that includes a lot of open space. Why does water freeze this way?



In ice, water molecules are connected by hydrogen bonds and arranged in an open hexagonal grid. This leaves a lot of open space. It is not the most efficient way to pack molecules. When water melts, this structure partially breaks down, and the molecules move closer together. Even though molecular mobility increases, density increases.

### Consider the Explain Question

**What does the molecular geometry of a chemical compound tell you about its properties?**

Go online to complete the scientific explanation.

dlc.com/ca9067s

### Check Your Understanding

**Go online to check your understanding of this concept's key ideas.**

dlc.com/ca9068s

## Applying Lewis Structures and Molecular Geometry

Different chemical compounds with the same chemical formula are considered isomers of each other. Isomers have the same chemical formula but different structures and therefore different properties. In other words, they are made of the same materials, but those materials are arranged differently in each **isomer**. For example, there are six isomers of butene, which has a chemical formula of $C_4H_8$. One of butene's isomers, 1-butene, readily forms polymers with other compounds. One such polymer, polyethylene, is used to make plastics, such as bags, buckets, and toys.

**PLASTIC BRICKS**

Plastics are polymers. What are polymers and what types of molecules are often polymerized to make plastics?

Another isomer of butene, isobutylene, has completely different applications due to its properties. Rather than form polymers, this compound reacts with methanol and ethanol to form compounds used as a fuel additive. Although the chemical composition of the isomers is the same, their different structures result in different properties, which makes them useful in different ways.

**VSEPR theory** can be applied to these structural formulas, or Lewis structures, to show the bends in the carbon chains. Common types of isomers include *cis-*, *trans-*, and *cyclo-*. *Cis-* and *trans-* are used to describe the location of similar atoms within the isomer. *Cis-* means "on the same side of the double bond" while *trans-* means "on the opposite side of the double bond." *Cyclo-* means "circular."

**TRANS-2-BUTENE**

Trans-2-butene is an isomer. How can you tell that this compound is a trans-isomer?

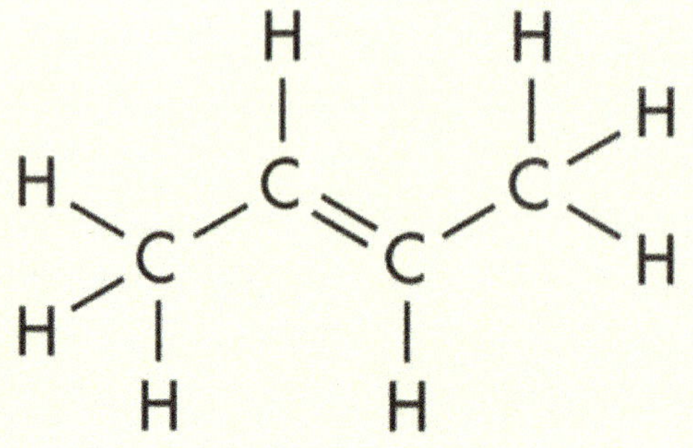

**CIS-2-BUTENE**

Cis-2-butene is also an isomer. How can you tell that this compound is a cis-isomer?

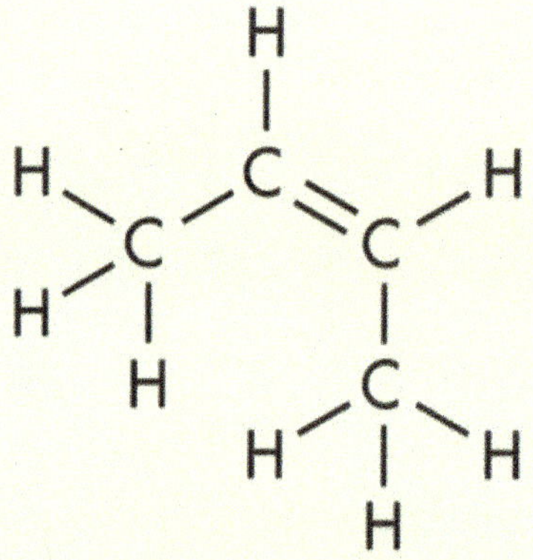

**CYCLOBUTENE**

Cyclobutene is a third isomer. If you drew an outline around this molecule, what would be its shape?

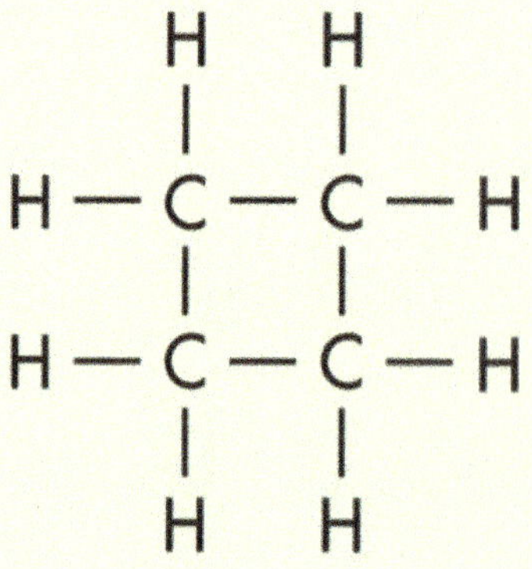

Although isomers have the same chemical composition, they differ in physical and chemical properties. Chemists working to develop new fuels or other materials might be especially interested in the study of isomers, their molecular structures, and their properties.

Why would a chemist look to isomers for new fuel sources? A stable burn is the goal when burning fuel, and some chemical isomers accomplish this better than others. Octane, a chemical component of gasoline, has several isomers, which are categorized as either linear or branched. Linear isomers of octane burn more quickly than branched isomers. Thus, a branched octane isomer, which provides a slow, even burn, is a more efficient fuel. The octane rating you see at the gas pump is the percentage of octane present in the fuel mixture.

The property differences between isomers are also applied in the food and drug industry. Do you like the taste of mint in your food, or enjoy using caraway seeds as a type of seasoning in rye bread? While mint gives a refreshing flavor and caraway seeds have a peppery taste, this difference in flavor is due to the isomers of a compound called carvone. Carvone's isomer-R form is found in mint leaves; it is the main contributor to the mint aroma. In its isomer-S form, carvone is found in caraway seeds contributing to a sharp, bitter, and very distinct flavor.

**MINT LEAVES**

Many compounds provide flavor or aroma. Such compounds may only have these properties if they are a specific isomer. Different isomers may smell or taste differently.

In the 1950s, thousands of pregnant women were prescribed a medicine called thalidomide, which was used to treat morning sickness. At the time, it was not known that human embryos would experience deformities due to the body's ability to transform this drug into its isomer-S form. Had the drug remained in its R form, it would have posed no health risk to the embryo. Thalidomide and its isomer-dependent side effects emphasize the need to place proposed drugs through years of rigorous testing before being administered to patients.

**ISOMER OF OCTANE**

Octane is a chemical compound found in the fuel used at a gas pump. What is the chemical formula of octane?

```
      H   H H   H H   H H   H
       \ /   \ /   \ /   \ /
H       C     C     C     C
 \     / \   / \   / \   / \
  C   C     C     C     C     H
 / \       / \   / \   / \
H   H H   H H   H H   H   H
```

## STEM and Molecular Geometry

The various unique properties of **water** can make this substance both useful and challenging. Scientists and engineers must find solutions to the problems caused by water. For example, expanding water can crack the strongest water pipes. Why does water behave this way when nearly every other substance on Earth becomes denser when it freezes? This phenomenon is due to the molecular structure of water. A water molecule is composed of an oxygen atom connected to two hydrogen atoms. The oxygen atom is slightly negative while the hydrogen atoms are slightly positive. This difference in charge causes the hydrogen bonds between water molecules to form a crystal lattice as the water cools in order to keep the oxygen atoms as far apart from each other as possible. The "empty" spaces in the lattice contribute to water becoming less dense as it freezes.

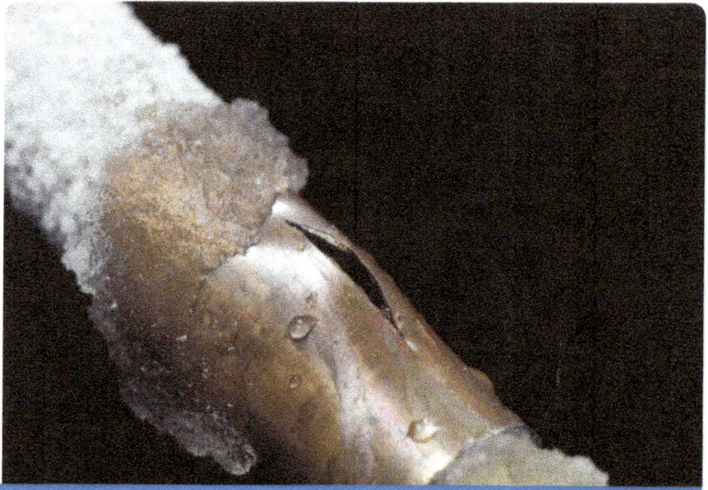

**CRACKED WATER PIPES**

What property of water caused this pipe to break?

When designing buildings or infrastructures, engineers must make sure water pipes are buried underground or are insulated if placed above ground. In very cold weather, people sometimes prevent pipe damage by allowing a slow drip of water from faucets. Why do you think this is helpful?

Engineers must also consider the effects of freezing and thawing water when designing other structures such as roads and bridges. Materials that repel water, such as asphalt, are useful. The shape of the road and the drainage of the surrounding area are also considered during design. Ditches are often dug along both sides of highways. This functions to drain water away from the pavement. Also, roads are typically shaped so that water runs off them easily.

**HIGHWAY SURVEYOR**

How is this highway engineered to minimize water-related damage?

CONCEPT
3.5

# Chemical Formulas



## LESSON OVERVIEW

dlc.com/ca9069s

### Lesson Questions

- How are the formulas for ionic compounds written?
- How are the formulas for molecular compounds written?
- How are empirical and molecular formulas distinguished?

### Lesson Objectives

By the end of the lesson, you should be able to:

- Compose and write formulas for ionic compounds.
- Compose and write formulas for molecular compounds.
- Distinguish between empirical and molecular formulas.

### Key Vocabulary

Which terms do you already know?

- ☐ chemical formula
- ☐ chemical symbol
- ☐ crystal lattice
- ☐ electronegativity
- ☐ empirical formula
- ☐ formula unit
- ☐ ion
- ☐ molecular formula
- ☐ molecule
- ☐ nonmetals
- ☐ polyatomic ion
- ☐ prefix
- ☐ subscript
- ☐ transition metals
- ☐ valence electron

## Deriving Chemical Formulas

**At some point in your life, your doctor will likely prescribe you a form of penicillin. The drug Penicillin I is a powerful antibiotic medicine used to cure bacterial infections. How did the scientists identify the formula of penicillin? What steps did they take to derive this formula?**

dlc.com/ca9070s

**EXPLAIN QUESTION**

**How are the formulas for ionic and covalent compounds determined?**

**CHEMICAL FORMULA OF WATER**

Based on the photo shown, how many hydrogen atoms are present in a molecule of water?

# How Are the Formulas for Ionic Compounds Written?

## Chemical Symbols and Subscripts

A **chemical formula** tells what elements make up a compound and how many ions or atoms of each element are present.

An ionic compound consists of cations and anions held together by the attraction between their opposite charges. The chemical formula of an ionic compound consists of two pieces of information. The first is a **chemical symbol**. A chemical symbol uses one to three letters of the alphabet to represent an element. Each element has its own symbol, as approved by the International Union of Pure and Applied Chemistry (IUPAC).

In an ionic compound, the symbol for the positively charged **ion** always appears on the left side of a chemical formula. The symbol for the negatively charged ion always appears on the right. The negative and positive ions in neutral ionic compounds are always electrically balanced. For example, consider the formula for calcium fluoride. This ionic compound contains the elements calcium and fluorine. An ion of fluorine is called fluoride. In its chemical formula, the symbol for calcium (Ca) will appear on the left because the calcium ion is positively charged. The symbol for fluorine (F) will appear on the right since the fluoride ion is negatively charged.

The second piece of information included in a chemical formula is the **subscript**, which follows each ion's chemical symbol. The subscript indicates the number of ions of each element present in each formula unit of the ionic compound. Because ionic compounds are composed of ions, they do not form discrete molecules. The ions in a sample of the compound are ordered in a repeating pattern, forming a **crystal lattice**. The smallest ratio of ions in any sample of the compound is the **formula unit**. For example, the compound calcium fluoride consists of one calcium ion bonded to two fluorine ions. Therefore, the formula for calcium fluoride is written as:

$$CaF_2$$

Note that the numeral "1" is not written as a subscript after the symbol Ca. When an ionic compound contains only one ion of a certain element, no subscript follows that ion's chemical symbol.

Consider another example. The ionic formula for common table salt, or sodium chloride, is NaCl. Since no subscripts follow the symbols for sodium (Na) or chlorine (Cl), the formula shows that only one ion of sodium is bonded to only one ion of chlorine.

**WATER MOLECULE: $H_2O$**

Numerical subscripts detail the ratios of elements in ionic and molecular compounds. What can you tell about the composition of water from its chemical formula?

## Determining Subscripts

How are the subscripts in the formula of an ionic compound determined? Remember, all ionic compounds are electrically balanced. The total positive charge of the cations in a formula unit is equal to the total negative charge of the anions in the formula unit. Consider the ionic compound calcium chloride.

The calcium ion has a charge of $2^+$ ($Ca^{2+}$). The chloride ion has a charge of negative 1, or $1^-$ ($Cl^-$). For the charges in calcium chloride to balance, the compound must contain two chloride ions for every calcium ion. This means that a balanced chemical formula for calcium fluoride could be written as $Ca_2Cl_4$. However, when writing chemical formulas for ionic compounds, the simplified ratio of ions is used. The simplest way to describe the ratio of calcium to chloride ions in calcium chloride is 1:2. Thus, the chemical formula is written:

$$CaCl_2$$

This ratio is due to calcium having two valence electrons. Chlorine can only receive one electron per atom; therefore, it takes two chlorine atoms to bond with one Ca atom.

## Subscripts for Transition Metal Cations

The process is a little trickier for cations formed by **transition metals** such as copper (Cu), iron (Fe), cobalt (Co), chromium (Cr), and mercury (Hg). Many transition metals can have multiple oxidation numbers and therefore can have more than one correct formula. Thus, most transition metals can form more than one ion. For example, copper can form compounds with either copper(I) ions ($Cu^{+}$) or copper(II) ions ($Cu^{2+}$).

To write a formula with a transition metal, the charge of the cation is needed. The name of the compound will contain a Roman numeral in parentheses to indicate the charge, immediately following the transition metal's name.

Using the correct charge, the same steps are used to write a formula that will have a ratio of cations to anions that results in a neutral compound. In the case of copper(II) chloride, the copper(II) ion has a charge of $2^{+}$ ($Cu^{2+}$), and the chloride ion has a charge of $1^{-}$ ($Cl^{-}$). For the charges to balance, the compound must contain one copper(II) ion for every two chloride ions. This 1:2 ratio is used to write the formula:

$$CuCl_2$$

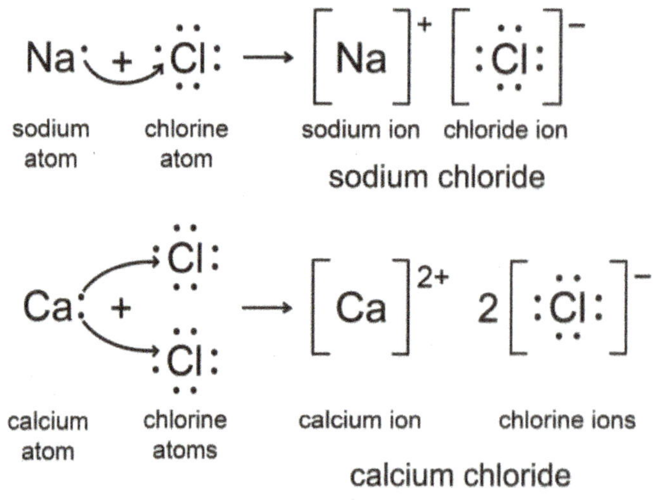

sodium atom | chlorine atom | sodium ion | chloride ion

sodium chloride

calcium atom | chlorine atoms | calcium ion | chlorine ions

calcium chloride

**IONIC COMPOUND FORMATION**

This image shows the formation of sodium chloride and calcium chloride. Why does the calcium ion have a charge of $2^{+}$?

## Polyatomic Ions

Many ions consist of single atoms that have lost or gained electrons. Ions may also form, however, when a group of two or more bonded atoms loses or gains an electron and thus carries a charge. Such groups of atoms are called polyatomic ions.

The atoms in a polyatomic ion are covalently bonded. Their bonds consist of pairs of shared valence electrons. However, the total number of electrons in a polyatomic ion differs from the total number of its protons.

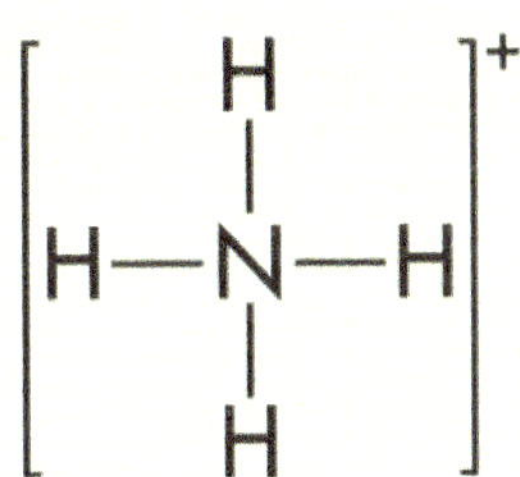


**POLYATOMIC ION: AMMONIUM**

Polyatomic ions are formed by the sharing of electron pairs between two nonmetal atoms. Why does the ammonium ion have a positive charge?

This is true because the polyatomic ion has gained or lost at least one valence electron. The electrons may have been transferred to or from another atom or group of atoms. The result is that the polyatomic ion possesses an overall positive or negative charge.

| Polyatomic Cations | | Polyatomic Anions | |
|---|---|---|---|
| $NH_4^+$ | ammonium | $CN^-$ | cyanide |
| $H_3O^+$ | hydronium | $OH^-$ | hydroxide |
| | | $SO_3^{2-}$ | sulfite |
| | | $SO_4^{2-}$ | sulfate |
| | | $NO_2^-$ | nitrite |
| | | $NO_3^-$ | nitrate |
| | | $ClO^-$ | hypochlorite |
| | | $ClO_2^-$ | chlorite |
| | | $ClO_3^-$ | chlorate |
| | | $ClO_4^-$ | perchlorate |

**SOME COMMON POLYATOMIC IONS**

Polyatomic ions contain covalently bonded atoms that, as a group, have a positive or negative charge. How can the charge be used to predict chemical formulas?

# How Are the Formulas for Molecular Compounds Written?

## Chemical Symbols and Subscripts

The ions in an ionic compound form a cluster of charged atoms or groups rather than a discrete **molecule**. Although attracted to one another, the ions are somewhat independent. This is not the case, however, in a molecular compound. Molecular compounds are also called covalent compounds.

In covalent compounds, valence electrons are not gained or lost by the compound's atoms. Instead, a covalent bond forms from a shared pair of electrons. This is because the **electronegativity** difference between the **nonmetals** in the bonds are not great enough to cause a transfer of electrons and therefore results in sharing. This is the reason that these groups of atoms form discrete molecules rather than ordered clusters.

Like the formulas for ionic compounds, the formulas for covalent compounds consist of chemical symbols and subscripts. A **molecular formula** describes the components of a molecule, including the number of atoms of each element that are present in one molecule of the compound.

A molecular formula may describe a molecule of a single element. For example, $H_2$ is the molecular formula for hydrogen gas. This means that in the gaseous state, hydrogen forms a molecule consisting of two covalently bonded atoms of hydrogen. Similarly, oxygen (with a molecular formula of $O_2$), fluorine ($F_2$), and chlorine ($Cl_2$) do the same. Note that these differ from the chemical symbols of these elements because they contain subscripts. Elements that form diatomic molecules are called diatomic elements.

A molecular formula may also describe a molecule that contains atoms of two or more elements. For example, the molecular formula of the compound hydrogen chloride is HCl. The molecular formula of the gas methane is $CH_4$. The molecular formula for a molecule of the sugar glucose is $C_6H_{12}O_6$. In each case, the molecular formula describes the element (the chemical symbols) and quantity of the atoms (the subscripts) that make up a discrete molecule.

**SUGAR OR SUCROSE**

Sugar, like glucose is formed from carbon, hydrogen and oxygen. Each molecule of sucrose contains six carbons. Can you predict how many hydrogen and oxygen atoms each molecule contains?



## Naming Molecular Compounds

Molecular formulas are a convenient way to describe the elements in a compound and the ratios of the ions or atoms present. The chemical name of a molecular compound also often reveals the elements it contains. In some cases, the name may also describe the amounts of each element present.

For example, the name *hydrogen chloride* (HCl) reveals that the molecule of this substance contains atoms of the elements hydrogen and chlorine. The name *hydrogen bromide* (HBr) reveals that a molecule of this substance contains atoms of the elements hydrogen and bromine.

What, however, do the names *carbon dioxide* and *carbon tetrachloride* reveal about the components of these molecules? The answer to this question lies in an analysis of the words and word parts in each name.

In the case of carbon dioxide, the names of two elements are mentioned. These are carbon and oxygen. It is clear that carbon dioxide molecules are made up of atoms of carbon and oxygen atoms.

Additionally, the name reveals something more, a quantitative fact. The **prefix** *di-* means "two." The name *carbon dioxide* reveals that a molecule of this compound contains two oxygen atoms. This is shown in the **chemical formula** for carbon dioxide, $CO_2$.

What about carbon tetrachloride? What does its name reveal about the components of its molecule? A clue to the answer lies in the prefix *tetra-*, which means "four." Therefore, a molecule of carbon tetrachloride contains four chlorine atoms. The molecule's chemical formula is $CCl_4$. Molecular compounds containing three or more elements have complex rules.

**COMMON PREFIXES USED IN CHEMISTRY**

Common Greek prefixes are often used to describe the quantity of atoms of each element in a molecular compound. What formula would you predict for the compound silicon tetrafluoride?

| Prefix | Meaning | Example | Formula |
|---|---|---|---|
| mono-, mon- | one | carbon monoxide | $CO$ |
| di- | two | carbon dioxide | $CO_2$ |
| tri- | three | phosphorous trichloride | $PCl_3$ |
| tetra- | four | carbon tetrachloride | $CCl_4$ |
| penta- | five | dinitrogen pentoxide | $N_2O_5$ |



## How Are Empirical and Molecular Formulas Distinguished?

### Empirical Formulas

A **molecular formula** describes the structure of a molecular compound. It describes the elements and the number of atoms of each present in one **molecule** of the compound. Ionic compounds, in contrast, are often composed of ordered clusters of the charged ions. Since these clusters can vary in size, the chemical formulas for ionic compounds are written as empirical formulas.

The **empirical formula** describes the simplest whole-number ratio of the atoms of each element in a compound. When trying to determine the **chemical formula** of a newly discovered molecule, scientists can often determine experimentally the empirical formula and the molecular weight. They then use that information to deduce the molecular formula.

| Compound | Molecular Formula | Empirical Formula |
|---|---|---|
| Benzene | $C_6H_6$ | CH |
| Acetylene | $C_2H_2$ | CH |
| Hydrogen peroxide | $H_2O_2$ | HO |

**MOLECULAR AND EMPIRICAL FORMULAS**

Empirical formulas are usually much simpler than molecular formulas. Why would chemists want to use an empirical formula?

For example, the molecular formula of the compound hydrogen peroxide is $H_2O_2$. The ratio of hydrogen to oxygen in the molecular formula is 2:2. The empirical formula of the same compound is HO, with a simplified ratio of 1:1. Similarly, the molecular formula of benzene is $C_6H_6$ and its empirical formula is CH.

| $C_6H_{12}O_6$ |
|---|
| *Molecular Formula of Glucose* |
| $CH_2O$ |
| *Empirical Formula of Glucose* |

**EMPIRICAL FORMULA OF GLUCOSE**

The empirical formula of glucose contains the same chemical symbols as the molecular formula, but the ratio of atoms is simplified. Can you contrast the appearance of these two formulas?



### Consider the Explain Question

**How are the formulas for ionic and covalent compounds determined?**

Go online to complete the scientific explanation.

dlc.com/ca9071s

### Check Your Understanding

**Why can an empirical formula, alone, not be used to derive the actual molecular formula for a given compound?**

dlc.com/ca9072s

# STEM in Action

## Applying Chemical Formulas

A seemingly slight change in the structure of a chemical compound, and therefore in its **molecular formula**, can significantly change the properties and applications of the compound. For example, methane, $CH_4$, is a combustible gas. It is also the main ingredient in natural gas, which is used as a fuel to heat homes and cook food. It is also used in some power plants to help generate electricity.

**NATURAL GAS METHANE USED AS A FUEL SOURCE**

Natural gas, which largely consists of the gas methane, is a fuel used for many purposes. If you change its structure and molecular formula, will methane continue to function as a fuel source?



If three chlorine atoms are substituted for three hydrogen atoms in methane, a new compound, chloroform, is formed. With the molecular formula $CHCl_3$ chloroform is a liquid that smells sweet. It has many current uses such as a building block for the production of other important chemicals. It also functions as a solvent, which is useful for dissolving other molecules.

By far, its most dramatic use was as an anesthetic in the 19th and early 20th centuries. When inhaled, vapors from this compound kept patients from feeling pain during operations or childbirth. Unfortunately, chloroform was found to have dangerous side effects, and its use as an anesthetic stopped in the early 20th century.

By substituting another chlorine atom for the last hydrogen atom, chloroform becomes carbon tetrachloride. The molecular formula for carbon tetrachloride is $CCl_4$. Carbon tetrachloride was once widely used to clean clothes in dry cleaning shops, in fire extinguishers, and as a pesticide. However, the compound was discovered to be a dangerous poison. In 1970, the United States banned its use in all consumer products.

**PESTICIDE SPRAYING ON FARMLAND**

The compound carbon tetrachloride was a chemical component in pesticide spray. Why is carbon tetrachloride no longer used in the chemical formulation of pesticide sprays?

The atoms of other elements, especially those of the halogens (fluorine, bromine, and iodine) can be substituted for one or more of the hydrogen atoms in methane. In each case, a new compound with a new molecular formula, new properties, and new applications is formed.

## STEM and Chemical Formulas

By the 1930s, rubber had become a vital resource. Because many military machines and civilian vehicles required rubber in order to function, a shortage of rubber would have caused serious consequences. As World War II approached, some nations far from sources of natural rubber in Southeast Asia realized this could impact their military success.

**HARVESTING RUBBER**

Natural rubber in the form of white sap can be collected and used to make latex. Can you think of a product made from latex rubber?

Part of the answer was found in butadiene, a substance with a molecular formula of $C_4H_6$. Discovered by scientists in 1863, the structure of butadiene was deciphered in 1886. Butadiene was a curious substance, but lacked practical applications. This all changed when, in 1910, a Russian chemist by the name of Sergei Lebedev found a way to link together many molecules of butadiene to form a chemical chain called a polymer.

The butadiene polymer has an intriguing property. It behaves in many ways like natural rubber. Unfortunately, the polymer is much softer than natural rubber and has a consistency that is nearly liquid.

As the years passed, organic chemists discovered that they could add molecules of other compounds to the butadiene polymer. One of the compounds they added was called styrene, whose molecular formula is $C_8H_8$.

Polymers made from combinations of butadiene and styrene molecules produce substances with properties very similar to those of natural rubber. This synthetic rubber became a substitute for natural rubber.

The invention of synthetic rubber contributed to the success of the United States in World War II. Today, this form of synthetic rubber is a key ingredient in car and truck tires and other rubber products used throughout the world.

It is estimated that more than seven million tons of synthetic rubber is produced each year. Synthetic rubber is used to make a wide range of products, such as automobile tires and flooring for a building.

CONCEPT
4.1

# Chemical Reactions and Equations

dlc.com/ca9073s

## LESSON OVERVIEW

### Lesson Questions

- What are the five types of chemical reactions?
- How can you predict the products of each of the five types of chemical reactions?
- How can you represent chemical reactions using chemical equations?

### Lesson Objectives

By the end of the lesson, you should be able to:

- Distinguish between five types of chemical reactions.
- Predict the products of a chemical reaction.
- Write balanced equations that represent chemical reactions.
- Write the net ionic equation and identify spectator ions for a chemical reaction.

### Key Vocabulary

Which terms do you already know?

- ☐ anion
- ☐ balanced chemical equation
- ☐ cation
- ☐ coefficient
- ☐ combustion reaction
- ☐ decomposition reaction
- ☐ double-displacement reaction
- ☐ double-replacement reaction
- ☐ ion
- ☐ law of conservation of matter
- ☐ net ionic equation
- ☐ oxidation-reduction reaction
- ☐ phase of matter
- ☐ product
- ☐ reactant
- ☐ reactivity series
- ☐ redox reaction
- ☐ single-displacement reaction
- ☐ single-replacement reaction
- ☐ solubility
- ☐ synthesis reaction

# Preparing for Chemical Reactions and Equations

**Why are fireworks different colors? What makes them explode? Why do they have to be lit on fire to explode?**

dlc.com/ca9074s

## EXPLAIN QUESTION

**What are the five main types of chemical reactions, and how can chemical equations help you predict the products of each of the five types of reactions?**

### FIREWORKS

Chemical reactions make fireworks exciting to watch and dangerous to make. What are some of the chemicals that react explosively to make fireworks?

# What Are Five Types of Chemical Reactions?

## Classifying Reactions into Five Types

Even for relatively mundane purposes, we might want to predict the reactions that occur when two substances are put together. We know that it is safe to combine bleach and detergent in the washing machine, for instance. But is it safe to mix bleach with an ammonia-based cleaning **product**? (No, it is not.) We know that adding baking soda to batter will help give a cake the shape and texture we want. Rather than try to memorize all the chemical reactions that are possible, it makes sense to find patterns that help us predict reactions.

Chemists classify reactions into categories based on the behaviors of the chemicals involved. By understanding the different types of reaction, scientists can predict the outcome of almost any combination of chemical substances.

## Synthesis Reactions

A **synthesis reaction** occurs when two or more reactants combine to form a single product. This type of reaction is represented by the following general equation:

$$A + X \rightarrow AX$$

A and X can be elements or compounds. AX is a compound.

To demonstrate a synthesis reaction, you can use a balloon filled with a mixture of hydrogen gas and oxygen gas. When a flame comes in contact with the balloon, the balloon bursts. However, the bang is much louder than that of just a popped balloon. The hydrogen and oxygen in the balloon react to form water. In this reaction, hydrogen and oxygen are the reactants, or the starting materials. Water is the product. This chemical reaction is represented by the following equation:

$$2H_2(g) + O_2(g) \rightarrow 2H_2O(g)$$

## Decomposition Reactions

In a **decomposition reaction**, a single compound undergoes a reaction that produces two or more simpler substances. The products might be elements, less complex compounds, or a combination of elements and compounds. This type of reaction is represented by the following general equation:

$$AX = A + X$$

A and X can be elements or compounds. AX is a compound.

To release hydrogen for use as a fuel, chemists use electrical energy to break down water into its constituents, hydrogen and oxygen. This reaction, the reverse of the synthesis reaction demonstrated with the balloon, is represented by the following equation.

$$2H_2O(g) \rightarrow 2H_2(g) + O_2(g)$$

Most decomposition reactions only take place when energy, such as heat or electricity, is added. The process of using an electric current to initiate and drive a reaction is called electrolysis.

## Combustion

During a **combustion reaction**, an element or compound rapidly combines with oxygen to form an oxide. The reaction releases heat and light.

The flame of a laboratory burner represents a combustion reaction. Methane, $CH_4$, supplied to the burner reacts with oxygen, and carbon dioxide and water are produced. Each of these products is an oxide. Combustion reactions are commonly used in heating, cooking, and transportation. This reaction is represented by the following equation:

$$CH_4(g) + 2O_2(g) \rightarrow CO_2(g) + 2H_2O(g)$$

**CHEMICAL REACTION FIRE**

A fire, like the one you might make in a fireplace, is the result of a certain type of chemical reaction. What type of reaction is the combustion of wood?

Some synthesis reactions can also be categorized as combustion reactions. The combustion reaction between hydrogen and oxygen is also a synthesis reaction. Hydrogen combines with oxygen to form an oxide (water) and the reaction releases thermal energy.

## Single-Displacement Reactions

The final two types of reactions are displacement reactions. They are also called replacement reactions because atoms or ions in the reactants are replaced with other elements.

A single-displacement reaction is a type of reaction that occurs when an atom or **ion** in a compound is replaced by a similar atom or ion. Single-displacement reactions can be represented by the following general equations:

$$A + BX \rightarrow AX + B$$

or

$$Y + BX \rightarrow BY + X$$

A, B, X, and Y are elements. AX, BX, and BY are compounds.

| Reactivity Series of Metals | |
|---|---|
| K | Most reactive |
| Ca | |
| Na | |
| Mg | |
| Al | |
| Zn | |
| Fe | |
| Ni | |
| Sn | |
| Pb | |
| Cu | |
| Ag | |
| Au | Least reactive |

**REACTIVITY SERIES OF METALS**

A list of metals in order of reactivity is used to predict the products of chemical reactions. Why are some metals more reactive than others?

For metals, a single-displacement reaction will occur if the element being replaced is lower in the **reactivity series** than the element that will take its place. When copper is placed in an aqueous solution of silver nitrate, two visible changes occur. The copper metal becomes covered with crystals of silver metal, and the solution becomes blue. This reaction is an example of a single-displacement reaction. In this case, the more active metal, copper, is replacing the less active metal, silver, in solution. This reaction is represented by the following equation:

$$Cu(s) + AgNO_3(aq) \rightarrow CuNO_3(aq) + Ag(s)$$

By conducting tests using combinations of elements and compounds, chemists have developed a list to predict when single-displacement reactions will happen. A reactivity series, sometimes referred to as an activity series, lists elements from the most reactive to the least reactive.

**COPPER AND SILVER NITRATE**

These two flasks contain copper and silver nitrate. Why do the contents of the flasks differ in appearance?

In the example above, the silver cations in the solution are reduced. They gain electrons and become silver atoms. The blue color of the solution is caused by the presence of copper(II) cations. Some copper atoms in the metal are oxidized. They give up electrons and become copper(II) ions. Silver (Ag) is located below copper (Cu) in the reactivity series. In this single-displacement reaction, the copper atoms replace the silver ions in silver nitrate.

The most active metals, such as those in Group 1, react vigorously with water to replace hydrogen in water. Less active metals, such as zinc, require the water to be in the form of steam for this reaction to take place. These reaction types are represented by the following equations:

$$2K(s) + 2H_2O(l) \rightarrow 2KOH(aq) + H_2(g)$$
$$Zn(s) + H_2O(g) \rightarrow ZnO(s) + H_2(g)$$

In a different type of **single-displacement reaction**, one halogen replaces another halogen in a compound. Fluorine is the most active halogen. It will replace all other halogens in their compounds. Each halogen is less active than any halogen above it on the periodic table. This reaction type is represented by the following equations:

$$Cl_2(g) + 2KBr(aq) \rightarrow 2KCl(aq) + Br_2(l)$$
$$Br_2(l) + 2KCl(aq) \rightarrow \text{no reaction}$$

## Double-Displacement Reactions

In a double-displacement reaction, the ions of two compounds exchange places and form a new compound. This means that the cations and anions trade places. Double-displacement reactions can be represented by the following general equation:

$$AX + BY = AY + BX$$

A, B, X, and Y in the reactants represent ions. AY and BX are ionic or molecular compounds.

For example, a sodium chloride solution is composed of sodium ions and chloride ions dissolved in water. A silver nitrate solution is composed of silver ions and nitrate ions dissolved in water. When the two solutions mix, the ions collide with each other. Sodium ions and nitrate ions remain dissolved because sodium nitrate is soluble. However, when silver and chloride ions collide, they bond together and form a precipitate. This reaction is represented by the following equation:

$$NaCl(s) + AgNO_3(aq) \rightarrow AgCl(s) + NO_3^{-1}(aq) + Na^{+1}(aq)$$

| Solubility Rules | |
|---|---|
| **Compounds of these ions tend to be soluble** | **Exceptions** |
| Alkali metals | |
| Ammonium | |
| Nitrate | |
| Acetate | |
| Chloride, Bromide | $Pb^{2+}$, $Hg_2^{2+}$, $Ag^+$ |
| Iodide | $Hg^{2+}$, $Pb^{2+}$, $Hg_2^{2+}$, $Ag^+$ |
| Sulfate | $Ca^{2+}$, $Sr^{2+}$, $Ba^{2+}$, $Pb^{2+}$, and $Hg_2^{2+}$ |
| **Compounds of these ions tend to be insoluble** | **Exceptions** |
| Carbonate | Alkali metal and ammonium |
| Chromate | Alkali metal and ammonium |
| Phosphate | Alkali metal and ammonium |

**SOLUBILITY RULES**

The solubility of ions is used to predict the results of double-displacement reactions. Why do the ions need to be soluble in order for a double-displacement reaction to occur?



**Solubility** describes how readily a substance dissolves in water. Solubility rules help scientists determine whether an ionic compound will be soluble or insoluble. The formation of a precipitate when two solutions are mixed is often a sign that a double-displacement reaction has occurred. The formation of the compound, water, or a gas are also signs that a double-displacement reaction has occurred.

Reactions between acids and bases are a type of double-displacement reaction. When acids and bases react, the products are usually water and a salt. For example, when hydrochloric acid reacts with sodium hydroxide, the products are the salt sodium chloride and water.

$$HCl(aq) + NaOH(aq) \rightarrow NaCl(aq) + H_2O(aq)$$

## Oxidation-Reduction Reactions

Oxidation-reduction reactions, also known as redox reactions, are reactions in which one element loses electrons, or is oxidized, while another gains electrons, or is reduced.

**FIZZING TABLETS**
Some medicines fizz when added to water. What type of reaction produces the carbon dioxide that causes the fizz?



This is a different way of classifying reactions, so many reactions can be classified as both a **redox reaction** and another type of reaction. For example, single-displacement reactions and combustion reactions are also redox reactions. Many synthesis and decomposition reactions are also classified as redox reactions.

## Chemical Bonding and Chemical Reactions

What happens during a chemical reaction? To answer this question, we must observe the action taking place at the atomic level. At this level, we would see that the chemical bonds linking atoms together in a **reactant** molecule are broken, and that the atoms reorganize in new arrangements that allow new bonds to form between new pairs of atoms. The result is a new substance—the product.

Chemical reactions involve the breaking of one set of bonds and the forming of a new set of bonds. The number and type of atoms remains the same throughout this process, so that only the bonding patterns change. This is an important feature of all chemical reactions—they obey the **law of conservation of matter**. This law, sometimes referred to as the law of conservation of mass, states that matter cannot be gained or lost in a chemical reaction.

## The Forces behind Chemical Bonding

What is involved as chemical bonds in reactants are broken and then new chemical bonds are formed in products? First, the reactant molecules must collide to be able to exchange atoms. Think about the reaction between hydrogen and oxygen:

$$2H_2 + O_2 \rightarrow 2H_2O$$

**A STABLE ARRANGEMENT**
When liquid water freezes, its molecules form a hexagonal structure. Why are molecules of ice stable?

Even though the molecules bump into each other, a mixture of hydrogen and oxygen gases is stable under ordinary conditions. But when a flame is brought into a mixture of $H_2$ and $O_2$, BOOM! The reaction that ensues is explosive. We can understand why the flame causes this reaction when we consider the different forces involved.

$H_2$ and $O_2$ molecules are normally stable because the molecules have less energy than the separate atoms. Within each molecule, there is a balance of electromagnetic forces operating on the two atoms bonded in each molecule. The two atoms are pushed apart as the positive charges of the atoms' nuclei repel one another. The two atoms are also pulled together as the negatively charged electrons of one atom attract the positive nucleus of the other atom.

In order for a stable atom to break apart, molecules must collide with energy great enough to break the chemical bonds. In the case of the hydrogen-oxygen reaction, a flame provides this energy. The heated molecules have more kinetic energy and collide with each other with more force.

Once the bonds between atoms are broken, the atoms can recombine to form new stable molecules with less energy than the reactant molecules or the individual atoms. In the case of $H_2$ and $O_2$, the hydrogen atoms may form bonds with oxygen instead of with each other.

**HINDENBURG DISASTER**

In 1937, forces holding together the stable hydrogen atoms filling the airship Hindenburg were disturbed by an ignition event. What chemical formula describes the reaction that resulted in the explosion?



## The Energy Associated with Chemical Bonding

As you saw with the reaction between $H_2$ and $O_2$ to form $H_2O$, molecules need to collide with enough energy for atoms to rearrange and form new bonds. When new bonds form, atoms tend to form bonds with the lowest possible potential energy. Energy is released. Because the chemical energy in molecules tends to be at a minimum, we rarely observe individual atoms in nature; instead, we observe atoms joined together by bonds. The stability in the compounds we observe is due to the lower energy of the bonded atoms.

Every reactant and every product has a certain amount of energy associated with its bonds. This energy is called bond energy. The bond energy of the O—H bond is higher than that of an H—H bond, but lower than that of an O—O bond. You can find the total bond energy of a molecule by adding up all of the bond energies for that molecule.

Because energy is never created or destroyed, the sum of all the energy present before a chemical reaction has to be equal to the sum of all the energy present after a chemical reaction. We can write an equation to show this:

Sum of bond energies of reactants + energy input
= Sum of bond energies of products + energy output

The minimum amount of energy needed for a reaction to begin is called the activation energy. This input of energy occurs in both exothermic and endothermic reactions.

In the case of the reaction between $H_2$ and $O_2$, the sum of the bond energies of the reactants was large compared to the sum of the bond energies of the product. A small amount of energy was added in the ignition flame (energy input), and a large amount of energy was released as the products formed (energy output). This reaction is said to be exothermic because energy is released as light and heat.

An endothermic reaction is one in which the sum of all the bond energies of the products is greater than that of the reactants.



## How Can You Predict the Products of Each of the Five Types of Chemical Reactions?

### Predicting the Products of a Reaction

During a **synthesis reaction**, two reactants form a single **product**. When the reactants are elements, the product is usually a binary compound composed of those elements. For example, the reaction of a metal with a nonmetal can produce a binary ionic compound. The formula of the product can be predicted from the ions that are likely to form. A bright white light is given off when magnesium reacts with oxygen. The product of this dramatic reaction between a metal and a nonmetal is the ionic compound magnesium oxide, MgO. When aluminum reacts with chlorine, the ionic compound aluminum chloride, $AlCl_3$, is the product.

Many decomposition reactions are the reverse of synthesis reactions. Therefore, in most cases, we can predict that a binary compound will decompose into its elements. Thus, sodium chloride can be decomposed into the elements sodium and chlorine. Water decomposes to form the elements hydrogen and oxygen.

Some compounds that are composed of polyatomic ions decompose in predictable ways. For example, metal carbonates break down to produce carbon dioxide and an oxide. Calcium carbonate in limestone decomposes upon heating. The products are carbon dioxide gas and calcium oxide.

$$CaCO_3(s) \rightarrow CaO(s) + CO_2(g)$$

**INSIDE AN OLD HOT WATER PIPE**

What is dissolved in water that can lead to hot water pipes becoming blocked?



Many acids decompose into nonmetal oxides and water. The breakdown of carbonic acid, $H_2CO_3$, in a carbonated drink produces bubbles of carbon dioxide gas. The other product of this decomposition is water, an oxide of hydrogen.

$$H_2CO_3 \rightarrow CO_2(g) + H_2O(g)$$

If the hydroxide **ion** is present in a compound, the products of the decomposition of the compound are water and a metal oxide. So, the decomposition of sodium hydroxide produces water and sodium oxide.

Because combustion reactions always involve a reaction with oxygen, the products are always oxides. Combustion reactions that are also synthesis reactions will produce a binary compound. For example, if a hydrocarbon such as methane is combined with excess oxygen, the products are carbon dioxide and water.

In a single-displacement reaction, an element and a compound react. The products are a different element and a new compound. To predict exactly what the products will be, remember that the replacement occurs between elements that have similar properties. Thus, the single-displacement reaction between copper and silver nitrate produces silver and copper(II) nitrate. In this case, metallic copper replaces metallic silver.

**NATIVE METALS**

Some metals exist in pure form in nature whereas others do not. What are the chemical characteristics of metals that exist in pure form in nature?



What will happen if silver metal is placed in a copper(II) nitrate solution? Nothing! The silver atoms cannot replace the copper(II) ions. This agrees with the information in the **reactivity series**: a single-displacement reaction will occur if the element being replaced is lower in the series than the element that would take its place. Copper, being above silver in the series (more reactive than silver), cannot be replaced by silver. However, if copper metal is placed in a silver nitrate solution, the reactants produce silver and copper nitrate, caused by a single-displacement reaction where copper replaces silver to produce copper nitrate.

In a double-displacement reaction, the reactants are two compounds, and the products are two different compounds. The products can be predicted by switching the ions in the compounds. Therefore, when silver nitrate and sodium chloride react, the products are sodium nitrate and silver chloride. This can be seen as either switching the cations or switching the anions.

Not all pairs of compounds will undergo a double-displacement reaction. For example, what products form when solutions of sodium chloride and potassium nitrate are combined? It might seem likely that the products would be potassium chloride and sodium nitrate. But when sodium chloride and potassium nitrate are actually combined, there is no reaction. Predictions about a double-displacement reaction must be checked against the **solubility** rules. If neither product is insoluble, then a reaction will not occur. No precipitate will form, and the ions will all remain in the solution.

### Predicting the Products of a Reaction: Sample Problem

What substance forms when aluminum reacts with sulfur?

**Solution**

The reactants, aluminum and sulfur, are two elements. The type of reaction that occurs is a synthesis reaction. The synthesis reaction between a metal and a nonmetal produces a single, binary ionic compound. The product is aluminum sulfide, $Al_2S_3$.



## How Can You Represent Chemical Reactions Using Chemical Equations?

### Writing and Balancing Chemical Equations

Scientists use chemical symbols to represent the elements. They use chemical formulas to represent compounds. In a similar way, they use chemical equations to represent chemical reactions.

One way to represent a chemical reaction is by writing a word equation. A word equation separates the names of the reactants from the names of the products using an arrow. The arrow can be read as "yields" or "produces." If the reaction involves more than one **reactant** or **product**, a plus sign is placed between the corresponding reactants or products. The word equation that represents the synthesis of water from hydrogen and oxygen is:

$$\text{hydrogen} + \text{oxygen} \rightarrow \text{water}$$

A formula equation is written by replacing each name with the chemical symbol or formula that represents that substance. Hydrogen and oxygen are both elements that exist at room temperature as molecules rather than single atoms. The formulas $H_2$ (hydrogen gas) and $O_2$ (oxygen gas) are therefore used instead of just H and O in the equation.

The formula equation is shorter than the descriptive word sentence. It also provides a visual description of the chemicals involved.

$$H_2 + O_2 \rightarrow H_2O$$

The formula equation is shorter than the descriptive word sentence. It also provides a visual description of the chemicals involved.

A key principle in writing formula equations is to ensure they are balanced. The formula equation above is still incomplete. Notice that there is one less oxygen atom on the product side of the equation than on the reactant side. We know that the oxygen atom did not simply disappear, so it must be accounted for in the equation.



During the synthesis of water, the chemical bonds in the hydrogen and oxygen molecules break. The atoms form new bonds as they reorganize to form water molecules. Each atom from the reactants becomes part of the product. To make sure that all of the atoms in the reactants and the products are accounted for, the formula equation must be balanced. A balanced reaction shows the same number of atoms of each element in both the reactants and the products. To balance a chemical equation, coefficients are placed before chemical formulas.

For the synthesis of water, a **coefficient** is needed before the water molecule. This balances the oxygen atoms:

$$H_2 + O_2 \rightarrow 2H_2O$$

Notice that the subscript for oxygen in $H_2O$ is not changed. Subscripts should never be changed to balance an equation. Changing a subscript would change a chemical formula, and the equation would no longer represent the same chemical reaction.

**RUSTED BIKE GEARS**

Rust formed on the gears of this bike when it was left out in the rain. What chemical equation represents this reaction?



Now we know two water molecules have formed for each molecule of reactant, but the equation is still not balanced. Due to the coefficient, there are now four hydrogen atoms on the product side of the arrow, but only two on the reactant side. Two more hydrogen atoms are needed in the reactants. A coefficient placed in front of $H_2$ balances the hydrogen atoms:

$$2H_2 + O_2 \rightarrow 2H_2O$$

## Writing and Balancing Chemical Equations: Sample Problem

Write a **balanced chemical equation** for the formation of potassium chloride and oxygen from the decomposition of potassium chlorate.

### Solution

The reactant for this **decomposition reaction** is potassium chlorate. The products are potassium chloride and oxygen. The word equation for the reaction is:

potassium chlorate → potassium chloride + oxygen

The formula equation for the reaction is:

$$KClO_3 \rightarrow KCl + O_2$$

There is one atom of potassium on each side of the equation. There is also one atom of chlorine on each side of the equation. However, there are three oxygen atoms in the reactants and only two in the products. The least common multiple of two and three is six. Balance the oxygen atoms with coefficients so that six atoms of oxygen are on each side of the equation:

$$2KClO_3 \rightarrow KCl + 3O_2$$

Now the oxygen atoms are balanced. However, now the potassium and chlorine atoms are not. To resolve this, place a coefficient in front of KCl:

$$2KClO_3 \rightarrow 2KCl + 3O_2$$



## Conservation in Chemical Reactions

The equations for chemical reactions always balance because they follow the **law of conservation of matter**. During a chemical reaction, matter is neither created nor destroyed. Each atom from the reactants is used in the products. Each atom has mass. A balanced equation, therefore, also shows that the mass of the reactants equals the mass of the products. However, the number of molecules is not always conserved during a reaction. This is evident from the equations for synthesis and decomposition reactions.

## Writing Net Ionic Equations

A balanced equation provides information about a chemical reaction. One important piece of information is the matter state of each substance involved. By convention, the symbols (s), (l), (g), and (aq) are used to distinguish these states.

Displacement reactions often involve ionic compounds in water, or aqueous solution. Solid ionic compounds, however, may not be reactive at all. When dry, crystals of table salt, for example, do not react with other compounds. When an ionic compound is dissolved in water, however, its ions can move throughout the solution and interact with other ions.

If sodium sulfate and barium nitrate are both dissolved in water and then combined, the reaction is described by the equation:

$$Na_2SO_4(aq) + Ba(NO_3)_2(aq) \rightarrow 2NaNO_3(aq) + BaSO_4(s)$$

One of the products is in the solid state because it has formed as a precipitate. The sodium nitrate remains dissolved in water. The **solubility** rules identify barium sulfate as insoluble.

It seems as though two compounds should be formed by this reaction. Remember, however, that when ionic compounds are dissolved their ions move freely through the water. The reaction of sodium sulfate and barium nitrate could also be described with the equation:

$$2Na^+(aq) + SO_4^{2-}(aq) + Ba^{2+}(aq) + 2NO_3^-(aq) \rightarrow 2NA^+(aq) + 2NO_3^-(aq) + BaSO_4(s)$$

The sodium ions and nitrate ions have not changed at all. Each **ion** is dissolved in water before and after the reaction. Ions that are unchanged during a displacement reaction are known as spectator ions. These ions “watch” the action, but do not participate. Since they are not involved in the reaction, the spectator ions can be removed from the chemical equation. The resulting **net ionic equation** shows only the substances that take part in a reaction.

$$SO_4^{2-}(aq) + Ba^{2+}(aq) \rightarrow BaSO_4(s)$$

The two ions in the solution react to form a precipitate.

## Writing Net Ionic Equations: Sample Problem

Write the net ionic equation for the reaction that occurs when a piece of zinc metal is placed into an aqueous solution of copper(II) sulfate.

**Solution**

The reactants include an element and an ionic compound in the solution. This reaction is a single-displacement reaction. Zinc is above copper in the **reactivity series**, so a reaction will occur. The zinc replaces copper to form zinc sulfate:

$$Zn(s) + CuSO_4(aq) \rightarrow ZnSO_4(aq) + Cu(s)$$

The total ionic equation is:

$$Zn(s) + Cu^{2+}(aq) + SO_4^{2-}(aq) \rightarrow Zn^{2+}(aq) + SO_4^{2-}(aq) + Cu(s)$$

The sulfate ion is unchanged during the reaction. Remove this spectator ion to find the net ionic equation:

$$Zn(s) + Cu^{2+}(aq) \rightarrow Zn^{2+}(aq) + Cu(s)$$

## Consider the Explain Question

**What are the five main types of chemical reactions, and how can chemical equations help you predict the products of each of the five types of reactions?**

dlc.com/ca9075s

Go online to complete the scientific explanation.

## Check Your Understanding

**How does a balanced chemical equation show that the law of conservation of mass is obeyed?**

dlc.com/ca9076s

## Applying Chemical Reactions and Equations

Decomposition reactions have long been applied in mining, construction, and demolition, as well as in warfare. Demolition often involves using sticks of dynamite. Dynamite is constructed by soaking an absorbent material in the chemical nitroglycerine. Nitroglycerine, $C_3H_5N_3O_9$, is a relatively simple molecule, but it decomposes so quickly that the explosion is violent. Another common explosive is trinitrotoluene—better known as TNT. The explosive release of energy from TNT also occurs during its decomposition.

**CONTROLLED EXPLOSION TO DEMOLISH AN OLD BUILDING**

Explosions are often used to bring down old buildings to make room for new ones. Great care must be taken to avoid damaging surrounding structures. How can chemical equations help the explosive experts predict how the explosion will occur?



The balanced equation that represents a TNT chemical reaction is:

$$2C_7H_5N_3O_6(s) \rightarrow 12CO(g) + 3N_2(g) + 5H_2(g) + 2C(s)$$

This reaction gives off thermal energy (heat). The force behind the blast is the formation of several gaseous products. A substance occupies a much larger volume in its gas phase than in its solid or liquid phases. When the reaction occurs, the released thermal energy heats the gases produced by the reaction. The gases expand outward and exert a large force on the surrounding material.

The quick expansion of gases from a high explosive is sometimes used to stop another type of chemical reaction: combustion. The shock wave from the explosion removes oxygen from the site of a fire. Oxygen is needed to keep the **combustion reaction** going; without it, the fire stops.

## STEM and Chemical Reactions and Equations

Starting in the late 1700s and early 1800s, a large portion of the manufacturing processes in the United States became dependent on machines. This change, called the Industrial Revolution, had enormous social benefits, but over time, it became apparent that there were also environmental drawbacks. A troubling environmental effect of the Industrial Revolution is known as acid precipitation, also known as acid rain. Acid precipitation is any form of precipitation that has an increased level of acidity. Certain chemicals released by factories and automobiles, as well as natural sources, can undergo synthesis reactions with rainwater to produce acids. The combination of acids and rainwater is acid precipitation.

Excess carbon dioxide released into the atmosphere, for example, dissolves in and reacts with rainwater. The resulting **synthesis reaction** forms carbonic acid, which causes precipitation to be naturally acidic:

$$H_2O(l) + CO_2(g) \rightarrow H_2CO_3(aq)$$

Other compounds released into the atmosphere, such as nitrogen oxides and sulfur dioxide, can also undergo synthesis reactions with rainwater and result in precipitation that is many times more acidic than normal.

**TREES DAMAGED BY ACID PRECIPITATION**

Trees at high altitudes are at increased risk of damage from acid precipitation. Why do reactions of atmospheric gases result in an acidic product?

Atmospheric scientists monitor the levels of certain compounds that have been released in the air that result in acidic precipitation. Sulfur dioxide, for example, is a **product** of the combustion reaction of sulfur-containing fuels, such as coal or farm diesel fuel. The sulfur dioxide enters the atmosphere, where it reacts with oxygen in the following synthesis reaction.

$$2SO_2(g) + O_2(g) \rightarrow 2SO_3(g)$$

The sulfur trioxide then reacts with water to form sulfuric acid.

$$H_2O(l) + SO_3(g) \rightarrow H_2SO_4(aq)$$

Acid precipitation damages structures such as bridges and buildings. It also harms plants and animals. It can make lakes too acidic to support aquatic plants and animals. Acid precipitation leaches chemicals from soil and rocks. These chemicals, which may include lead and mercury, are carried to streams and lakes, where they can harm organisms.

# Balancing Chemical Equations

**Balance each of the chemical equations shown below.**

**1.** $NaClO_3 \rightarrow NaCl + O_2$

**2.** $Al + O_2 \rightarrow Al_2O_3$

**3.** $Ca(OH)_2 + HNO_3 \rightarrow Ca(NO_3)_2 + H_2O$

**4.** $Al + Sn(NO_3)_2 \rightarrow Al(NO_3)_3 + Sn$

**5.** $CH_4 + O_2 \rightarrow CO_2 + H_2O$

CONCEPT **4.2**

# Mathematics of Formulas and Equations

dlc.com/ca9077s

## LESSON OVERVIEW

### Lesson Questions

- How is molar mass calculated and why is it useful?
- How are percent composition of a compound, empirical formula of a compound, and molecular formula of a compound calculated?
- How are the principles of stoichiometry used to calculate quantities of reactants or products in a chemical reaction?

### Lesson Objectives

By the end of the lesson, you should be able to:

- Determine molar mass and formula mass.
- Determine the percent composition of a compound.
- Apply stoichiometric principles.

### Key Vocabulary

Which terms do you already know?

- ☐ atom
- ☐ average atomic mass
- ☐ Avogadro's number
- ☐ balanced chemical equation
- ☐ chemical reaction
- ☐ compound
- ☐ crystal lattice
- ☐ element
- ☐ empirical formula
- ☐ formula mass
- ☐ formula unit
- ☐ limiting reagent
- ☐ mass
- ☐ molar mass
- ☐ mole
- ☐ molecular formula
- ☐ molecular mass
- ☐ molecule
- ☐ percent composition
- ☐ percent yield
- ☐ periodic table of the elements
- ☐ product
- ☐ reactant
- ☐ standard temperature and pressure (STP)
- ☐ stoichiometry

# Understanding the Importance of Mathematics of Formulas and Equations

**Can you imagine an explosion saving a person's life? That is exactly what can happen when airbags inflate during automobile accidents.**

dlc.com/ca9078s



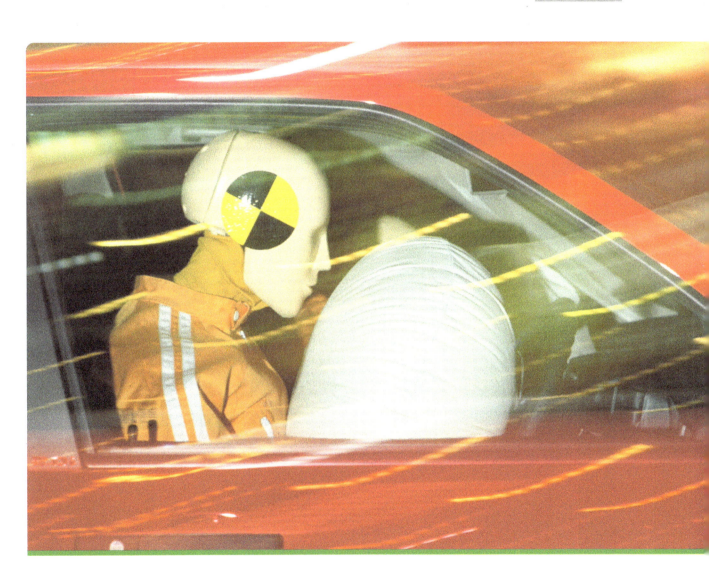

## EXPLAIN QUESTION

**How can you use average atomic masses from the periodic table to find the formula mass and the molar mass of ammonia, $NH_3$, and the percent composition of each element in ammonia?**

### AIRBAG EXPLOSION

A collision triggers a chemical reaction inside an airbag. How would a chemist use chemical formulas and equations to predict what will happen inside the airbag?

# How Is Molar Mass Calculated and Why Is It Useful?

## The Mole

In chemistry, it is rare that anyone works with single atoms or molecules. More often, chemists work with amounts of matter that are easily observed. Such matter is measured in gram or kilogram quantities. These quantities contain billions of atoms or molecules.

Recall that a **mole** is defined as $6.022 \times 10^{23}$ atoms or molecules. A mole is abbreviated "**mol**" in calculations. This number is called **Avogadro's number** in honor of Amedeo Avogadro, the Italian scientist who showed that one mole of any gas has the same number of particles as one mole of any other gas.

Because one mole contains so many particles, it offers a convenient unit for expressing the gram and kilogram quantities of substances that chemists tend to work with. Instead of trying to talk about millions of atoms or molecules, they can talk about moles of atoms or molecules instead.

## Molar Mass of an Element

Since Avogadro's time, much evidence has been gathered to support the mole concept. Chemists now use this idea often and have extended it to include all substances. One mole of any pure substance has the same number of particles as one mole of any other pure substance.

However, even though one-mole quantities of two substances contain the same number of particles, they do not have the same **mass**. Mass can vary depending on the identities of the elements present in each substance.

Suppose a chemist has one mole of helium gas and one mole of neon gas. An **atom** of neon is heavier than an atom of helium. This means that a mole of neon will be heavier than a mole of helium.

The mass of one mole of an **element** is easily determined from the **average atomic mass** of that element. Locate the average atomic mass of the element listed on the periodic table. The value shown is equal to the mass in grams of one mole of that element.

Using this approach, the masses of one atom of helium, one atom of neon, and one mole of each of these elements can be summarized as shown below.

| | Mass of one atom | Mass of one mole |
|---|---|---|
| Helium | 4.002 amu | 4.002 g |
| Neon | 20.180 amu | 20.180 g |

**COMPARING MASS TABLE**

This table shows the relationship between the atomic mass units (*amu*) and gram weight of atoms. What is the molar mass of hydrogen, $H_2$, if its atomic mass is 1.008 amu?

The **molar mass** of an element is defined as the mass of one mole of that element. The units for molar mass are grams per mole, so neon's molar mass is written as 20.180 g/mol. The molar mass of each element can be found on the periodic table.

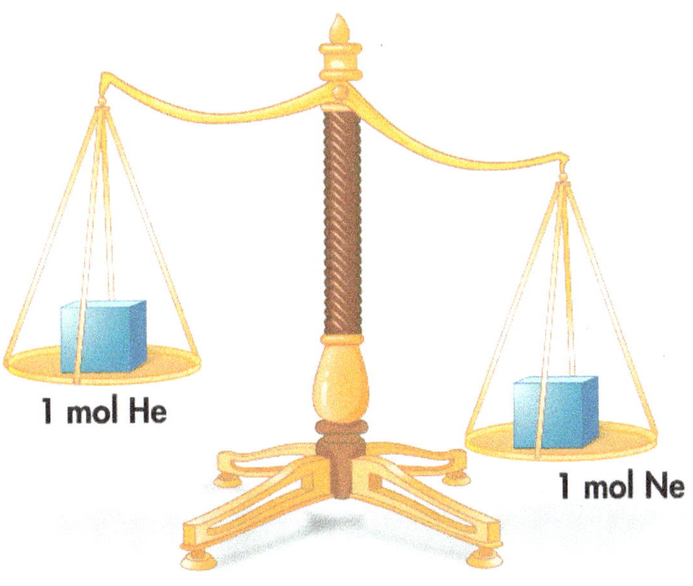




**COMPARISON OF HELIUM AND NEON MASSES**

One mole of helium atoms contains the same number of atoms as one mole of neon atoms, but has less mass. Why isn't the mass the same?

## Molar Mass of a Compound

Recall that compounds are composed of two or more elements whose atoms are bonded together. The molar mass of a **compound** can be found using the molar masses of its elements.

Consider the compound methane for example. Methane's chemical formula is $CH_4$. Its molar mass is determined by adding together the molar masses of each carbon and hydrogen atom present in one **molecule**.

One mole of $CH_4$ contains one mole of carbon atoms and four moles of hydrogen atoms. Its molar mass is the sum of the molar masses of these elements:

1 mole of carbon: 12.011 g/mol

4 moles of hydrogen 4(1.008) g/mol = 4.032 g/mol

Total mass = 12.011 g/mol + 4.032 g/mol = 16.043 g/mol

The compound methane has a molar mass of 16.043 grams per mole.

## Using Molar Mass in Conversions

It is not easy to measure moles in the lab. It is, however, easy to measure mass in the lab. Electronic balances are standard pieces of equipment in most chemistry labs.

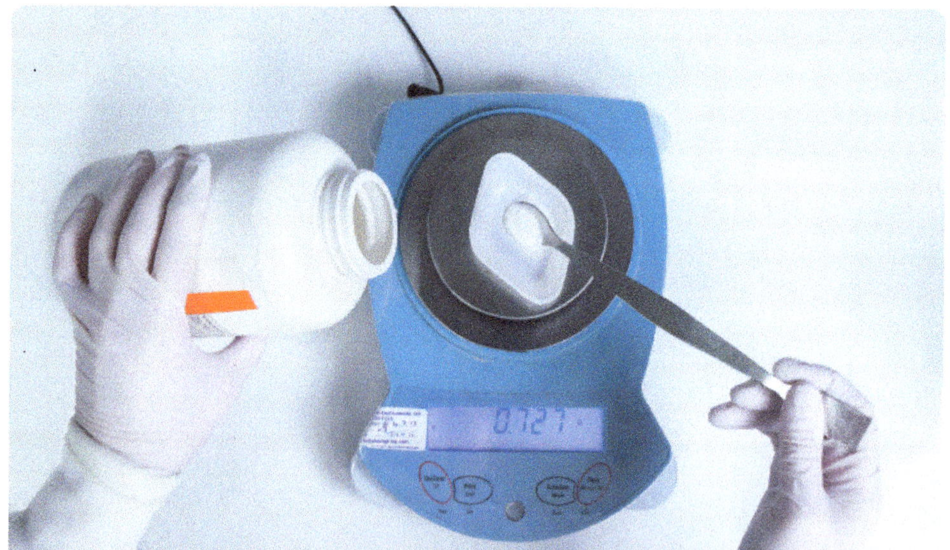

**MEASURING MASS**

This digital scale is used to measure the mass of a substance. What are the advantages of using digital scales in the laboratory?



Often, chemists perform calculations to determine the number of moles of a substance needed for a particular reaction. They then do another calculation to convert moles to grams. Knowing the number of grams allows them a means for accurately measuring out the needed quantity of the substance.

To do this type of calculation, chemists use the molar mass as a conversion factor. In general, the following scheme summarizes the types of conversions that are possible:

$$\text{atoms or molecules} \underset{\textit{use Avogadro's number}}{\longleftrightarrow} \text{moles} \underset{\textit{use molar mass}}{\longleftrightarrow} \text{grams}$$

Numbers of atoms or molecules can be converted to moles using Avogadro's number. Moles can be converted to grams using molar mass.

These conversions can also be carried out in reverse. Grams can be converted to moles using molar mass. Moles can be converted to numbers of atoms or molecules using Avogadro's number.

## Using Molar Mass in Conversions: Sample Problem

Calculate the number of moles and the mass in grams of $2.57 \times 10^{18}$ atoms of sodium.

**Solution**

The number of atoms of sodium is given. The first step will involve a conversion of this given value to the next logical value. In the following scheme, Avogadro's number can be used to convert numbers of atoms to moles:

$$\text{atoms or molecules} \underset{\textit{use Avogadro's number}}{\longleftrightarrow} \text{moles} \underset{\textit{use molar mass}}{\longleftrightarrow} \text{grams}$$

To convert numbers of atoms to moles, use the conversion factor based on Avogadro's number:

$$1 \text{ mole} = 6.022 \times 10^{23} \text{ atoms}$$

Set up an equation in which the number of atoms of sodium is multiplied by a conversion factor based on Avogadro's number. Be sure that the conversion factor is written so that like units (atoms) cancel and only the units of moles will be left in the final answer:

$$(2.57 \times 10^{18} \cancel{\text{sodium atoms}}) \left( \frac{1 \text{ mol sodium}}{6.022 \times 10^{23} \cancel{\text{sodium atoms}}} \right)$$
$$= 4.27 \times 10^{-6} \text{ mol sodium}$$

There are $4.27 \times 10^{-6}$ moles of sodium present.

Once moles of sodium have been calculated, this value can be converted to grams of sodium. This conversion requires the molar mass of sodium. From the periodic table, the molar mass of sodium is 22.990 grams per mole.

Set up an equation in which the moles of sodium are multiplied by a conversion factor based on sodium's molar mass. Be sure that the conversion factor is written so that like units (moles) cancel and only the units of grams will be left in the final answer:

$$(4.27 \times 10^{-6} \text{ \sout{mol sodium}})\left(\frac{22.990 \text{ g sodium}}{1 \text{ \sout{mol sodium}}}\right) = 9.81 \times 10^{-5} \text{ g sodium}$$

There are $9.81 \times 10^{-5}$ grams of sodium present.

## How Are Percent Composition of a Compound, Empirical Formula of a Compound, and Molecular Formula of a Compound Calculated?

### Stoichiometry

The term **stoichiometry** is used to describe numerical relationships in chemistry. Stoichiometric relationships based on the **mole** are always whole number ratios. There are two types of stoichiometry:

- Composition stoichiometry is related to numerical relationships between elements in chemical compounds. For example, a water **molecule** contains a 2:1 molar ratio of hydrogen atoms to oxygen atoms.
- Reaction stoichiometry is related to molar ratios of reactants and products in chemical reactions. For example, in the reaction of hydrogen with oxygen to form water, there is a 1:1 molar ratio of $H_2$ **reactant** to $H_2O$ **product**.

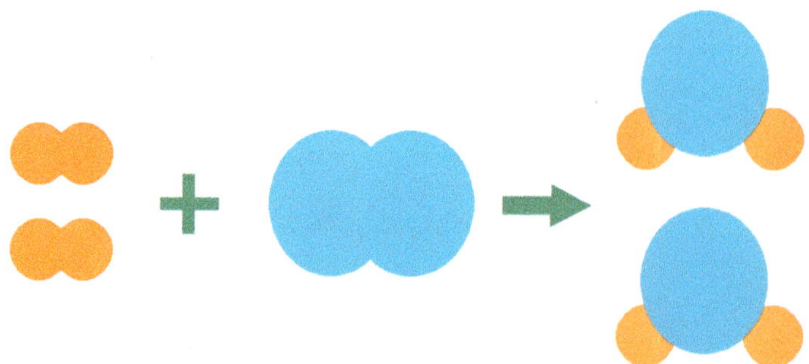

**COMBUSTION OF HYDROGEN**

Both composition stoichiometry and reaction stoichiometry can be used to analyze the reaction of hydrogen with oxygen to produce water. What type of stoichiometry could be used to calculate the percentage of each element that makes up the mass of the water molecule?

## Examples of Composition Stoichiometry

Composition stoichiometry is related to the numerical relationships between elements in chemical compounds. The different ways that compounds can be described include the following:

- percent water of a hydrate
- percent composition
- empirical formula
- molecular formula

The first example listed is restricted to compounds that form hydrates, while the other three apply to all compounds. Each of these will be discussed in the order shown.

**THE STOICHIOMETRY OF WATER**

The composition stoichiometry of water is based on its hydrogen and oxygen content. What is the ratio of hydrogen to oxygen in water?

## Calculating the Percent Water in a Hydrate

Hydrates are compounds that bind water molecules in specific ratios when they crystallize. An example is copper(II) sulfate pentahydrate.

$CuSO_4 \cdot 5H_2O$ is bright blue in color. Within the crystal lattice of this compound are five water molecules for every one copper(II) sulfate formula unit. When the compound is heated, the water molecules can be evaporated leaving a white powder, $CuSO_4$.

The percent water in this hydrate is found by the following equation:

$$\% \text{ water} = \left(\frac{\text{grams } H_2O}{\text{grams } CuSO_4 \cdot 5H_2O}\right) \times 100$$

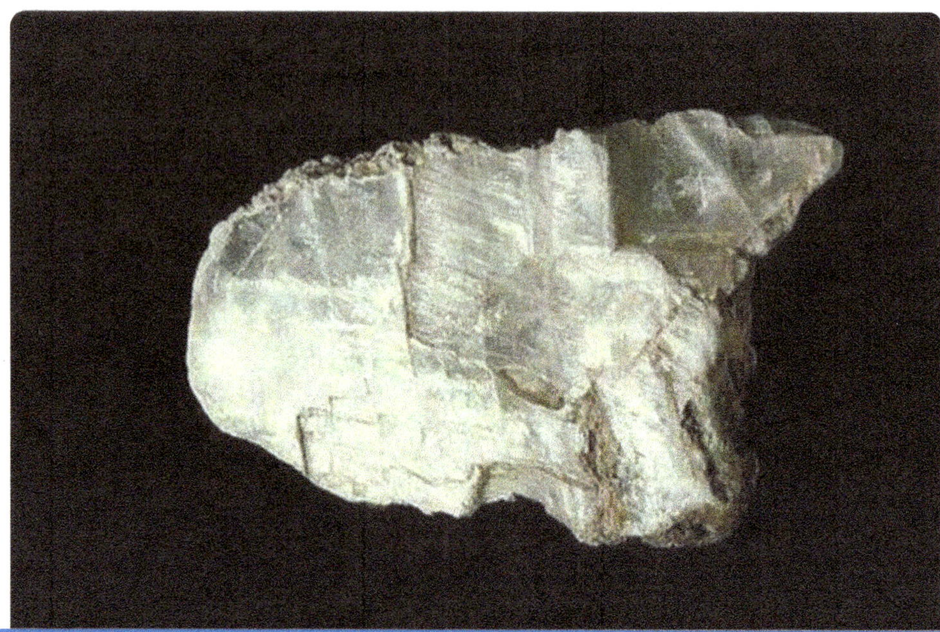

**WATER IN GYPSUM**
Gypsum is a naturally occurring mineral consisting of hydrated calcium sulfate. How could you determine the percent water in this sample of gypsum?



## Calculating Percent Composition

Percent composition is defined as the percent by **mass** of each **element** in the compound. These values can be determined if the molecular formula is known. Molar masses are used to do these calculations.

For each element, the percent composition is given by the equation shown below. It is useful to assume that 1 mole of the compound is being analyzed.

$$\% \text{ element} = \left(\frac{\text{mass of element}}{\text{mass of compound}}\right) \times 100$$

The percent by mass of every element in the compound is calculated in this way. When these calculations are complete, check them by adding them together. The sum should add up to 100%. This is the theoretical percent by mass of the compound. If the compound is separated into its element experimentally, the mass may be exactly 100% due to experimental error.

**MOLECULAR MODEL OF AMMONIA**

How can you determine the percent composition of nitrogen and hydrogen in ammonia?

## Calculating Percent Composition: Sample Problem

Calculate the percent composition of water, $H_2O$.

**Solution**

First, calculate the **molar mass** of water. One water molecule contains two hydrogen atoms and one oxygen **atom**.

1 mole of oxygen: 15.999 g/mol
2 moles of hydrogen: 2(1.008) g/mol = 2.016 g/mol
total mass = 15.999 g/mol + 2.016 g/mol = 18.015 g/mol

Next, assume that one mole of water is being analyzed. Calculate the mass of one mole of water using the molar mass of water.

$$\text{mass of } H_2O = 1\ \cancel{\text{mol } H_2O} \times \left(\frac{18.015 \text{ g } H_2O}{1\ \cancel{\text{mol } H_2O}}\right) = 18.015 \text{ g } H_2O$$

Calculate the mass of each element:

$$\text{mass of H} = 2\ \cancel{\text{mol H}} \times \left(\frac{1.008 \text{ g H}}{1\ \cancel{\text{mol H}}}\right) = 2.016 \text{ g H}$$

$$\text{mass of O} = 1\ \cancel{\text{mol O}} \times \left(\frac{15.999 \text{ g O}}{1\ \cancel{\text{mol O}}}\right) = 15.999 \text{ g O}$$

Finally, calculate the percent mass of each element by dividing the mass of the element by the mass of the compound. Multiply by 100 to obtain a percentage:

$$\%\text{ H} = \left(\frac{2.016 \text{ g H}}{18.015 \text{ g } H_2O}\right) \times 100 = 11.19\%$$

$$\%\text{ O} = \left(\frac{15.999 \text{ g H}}{18.015 \text{ g } H_2O}\right) \times 100 = 88.81\%$$

Check the results by adding the percentages to see if they add to give 100%:

$$\begin{array}{r} 11.19\% \\ +\ 88.81\% \\ \hline 100.00\% \end{array}$$

The answers appear to be reasonable.

### Molecular Formulas and Empirical Formulas

Recall that the molecular formula gives the number of atoms of each element found in one molecule of the compound. If these numbers are divisible by a common factor, they can be simplified to give the simplest ratio of elements in the compound. The simplest molar ratio of elements in a compound is the compound's empirical formula.

Nonmetallic elements often form several different compounds when they bond with another nonmetallic element. For example, sulfur and oxygen can bond to form sulfur monoxide, SO, or disulfur dioxide, $S_2O_2$. The empirical formula for both compounds is the same, SO, but their molecular formulas are different.

**A Crystal Lattice**

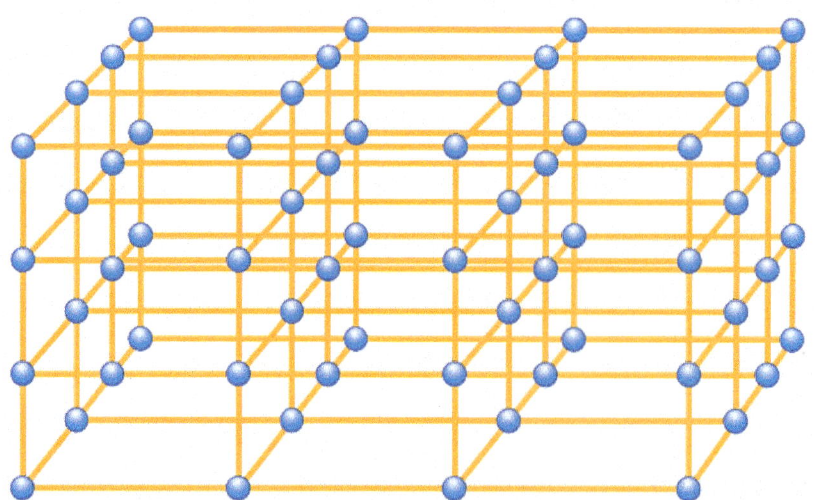

**CRYSTAL**

A solid ionic compound forms a crystal lattice. How does the crystal lattice structure relate to the concept of a molecule?

Some compounds do not have molecular formulas and can only be described by empirical formulas. Ionic compounds are composed of repeating units of ions that form a crystal rather than individual molecules. Since these compounds do not form discrete molecules, they cannot be said to have molecular formulas. Instead, they are described only by empirical formulas that give the simplest ratios of the elements in the compound.

| Compound | Molecular Formula | Empirical Formula |
|---|---|---|
| Benzene | $C_6H_6$ | CH |
| Acetylene | $C_2H_2$ | CH |
| Hydrogen peroxide | $H_2O_2$ | HO |

**EMPIRICAL AND MOLECULAR FORMULAS**

This table shows molecular and empirical formulas for three compounds. Why are the empirical formulas of benzene and acetylene the same even though their molecular formulas are different?

## Calculating Empirical Formula from Percent Composition

When chemists synthesize or discover a new compound, they often are able to more easily determine its percent composition than its chemical formula. However, they can calculate the empirical formula of the compound once they know its percent composition.

To do the calculations, it is helpful to assume that a 100-gram sample is being used. That way, the percentage values can be assumed to represent grams of each element. Then, follow these steps:

- Use the grams and the molar mass of each element to calculate the moles of each element.
- Compare the moles of all the elements to obtain molar ratios.
- Calculate the simplest ratios to find the empirical formula.

## Calculating Molecular Formula from Empirical Formula

The empirical formula gives the simplest molar ratio of elements in a compound. The molecular formula gives the exact amounts of each element in one molecule of the compound. Therefore, the empirical and molecular formulas are related by some whole number factor.

$$x(\text{empirical formula}) = \text{molecular formula}$$

The molecular formula can be determined if the empirical formula and the molar mass of the compound are given. Calculate the **formula mass** by adding together the molar masses of the elements in the empirical formula. Then divide the formula mass by the molar mass. They should be related by a factor that is a whole number.

$$\frac{\text{formula mass}}{\text{molar mass}} = x$$

This whole number can be used to multiply the molar ratios in the empirical formula to give the molecular formula.

## How Are the Principles of Stoichiometry Used to Calculate Quantities of Reactants or Products in a Chemical Reaction?

### Reaction Stoichiometry

Recall that there are two types of **stoichiometry**: composition stoichiometry and reaction stoichiometry. Composition stoichiometry deals with ratios of elements in chemical compounds. Reaction stoichiometry deals with ratios of all of the compounds involved in a **chemical reaction**.

In the reaction of hydrogen and oxygen to form water, the reaction stoichiometry is given by the balanced equation:

$$2H_2 + O_2 \rightarrow 2H_2O$$

The reaction stoichiometries for this reaction can be summarized as follows:

- There is a 1:2 molar ratio of $O_2$ to $H_2$.
- There is a 1:1 molar ratio of $H_2$ to $H_2O$.
- There is a 1:2 molar ratio of $O_2$ to $H_2O$.

SCHOOL BUS

The motor of a school bus burns fuel in a combustion reaction. How can stoichiometry be used to calculate the masses and ratios of reactants or products in the reaction?



## Calculating Reactant or Product Quantities Using Reaction Stoichiometry

Chemists use reaction stoichiometries to do calculations involving chemical reactions. For example, the **mass** of **product** expected from a reaction can be calculated using the mass of **reactant** supplied to run the reaction.

Use the following steps to carry out calculations involving reaction stoichiometries:

- Begin with the **balanced chemical equation** for the reaction.
- Determine the molar ratio of reactant to product from the balanced equation.
- Convert the mass of reactant to moles.
- Use the molar ratio of reactant to product as a conversion factor. Multiply the molar ratio by the moles of reactant to find the moles of product.
- Convert the moles of product to mass.

## Calculating Reactant or Product Quantities Using Reaction Stoichiometry: Sample Problem

Calculate the mass of aluminum oxide, $Al_2O_3$, formed when 2.00 g Al is reacted with oxygen from the air. The balanced chemical equation for this reaction is:

$$4Al + 3O_2 \rightarrow 2Al_2O_3$$

**Solution**

First, determine the molar ratio of Al to $Al_2O_3$ from the balanced equation.

$$4 \text{ mol Al} \rightarrow 2 \text{ mol } Al_2O_3$$

Next, convert the mass of reactant to moles.

$$2.00 \cancel{\text{g Al}} \times \left(\frac{1 \text{ mol Al}}{26.982 \cancel{\text{g Al}}}\right) = 7.41 \times 10^{-2} \text{ mol Al}$$

Use the molar ratio of reactant to product as a conversion factor. Multiply the molar ratio by the moles of reactant to find the moles of product.

$$7.41 \times 10^{-2} \cancel{\text{mol Al}} \times \left(\frac{2 \text{ mol } Al_2O_3}{4 \cancel{\text{mol Al}}}\right) = 1.48 \times 10^{-1} \text{ mol } Al_2O_3$$

Finally, convert the moles of product to mass.

$$1.48 \times 10^{-1} \cancel{\text{mol } Al_2O_3} \times \left(\frac{101.961 \text{ g } Al_2O_3}{1 \cancel{\text{mol } Al_2O_3}}\right) = 15.1 \text{ g } Al_2O_3$$

Thus, 2.00 grams Al will produce 15.1 g $Al_2O_3$.

## Stoichiometry of Gases: Volume-Volume Calculations: Sample Problem

Calculate the volume of nitrogen monoxide that will be produced in a reaction using 3,500 L of oxygen gas. The balanced chemical equation for this reaction is:

$$4NH_3(g) + 5O_2(g) \rightarrow 4NO(g) + 6H_2O(l)$$

**Solution**

First, the volume ratio is the same as the mole ratio, if the two gases are at the same temperature and pressure.

$$\text{Mole ratio of } NO/O_2 = 4/5$$
$$\text{Volume ratio } NO/O_2 = 4/5$$

Next, use the volume ratio of the reactant to product as a conversion factor. Multiply the volume of the reactant by the mole ratio of the product over the reactant to find the volume of product.

$$3{,}500 \text{ L } O_2 \times \left(\frac{4 \text{ L NO}}{5 \text{ L } O_2}\right) = 2{,}800 \text{ L NO}$$

## Stoichiometry of Gases: Mass-Mass: Sample Problem

If stoichiometric calculations of gases involve mass, convert the mass of the gas to moles using the molar mass.

How many grams of carbon monoxide gas are produced from the reaction of carbon and oxygen gas, if 12.0 grams of oxygen are consumed in the reaction? The balanced chemical equation for the reaction is:

$$2C(g) + O_2(g) \rightarrow 2CO(g)$$

**Solution**

First, convert the mass of oxygen to moles of oxygen using the molar mass.

$$12.0 \text{ g } O_2 \times \left(\frac{1 \text{ mol } O_2}{32.00 \text{ g } O_2}\right)$$

Second, use the mole ratio of the reactant to product as a conversion factor. Multiply the mole ratio by the moles of reactant to find the moles of product.

$$12.0 \text{ g } O_2 \times \left(\frac{1 \text{ mol } O_2}{32.00 \text{ g } O_2}\right) \times \left(\frac{2 \text{ mol CO}}{1 \text{ mol } O_2}\right)$$

Next, multiply the moles of carbon monoxide by the molar mass of carbon dioxide to calculate the mass of carbon monoxide produced.

$$12.0 \text{ g } O_2 \times \left(\frac{1 \text{ mol } O_2}{32.00 \text{ g } O_2}\right) \times \left(\frac{2 \text{ mol CO}}{1 \text{ mol } O_2}\right) \times \left(\frac{28.01 \text{ CO}}{1 \text{ mol CO}}\right) = 21.0 \text{ g CO}$$

## Limiting Reagent

When a chemical reaction involves two or more reactants, these reactants are not usually present in the exact molar ratio shown by the balanced equation. Usually one of the reactants is present in smaller amounts than the other(s). In this case, this reactant is said to be the limiting reactant or **limiting reagent**. It limits the amount of product that can form.

If the quantities of all reactants present at the start of the reaction are known, it is possible to use these values to calculate the quantity of product that could be produced from each. The reactant found to produce the least amount of product is the limiting reagent.

The maximum amount of product that will be formed during a chemical reaction can be found by using the limiting reagent. This amount of product is the theoretical yield of the reaction. The actual yield may be smaller than this. This may occur if the reaction does not proceed to completion.

## Limiting Reagent: Sample Problem

Calculate the mass of CaO that is possible if 4.8 grams Ca reacts with 10.0 grams $O_2$.

**Solution**

Step 1: Write a balanced equation for the reaction.

$$2Ca + O_2 \rightarrow 2CaO$$

Steps 2 and 3: Convert mass to moles and determine how much CaO is possible with each reactant.

$$4.8 \text{ g Ca} \times \frac{1 \text{ mol Ca}}{40.1 \text{ g Ca}} \times \frac{2.00 \text{ mol CaO}}{2.00 \text{ mol Ca}} \times \frac{56.1 \text{ g CaO}}{1.00 \text{ mol CaO}} = 6.7 \text{ g CaO}$$

$$10.0 \text{ g } O_2 \times \frac{1.00 \text{ mol } O_2}{32.0 \text{ g } O_2} \times \frac{2.00 \text{ mol CaO}}{2.00 \text{ mol } O_2} \times \frac{56.1 \text{ g CaO}}{1.00 \text{ mol CaO}} = 17.5 \text{ g CaO}$$

Step 4: The reactant that produces the smaller amount of product is the limiting reagent.

Ca produces less CaO than does $O_2$ (6.7 g MgO versus 17.5 g MgO). Therefore, Ca is the limiting reagent.

Step 5: The reactant that produces a larger amount of product is the excess reagent.

$O_2$ produces more CaO than Ca, therefore $O_2$ is the excess reagent in this reaction.

### Calculating Percent Yield

The maximum amount of product formed using the limiting reagent is the theoretical yield of the reaction. The actual yield may be smaller than this. Reasons for smaller than expected yield of product include:

- the reaction does not proceed to completion
- some product is lost during the steps following reaction in which the product is isolated
- some product decomposes or undergoes some other reaction that converts it to a different **compound**

The **percent yield** gives an indication of the success of isolating a product from a chemical reaction. Percent yield is calculated by dividing the actual yield by the theoretical yield and multiplying by 100.

$$\text{percent yield} = \left(\frac{\text{actual yield of product in grams}}{\text{theoretical yield of product in grams}}\right) \times 100$$

#### Consider the Explain Question

**How can you use average atomic masses from the periodic table to find the formula mass and the molar mass of ammonia, $NH_3$, and the percent composition of each element in ammonia?**

dlc.com/ca9079s

Go online to complete the scientific explanation.

#### Check Your Understanding

**How can molar mass be used to determine the number of molecules in a sample compound, if its mass and molecular formula is known?**

dlc.com/ca9080s

# Applying Mathematics of Formulas and Equations

Calculations involving quantities of substances are essential to many activities people do every day. How do you think chemistry calculations apply to brushing your teeth? Manufacturers of personal care products use stoichiometric calculations to determine how ingredients can be produced and combined. A scientist who develops personal care products is called a *formulation chemist.* If you look on the side of a tube of toothpaste, you might see ingredient listings such as "sodium fluoride (0.15%)" or "potassium nitrate (5%)." Sodium fluoride is added to make teeth more resistant to decay. Potassium nitrate may be added to toothpaste to reduce pain in sensitive teeth. How do you think formulation chemists determined the percentages of each ingredient? How could they have used the percentages to specify the masses of each **compound** that should be included in the toothpastes?

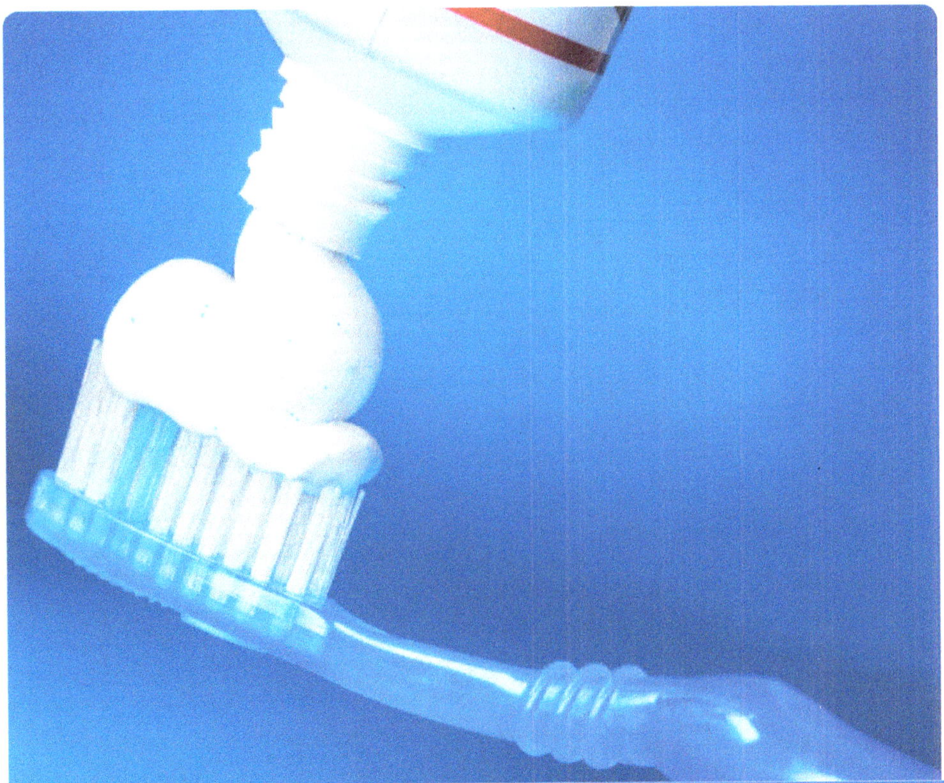

**TOOTHPASTE**
Stoichiometry plays a role in toothpaste manufacturing. How could you determine the molar mass of a compound listed on the side of a tube of toothpaste?

When calculating percentages of compounds to include in toothpaste, formulation chemists must use the chemical formulas of the compounds. The formula for sodium fluoride, for example, is NaF, and the formula for potassium nitrate is $KNO_3$. Chemists can use these formulas to calculate the molar masses of the compounds, and then based on experiments, they can determine the **mass** percentages of each compound that is most effective. A similar procedure is used for other personal care items, such as soap, shampoo, and deodorant. In each case, formula scientists use formulas and equations as they perform experiments to determine quantities of each compound to use in the **product**.

Even baking cookies requires the mathematics of formulas and equations. Mixing ingredients in the kitchen for cooking or baking often relies on calculating percentages. If a batch of 24 cookies requires two and three-quarter cups of flour, how much flour would be required to make 500 cookies? Professional bakers who produce large quantities of baked goods for sale must make calculations such as this to scale up their recipes. How is this similar to calculations performed in a chemistry laboratory? Whether you are dealing with recipes that involve units of tablespoons, cups, and ounces, or chemical equations that involve moles and atomic mass units, the stoichiometric calculations are similar. Both can include changing units, scaling values, and determining percentages.

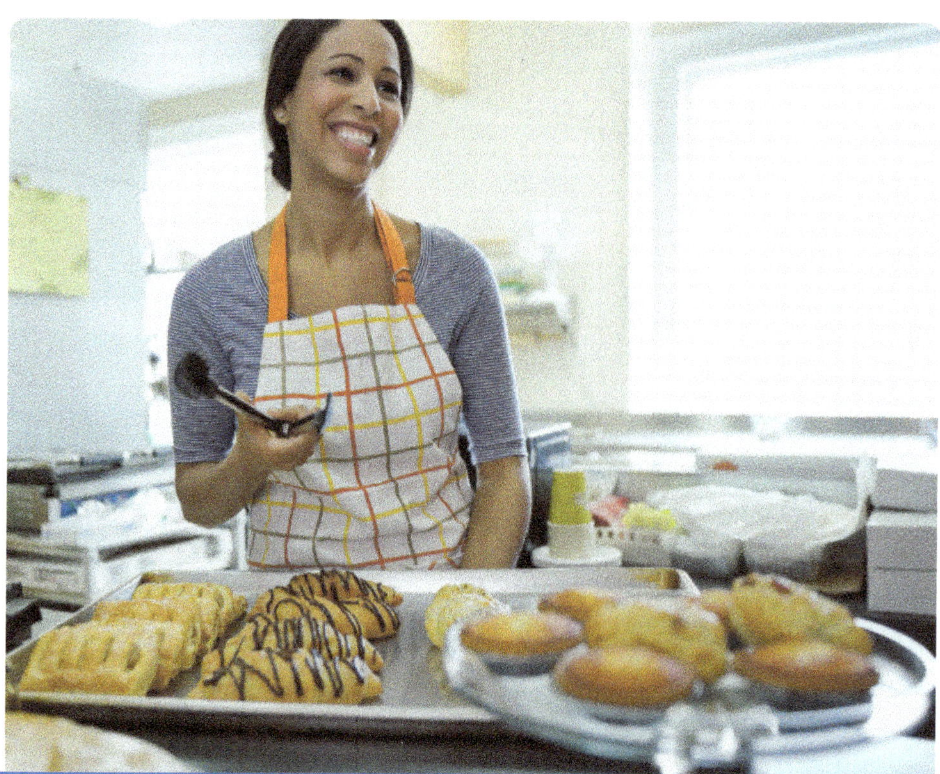

**BAKING PASTRIES**

Professional baking relies on precision. How would a professional baker use formulas and equations to determine the correct quantity of ingredients in a recipe?

## STEM and Mathematics of Formulas and Equations

People have come to depend on synthetic polymers such as polyethylene, polystyrene, and polyacrylate. A polymer is a compound consisting of large molecular chains of smaller molecules, called monomers, bonded together. Natural polymers include proteins in your body, but synthetic polymers can be found everywhere in modern civilization. For example, have you ever noticed the recycling symbol 1 in a triangle printed on a plastic container? That symbol means the container is produced from a polymer called polyethylene terephthalate (PET). Disposable water bottles and clear plastic containers used for strawberries or blueberries are often made of PET. The plastic recycling symbol 2 means the container is produced from high-density polyethylene. This type of plastic is used for grocery bags as well as milk and juice bottles. Polystyrene, designated by the recycling symbol 6, is used to make packing peanuts and egg cartons. Other recycling numbers refer to other types of polymers. Polymers have numerous familiar uses. Polyacrylate is a polymer used in paints and adhesives. Electrical insulation, toys, home siding, and upholstery are all made of polymers.

Synthetic polymers are developed in chemical laboratories. Much research is carried out to discover new polymers and modify existing ones. Because they are synthesized using chemical reactions, polymers can be described using **stoichiometry**. Polymer chemists use stoichiometric principles to plan and carry out reactions that they hope will produce new polymers.

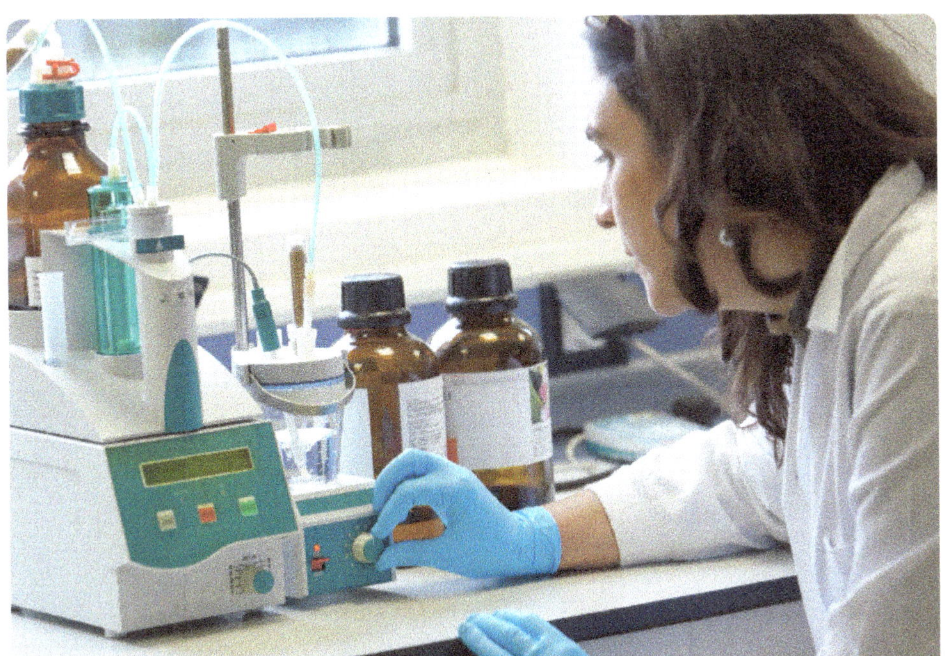

**ANALYZING A POLYMER**

This polymer chemist is performing an analysis to determine the composition of a polymer. How is stoichiometry useful in this process?

# Molar Mass Conversions

**Use molar mass as a conversion factor to solve the problems below.**

**1.** What is the mass in grams of 0.250 mol $CH_2Cl_2$?

**2.** Suppose you have 32.7 g $PbCrO_4$. How many moles does this represent?

**3.** Calculate the number of molecules present in $1.55 \times 10^{-3}$ g HCl.

**4.** What is the mass in grams of $4.61 \times 10^{18}$ molecules of $CO_2$?

## Percent Composition

**Use the indicated compound's molar mass and the molar masses of its elements to determine the percent composition in each case below.**

1. $C_6H_6$

2. $Mg_3N_2$

3. $NaH_2PO_4$

4. $Cr_2(SO_4)_3$

# Empirical and Molecular Formulae Practice Problems

**Use your knowledge of empirical and molecular formulae to answer the following questions. Show your calculations.**

**1.** What are the empirical formulae of the following molecular compositions?

**a.** 92.81% bromine, 7.19% phosphorus

**b.** 70% iron, 30% oxygen

**c.** 40% carbon, 6.7% hydrogen and 53.3% oxygen

**2.** Find the molecular formulae of the following:

**a.** A compound has the formula $C_5H_4$. Its molar mass is 126.16 g/mol. Find its molecular formula.

**b.** A compound contains 40.0% carbon, 5.7% hydrogen and 53.3% oxygen. It has an atomic mass of 175 g/mol. What is the molecular formula?

# Mole Conversions

**Manipulate chemical quantities using mathematical procedures, including scientific notation.**

1. How many formula units are present in 2.50 moles of $CaCl_2$?

2. A teacher needs 0.75 kg of aluminum for a laboratory experiment on single replacement reactions. How many atoms of aluminum will be used in the lesson?

3. $1.50 \times 10^{24}$ molecules of oxygen are burned in a combustion reaction. What is the mass of $O_2$ consumed in the reaction?

4. $3.15 \times 10^{23}$ molecules of water were collected in a distillation flask. How many moles of $H_2O$ were purified?

5. $5.60 \times 10^{-4}$ g of hydrogen gas are produced during the electrolysis of water. Calculate the number of hydrogen molecules that are produced in the reaction.

## Reaction Stoichiometry

**Use information from the balanced chemical equation in each case below to solve reaction stoichiometry problems.**

**1.** How many grams of water are required to form 9.00 mol $TiO_2$ according to the reaction described by the following equation?

$$TiCl_4 + 2H_2O \rightarrow TiO_2 + 4HCl$$

**2.** Iron reacts with sulfur to produce iron(II) sulfide according to the chemical equation shown below. What is the most iron(II) sulfide (in grams) that could be made using 0.500 mol sulfur? Use the equation $Fe + S \rightarrow FeS$.

**3.** What mass of $O_2$ will react with exactly 6.00 g P to form $P_2O_5$ in the following reaction?

$$4P + 5O_2 \rightarrow 2P_2O_5$$

# Gas Stoichiometry: Volume-Volume Calculations

**Use the balanced chemical equations given to calculate the mass of a gas. Assume all gases are under the same conditions of temperature and pressure.**

The combustion reaction of methane is represented by the following equation:

$$CH_4(g) + 2O_2(g) \rightarrow 2H_2O(g) + CO_2(g)$$

**1.** Calculate the volume of carbon dioxide gas produced when 50 L of methane are burned in the reaction.

**2.** How many liters of oxygen gas are required to produce 25 L of gaseous $H_2O$?



Methane gas is produced from the reaction of solid carbon and hydrogen gas.

$$C_{(s)} + 2H_{2(g)} \rightarrow CH_{4(g)}$$

**3.** How many liters of hydrogen gas at STP are required to produce 40 L of methane?

The synthesis reaction for water in the following equation is represented below:

$$2H_{2(g)} + O_{2(g)} \rightarrow 2\ H_2O_{(g)}$$

**4.** How many liters of hydrogen gas are needed to completely react with 10 L of oxygen gas?

**5.** How many liters of oxygen gas are required to produce 15 liters of water vapor?

# Gas Stoichiometry: Mass-Mass Calculations

**Use the balanced chemical equations given to calculate the mass of a gas. Assume all gases are under the same conditions of temperature and pressure.**

The combustion reaction of propane is represented by the following equation:

$$C_3H_{8(g)} + 5O_{2(g)} \rightarrow 4H_2O_{(g)} + 3CO_{2(g)}$$

**1.** If 120 g of propane, $C_3H_8$, is burned in excess oxygen, how many grams of water vapor are formed?

**2.** What is the mass of oxygen gas that is consumed in the reaction in problem 1?



The synthesis reaction for the production of water is represented in the following equation:

$$2H_{2(g)} + O_{2(g)} \rightarrow 2H_2O_{(g)}$$

**3.** What is the mass of oxygen gas required to completely react with 55 grams of hydrogen gas?

**4.** What is the mass of water vapor that is produced in the reaction in #3?

The balanced chemical equation for the synthesis of methane is:

$$C_{(s)} + 2H_{2(g)} \rightarrow CH_{4(g)}$$

**5.** How many grams of hydrogen gas at are required to produce 40 grams of methane? All gases are under the same conditions.

CONCEPT 4.3

# Oxidation-Reduction Reactions



## LESSON OVERVIEW

### Lesson Questions

- How are oxidation numbers determined for elements in a compound?
- How are the oxidized and reduced species identified in a redox reaction from the changes in oxidation number?
- How are oxidizing and reducing agents identified in a redox reaction?
- How is the number of electrons transferred in a redox reaction used to write a balanced equation?
- How can a reactivity series be used to determine the most easily oxidized elements?

### Key Vocabulary

Which terms do you already know?

- [ ] anion
- [ ] cation
- [ ] double-replacement reaction
- [ ] law of conservation of mass
- [ ] metals
- [ ] nonmetals
- [ ] oxidation
- [ ] oxidation number
- [ ] oxidation-reduction reaction
- [ ] oxidizing agent
- [ ] reactivity series
- [ ] redox reaction
- [ ] reducing agent
- [ ] reduction
- [ ] single-replacement reaction
- [ ] stoichiometry
- [ ] synthesis reaction

dlc.com/ca9081s

## Lesson Objectives

By the end of the lesson, you should be able to:

- Predict the oxidation number of an element in a compound.
- Evaluate changes in the oxidation number to identify the species oxidized and reduced in a redox reaction.
- Distinguish between oxidizing and reducing agents in redox reactions.
- Balance redox equations based on the number of electrons transferred.
- Compare and contrast elements in an activity series to identify elements most easily oxidized.
- Show how the result of a chemical reaction is related to atomic structure, trends in the periodic table, and patterns of chemical properties.
- Develop a logical argument using mathematical models to show that atoms, and therefore mass, are conserved during a chemical reaction.

# Thinking about Oxidation-Reduction Reactions

**Have you ever gazed into a glowing campfire? Have you felt its heat as it warms your hands or melts a marshmallow?**

dlc.com/ca9082s

## EXPLAIN QUESTION

**What happens during an oxidation-reduction reaction between copper metal and a silver nitrate solution?**

**BURNING FUEL**

The chemical reaction that causes wood to burn generates a large amount of heat as energy. How do you know that a chemical reaction is taking place as wood burns?

# How Are Oxidation Numbers Determined for Elements in a Compound?

## Assigning Oxidation Numbers

Chemical reactions involve the sharing or transfer of electrons between elements. Understanding the structure of atoms of different elements and the role of valence electrons that participate in reactions is key to predicting an element's chemical properties. Chemists have developed a method for keeping track of these electrons. The method uses **oxidation** numbers.

Oxidation numbers depend on the number of valence electrons in an atom of an element. Recall that the number of electrons present determines the charge of a particle. Oxidation numbers are therefore also related to charge. For a single ion, the **oxidation number** equals the charge of the ion.

In a neutral molecule, the oxidation numbers of all the atoms present add up to zero. Because atoms in their elemental form are always neutral, their oxidation number is always zero. Examples of the elemental forms of atoms are sodium, Na(s); chlorine, $Cl_2$(g); oxygen, $O_2$(g); and helium, He(g).

The periodic table of the elements can be used to find the oxidation numbers of some elements. For example, the group 1 **metals** all have one valence electron and usually form ions with a charge of 1+. Therefore, the oxidation number for metals from this group is +1, unless they are in pure elemental form. Similarly, group 2 metals have two valence electrons, form ions with a charge of 2+, and have an oxidation number of +2 when not in pure elemental form.

| Atom(s) | Oxidation Number | Compound Example |
|---|---|---|
| Li, Na, K | +1 | LiF, NaCl, KBr |
| Be, Mg, Ca | +2 | BeO, $MgCl_2$, $CaF_2$ |
| H (nonmetal)<br>H (metal) | +1<br>−1 | HCl, $H_2O$<br>LiH, NaH |
| O<br>O (peroxide) | −2<br>−1 | $CO_2$, $H_2O$, MgO<br>$H_2O_2$ |
| F | −1 | LiF, $CaF_2$, $NF_3$ |

**TABLE OF COMMON OXIDATION NUMBERS**
Some atoms always have the same oxidation number. What trends in the periodic table relate to the trends in oxidation number?

The elements listed in the Table of Common Oxidation Numbers always have the oxidation numbers listed. Fluorine, for example, always has an oxidation number of −1, because it is so highly electronegative, and therefore tends to add one electron to its valence shell when bonding. For some elements, however, the oxidation number varies depending on the other elements present. For example, hydrogen has the expected oxidation number of +1 when bonded to a nonmetal, but this changes to −1 when hydrogen is bonded to a metal. Oxygen has an oxidation number of −1 when forming peroxides, but it has an oxidation number of −2 in all other compounds. The element sulfur commonly has oxidation numbers of −2, +2, +4, and +6.

Keep in mind that oxidation numbers describe individual atoms or ions, not compounds. To find the oxidation number of each element in a compound, first assign numbers for those elements that never change, and then find the sum. The difference between that value and the overall charge of the ion or molecule will give the oxidation number of the remaining element.

How is the oxidation number of each atom in a polyatomic ion determined? Note that when chemists list the oxidation number of an element in a compound, the sign is placed before the number (−2 or +3). This is different from the signs that show the charge of an ion. When showing the charge of an ion, the number is always placed before the sign.

Consider the sulfate ion, $SO_4^{2-}$. As shown in the Table of Common Oxidation Numbers, each of the four oxygen atoms has an oxidation number of −2. The entire ion has a charge of 2−. The oxidation number of sulfur is calculated from the difference between the total charge of the oxygen atoms and the charge of the polyatomic ion.

$$\begin{aligned} 4(\mathrm{O}) + \mathrm{S} &= \mathrm{SO_4^{2-}} \\ 4(-2) + \mathrm{S} &= -2 \\ -8 + \mathrm{S} &= -2 \\ \mathrm{S} &= +6 \end{aligned}$$

Thus, in the sulfate ion, the oxidation number of sulfur is +6.

### Assigning Oxidation Numbers: Sample Problem

Assign the oxidation numbers for each atom in a neutral molecule of $HClO_2$.

**Solution**

In the compound, $HClO_2$, hydrogen is bonded to a polyatomic ion. Oxygen is present in the non-peroxide form. The following oxidation numbers can be assigned:

$$H: +1$$
$$O: -2$$

The total charge on the molecule is zero; therefore, the sum of all the oxidation numbers is zero.

$$\text{H} + \text{Cl} + 2(\text{O}) = 0$$

Substitute the oxidation numbers for hydrogen and oxygen to solve for the oxidation number of Cl.

$$+1 + \text{Cl} + 2(-2) = 0$$
$$+1 + \text{Cl} + -4 = 0$$
$$+1 + \text{Cl} = 4$$
$$\text{Cl} = +3$$

In $HClO_2$, hydrogen has an oxidation number of +1, oxygen has an oxidation number of −2, and chlorine has an oxidation number of +3.



## How Are the Oxidized and Reduced Species Identified in a Redox Reaction from the Changes in Oxidation Number?

### Describing Oxidation-Reduction Reactions

Many reactions that produce energy, such as burning fuel for energy, are **oxidation-reduction** reactions. During oxidation-reduction reactions, or redox reactions, electrons transfer between substances. Other common redox reactions include synthesis reactions, **single-replacement reactions**, and double-replacement reactions. All redox reactions can be written as two separate equations that together describe the step-by-step movement of electrons.

For example, when magnesium metal is placed into a beaker containing a solution of HCl, a gas forms. The balanced chemical equation that describes the formation of gaseous hydrogen is as follows:

$$Mg(s) + 2H^{+}(aq) \rightarrow Mg^{2+}(aq) + H_2(g)$$

In this chemical reaction, two separate processes occur at the same time—oxidation and reduction. Half-reactions are used to show the movement of electrons during each process.

**MAGNESIUM**

Solid magnesium is a shiny, silver metal. What happens to electrons in solid magnesium when it is immersed in a solution of hydrochloric acid?



Solid magnesium has an **oxidation number** of zero. It loses two electrons as the magnesium ion is formed.

$$Mg(s) \rightarrow Mg^{2+}(aq)$$

The number and type of atoms are balanced in this equation, but the charges are not balanced. To balance the charge on each side of the equation, two electrons are needed on the product side of the equation.

$$Mg(s) \rightarrow Mg^{2+}(aq) + 2e^{-}$$

During the reaction, the oxidation number of magnesium changes. The reactant, Mg(s), has an oxidation number of zero. The product, $Mg^{2+}$, has an oxidation number of +2. So, the magnesium goes from a lower (0) to a higher (+2) oxidation number. When the oxidation number of an atom increases, the atom becomes oxidized via the process of oxidation. This equation, where solid magnesium forms the magnesium ion, represents the oxidation half-reaction. The magnesium is oxidized, since its oxidation number increases during the reaction.

During the same reaction, each of the hydrogen ions gains an electron.

$$2H^{+}(aq) \rightarrow H_2(g)$$

This time, in order to balance the charge on each side of the equation, two electrons are added to the reactants.

$$2H^{+}(aq) + 2e^{-} \rightarrow H_2(g)$$

**CHANGING OXIDATION NUMBERS**

When a substance is oxidized, its oxidation number increases. Can you use the periodic table to predict how a substance's oxidation number changes when it is oxidized?

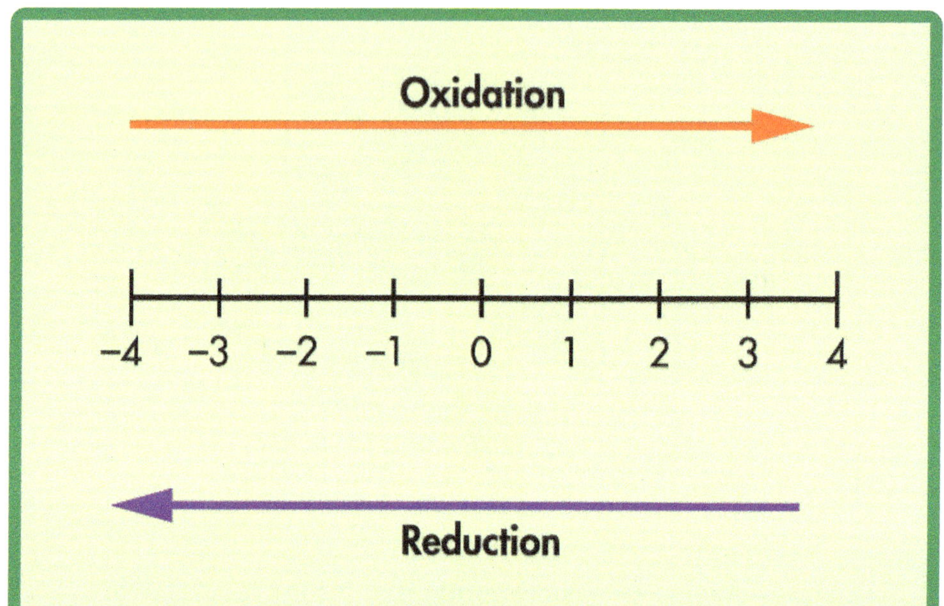


During the reaction, the oxidation number of the hydrogen atom changes. Each of the hydrogen ions in the reactants has an oxidation number of +1, while the hydrogen gas in the products has an oxidation number of 0. The oxidation number of the hydrogen ion decreases during the reaction, from +1 to 0. When the oxidation number of an atom decreases, the atom becomes reduced via the process of reduction. Thus, the reaction of hydrogen ions forming hydrogen gas represents the reduction half-reaction. The hydrogen ion is reduced since its oxidation number decreases. A way to remember this is the mnemonic, **LEO** the lion goes **GER**. **L**oss of **E**lectrons is **O**xidation and **G**ain of **E**lectrons is **R**eduction.

**LIONS**

Using the mnemonic, LEO the lion goes GER, can help you remember some oxidation and reduction rules. How might these rules apply to the principle that atoms, and therefore mass, are always conserved during a chemical reaction?



During a **redox reaction**, the oxidation and the reduction reactions both occur at the same time. If the oxidation and reduction reactions are combined, the following equation can be written:

$$Mg(s) + 2H^+(aq) + \cancel{2e^-} \rightarrow Mg^{2+}(aq) + \cancel{2e^-} + H_2(g)$$

The electrons on each side of the equation cancel each other out. This equation is the initial equation for the reaction.

$$Mg(s) + 2H^+(aq) \rightarrow Mg^{2+}(aq) + H_2(g)$$

Two electrons are produced in the oxidation half-reaction. This is equal to the two electrons used in the reduction half-reaction.

## How Are Oxidizing and Reducing Agents Identified in a Redox Reaction?

### Identifying Terms in Redox Reactions

In a **redox reaction**, half-reactions can be written to identify the substances being oxidized or reduced. Let's look again at the reaction between solid magnesium and the acid.

$$Mg(s) + 2H^+(aq) \rightarrow Mg^{2+}(aq) + H_2(g)$$

Solid magnesium, Mg(s), loses two electrons, and each hydrogen atom gains one electron.

The **oxidation** half-reaction identifies which substance loses, or donates, electrons in the reaction. In this example, solid magnesium is oxidized.

$$Mg(s) \rightarrow Mg^{2+} + 2e^{-} \text{ (reducing agent)}$$

Alternatively, the **reduction** half-reaction can be used to identify the substance that gains, or accepts, electrons. In this reaction, the hydrogen ion is reduced.

$$H^{+} + e^{-} \rightarrow H_2 \text{ (oxidizing agent)}$$

Another way to describe the components of a redox reaction is to identify the substance that causes oxidation. Since the hydrogen ion causes solid magnesium to oxidize, it is the **oxidizing agent** in the reaction. The oxidizing agent is the substance that is reduced in the reaction.

Conversely, in this reaction, the hydrogen atom is reduced. The substance that has caused the reduction to occur is solid magnesium. Thus, solid magnesium is the **reducing agent**. It has become oxidized in the reaction.

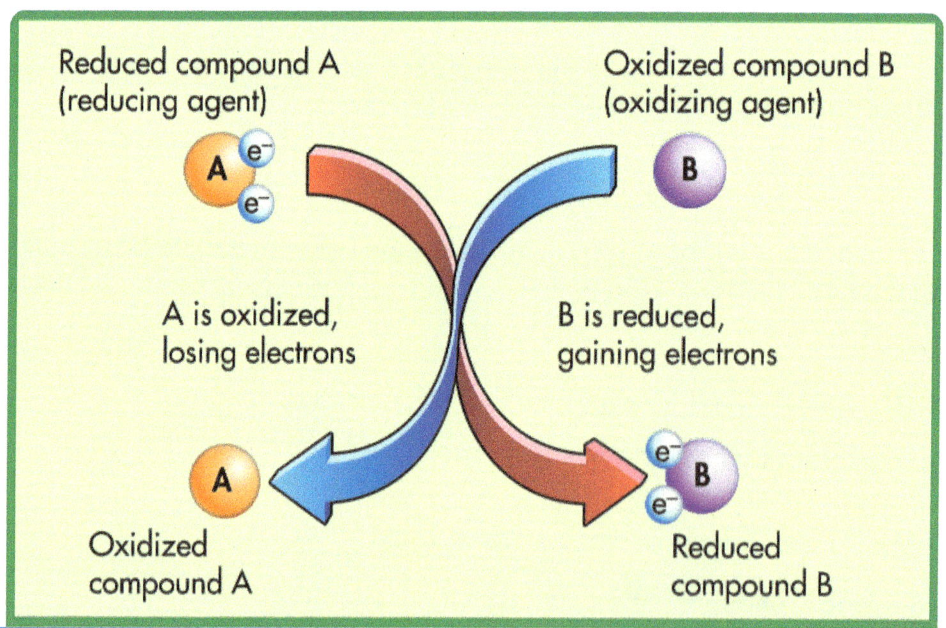


**IDENTIFYING REDUCING AND OXIDIZING AGENTS**
The reducing agent loses electrons and becomes oxidized, while the oxidizing agent gains electrons and becomes reduced. Can you infer the chemical equation for the oxidation half-reaction shown here?

# How Is the Number of Electrons Transferred in a Redox Reaction Used to Write a Balanced Equation?

## Balancing Redox Reactions

The **law of conservation of mass** states that matter cannot be created or destroyed. When balancing equations based on **stoichiometry**, this law is followed. In the same way, electrons cannot be created or destroyed during a chemical reaction. Therefore, the reaction must be balanced not only in terms of atoms, but also in terms of electrons.

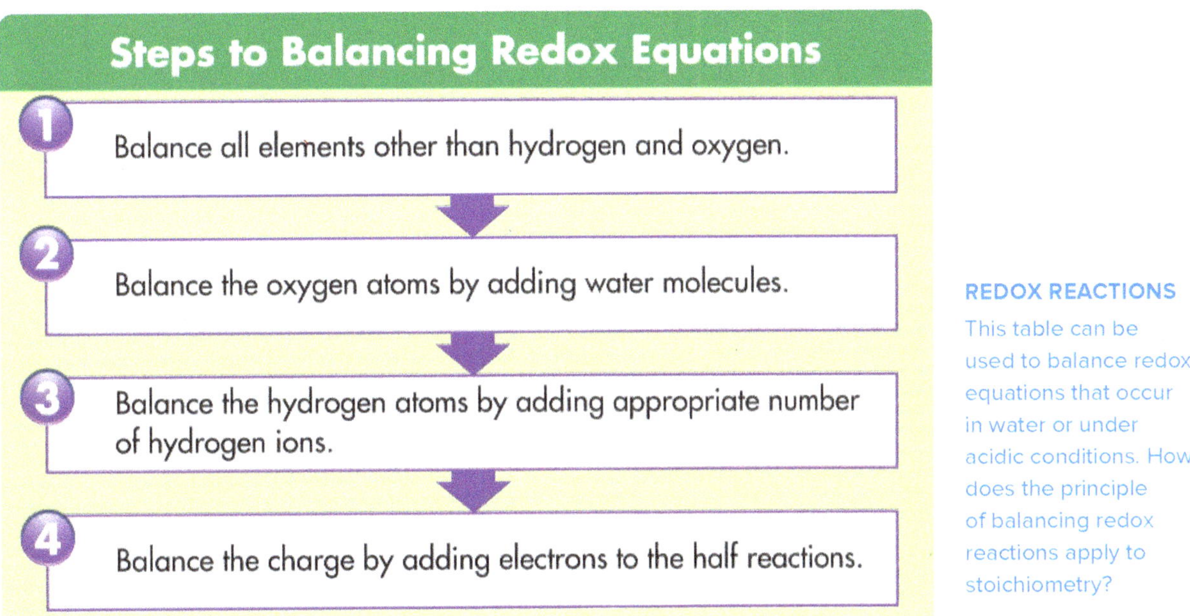

Steps to Balancing Redox Equations
1 Balance all elements other than hydrogen and oxygen.
2 Balance the oxygen atoms by adding water molecules.
3 Balance the hydrogen atoms by adding appropriate number of hydrogen ions.
4 Balance the charge by adding electrons to the half reactions.


**REDOX REACTIONS**
This table can be used to balance redox equations that occur in water or under acidic conditions. How does the principle of balancing redox reactions apply to stoichiometry?



The half-reaction method is often used to balance electrons in a **redox reaction**. This method was used previously to describe the reaction between magnesium and HCl. When writing **oxidation** and **reduction** half-reactions, the number of electrons lost or gained must be shown in the equation. The same number of electrons that are lost because of oxidation must be gained by reduction. Since electrons are matter and the law of conservation of matter states matter cannot be created nor destroyed through ordinary chemical or physical means, the number of electrons must be balanced.

The first step in using half-reactions to balance chemical equations is to identify the substances that are oxidized and reduced. In the following reaction, iron and tin ions react:

$$Fe^{3+}(aq) + Sn^{2+}(aq) \rightarrow Fe^{2+}(aq) + Sn^{4+}(aq)$$

Although it appears that the reaction may be balanced because the numbers of atoms balance, the reaction is not balanced based on charge. Adding up the charges on each side of the equation would result in a value of +5 for the reactants, and +6 for the products.

The iron(III) ion ($Fe^{3+}$) has an **oxidation number** of +3, and the tin(II) ion ($Sn^{2+}$) has an oxidation number of +2. However, during the reaction, the iron becomes reduced, as its oxidation number changes from +3 to +2. The reduction half-reaction is written as:

$$Fe^{3+}(aq) + e^- \rightarrow Fe^{2+}(aq)$$

At the same time, the tin ion becomes oxidized, as its oxidation number changes from +2 to +4 in the following half-reaction:

$$Sn^{2+}(aq) \rightarrow Sn^{4+}(aq) + 2e^-$$

To balance the charges, each half-reaction must transfer the same number of electrons. To do this, we multiply the entire reduction half-reaction by two so that the electrons cancel each other.

$$2[Fe^{3+}(aq) + e^- \rightarrow Fe^{2+}(aq)]$$

If both half-reactions are added, the electrons will cancel and the balanced reaction can be identified.

$$2Fe^{3+}(aq) + \cancel{2e^-} \rightarrow 2Fe^{2+}(aq)$$
$$Sn^{2+}(aq) \rightarrow Sn^{4+}(aq) + \cancel{2e^-}$$

The combined redox reaction is

$$2Fe^{3+}(aq) + Sn^{2+}(aq) \rightarrow 2Fe^{2+}(aq) + Sn^{4+}(aq)$$

As a check, both the reactant and products have an 8+ charge, so both the atoms and the charges are balanced.

When balancing redox reactions, be sure to transfer the coefficients in the half-reactions to the final balanced equation. In the reaction above between $Fe^{3+}$ and $Sn^{2+}$, the coefficient for each iron ion ensured the reaction was balanced for both mass and charge.

The majority of redox reactions take place in water (aqueous) or acidic conditions. When this happens, water molecules and hydrogen ions can be used to balance the redox equations.

**RUSTY IRON PIPES**

This rust began to form when a reaction passed electrons from iron to oxygen. How could stoichiometry be used to show whether the iron was reduced or oxidized?

## Balancing Redox Reactions: Sample Problem

Balance the following reaction in acid using the half-reaction method:

$$Cr^{3+}(aq) + Cl_2(g) \rightarrow Cr_2O_7^{2-}(aq) + Cl^-(aq)$$

**Solution**

Identify the atoms that are oxidized and reduced and write the corresponding half-reactions:

$Cr^{3+}$ goes from an oxidation state of +3 to +6 during the reaction. It becomes oxidized. To account for the oxygen atoms, assume that the reaction takes place in water, so add water to the reactants. $Cl_2$ begins with an oxidation number of zero and is reduced to −1.

Oxidation:

$$Cr^{3+}(aq) \rightarrow Cr_2O_7^{2-}(aq)$$

Reduction:

$$Cl_2(g) + \rightarrow 2Cl^-(aq)$$

Step 1: Balance all elements other than hydrogen and oxygen.

Oxidation:

$$2Cr^{3+}(aq) \rightarrow Cr_2O7_2^{-}(aq)$$

Reduction:

$$Cl_2(g) + \rightarrow 2Cl^{-}(aq)$$

Step 2: Balance the oxygen atoms by adding water molecules to the equation where it is needed to balance the oxygen atoms.

Oxidation:

$$2Cr^{3+}(aq) + 7H_2O(l) \rightarrow Cr_2O_7^{2-}(aq)$$

Reduction (oxygen is not present in this half-reaction, so it is not necessary for balancing the number of oxygen atoms):

$$Cl_2(g) \rightarrow 2Cl^{-}(aq)$$

Step 3: Balance the hydrogen atoms by adding the appropriate number of hydrogen atoms.

Oxidation:

$$2Cr^{3+}(aq) + 7H_2O(l) \rightarrow Cr_2O_7^{2-}(aq) + 14H^{+}(aq)$$

Reduction (hydrogen is not present in this half-reaction, so it is not necessary for balancing the number of hydrogen atoms):

$$Cl_2(g) \rightarrow 2Cl^{-}(aq)$$

Balance the charges by adding electrons to the half-reactions.

Oxidation: The charge on the reactants side is +6 and the charge on the products side is +12. Add electrons to the products to balance the charge.

$$2Cr^{3+}(aq) + 7H_2O(l) \rightarrow Cr_2O_7^{2-}(aq) + 14H^{+}(aq) + 6e^{-}$$

Left-hand side:

$$2Cr^{3+}(2 \times 3^{+}) = 6^{+}$$

Right-hand side:

$$Cr_2O_7^{2-}(1 \times 2^{-}) + 14H^{+}(14 \times 1^{+}) = 12^{+}$$

Difference:

$6+$ ($6e^-$ are needed on the right to balance the charges)

Reduction: The charge on the reactants side is 0 and the charge on the products side is $-2$. Add electrons to the reactants to balance the charge.

$$Cl_2(g) + 2e^- \rightarrow 2Cl^-(aq)$$

Finally, combine the oxidation and reduction reactions. Ensure that each side of the equation has the same number of electrons, and that the coefficients are transferred to the final reaction.

Oxidation:

$$2Cr^{3+}(aq) + 7H_2O(l) \rightarrow Cr_2O_7^{2-}(aq) + 14H^+(aq) + 6e^-$$

Reduction:

$$3[Cl_2(g) + 2e^- \rightarrow 2Cl^-(aq)]$$

Combined:

$$2Cr^{3+}(aq) + 7H_2O(l) \rightarrow Cr_2O_7^{2-}(aq) + 14H^+(aq) + 6e^-$$
$$3Cl_2(g) + 6e^- \rightarrow 6Cl^-(aq)$$

So the balanced reaction is:

$$2Cr^{3+}(aq) + 7H_2O(l) + 3Cl_2(g) \rightarrow Cr_2O_7^{2-}(aq) + 14H^+(aq) + 6Cl^-(aq)$$

# How Can a Reactivity Series Be Used to Determine the Most Easily Oxidized Elements?

## Metal Reactivity Series

Scientists organize **metals** according to their tendency to react with other substances. This organization is referred to as the **reactivity series** of metals. The most reactive species is found at the top of the table and the least reactive is at the bottom of the series.

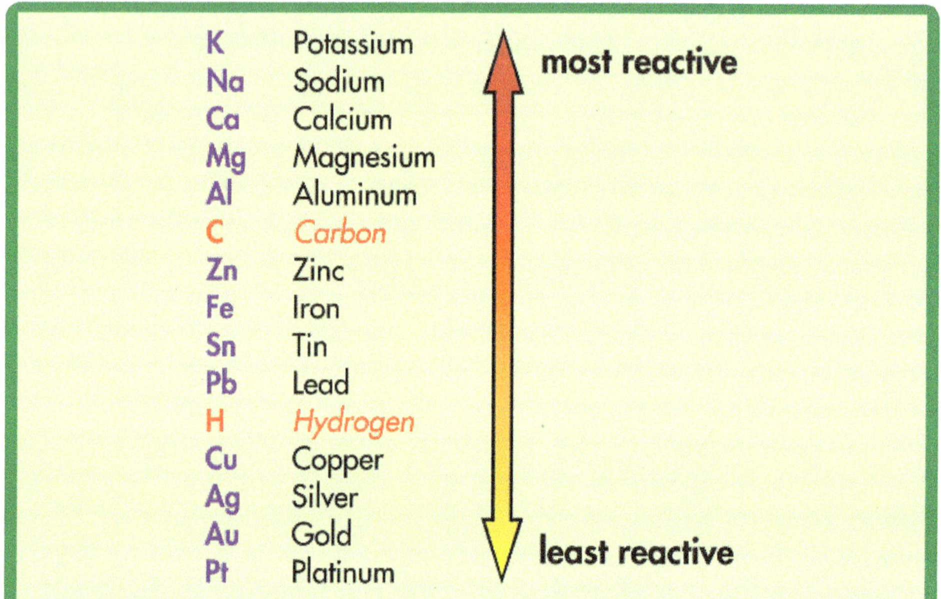


**REACTIVITY SERIES**

A reactivity series lists the most reactive metals at the top of the series and the least reactive metals at the bottom. How does the reactivity of carbon compare to that of hydrogen?

The reactivity of these metals can be illustrated with water. Sodium metal violently reacts with water, releasing hydrogen gas:

$$2Na(s) + 2H_2O(aq) \rightarrow 2Na^{+}(aq) + 2OH^{-}(aq) + H_2(g)$$

Yet, when solid iron is placed in water, the reaction occurs very slowly. The iron metal is lower in the reactivity series than sodium, and thus reacts more slowly with water.

There are other differences among the metals in the reactivity series. The metals at the top of the table are more likely to lose electrons and form cations. The metals at the top of the table are more corrosive, or tarnish easily. Also, the metals at the top of the table are stronger reducing agents and have a greater tendency to become oxidized than metals lower on the table.

**SODIUM IN WATER**

Sparks fly when sodium is combined with water. Why is this reaction so intense?

## Corrosion

Many metals react with oxygen in the presence of water and acid. One of the most common of these corrosion reactions is the **oxidation** of iron to produce rust. During this **redox reaction**, the oxidized species produced is iron oxide ($Fe_2O_3$). The first step of this reaction produces the iron(II) ion.

$$2Fe(s) + O_2(g) + H^+(aq) \rightarrow 2H_2O(l) + Fe^{2+}(aq)$$

The $Fe^{2+}$ ion then reacts with oxygen to form iron oxide. Iron oxide is a flaky substance and easily breaks away from the metal surface, allowing more metal to be exposed. This makes possible further oxidation of the iron metal.

Aluminum metal also forms an oxide ($Al_2O_3$) on its outer surface when oxidized. However, unlike iron oxide, aluminum oxide has a dense, stable structure. Because of this stability, the aluminum oxide protects the layers of aluminum underneath from corrosion. In both cases, the structure of the oxide product determines the rate and extent of corrosion.

### Consider the Explain Question

**What happens during an oxidation-reduction reaction between copper metal and a silver nitrate solution?**

Go online to complete the scientific explanation.

dlc.com/ca9083s

### Check Your Understanding

**Go online to check your understanding of this concept's key ideas.**

dlc.com/ca9084s

## Applying Oxidation-Reduction Reactions

Many metal and nonmetal objects used every day have undergone a process called electroplating. During this procedure, a metallic coating is applied to the surface of a metal in a **redox reaction**. Electroplating is applied to objects for decoration, to increase their functionality, and to inhibit corrosion.

Inexpensive jewelry with a fine coating of gold or silver is produced this way. Automobile manufacturers rely on chrome plating to coat many car parts, such as wheel rims, with chromium to protect against corrosion. Copper or brass electrical connectors are often plated with silver because silver has a higher conductivity and tarnishes much slower than the base **metals**. This results in strong connectors with high conductivity.

During electroplating, the metal object to be plated is given a negative charge when the object is placed into a salt solution. This solution contains the metal ion that will be plated onto the metal object. Finally, a positive charge is applied to the solution. This drives the metal ions onto the negatively charged metal. The metal ions become reduced, and "plate" onto the metal object. Here is the redox reaction for electroplating an object with nickel:

**Oxidation**: object− → object + $e^-$

**Reduction**: $Ni^+(aq) + e^- \rightarrow Ni(s)$

Common everyday objects that use electroplating include compact discs (CDs) and DVDs. Production of CDs or DVDs begins with electroplating a thin nickel coating on the glass master disc. The thickness of the nickel plating is tightly controlled to achieve optimal sound quality. The master disc is then used as a template to mass produce other discs on a plastic backing with a reflective metal layer on top.

In addition, scientists have discovered that the transfer of electrons plays a role in the transformation of many substances, such as the smelting of metals. Most metals are rarely found in their pure solid form, and instead are mined as metal ore. These ores are often metal oxide compounds. For example, aluminum oxide, $Al_2O_3$, is the metal ore of aluminum.

To obtain the pure form of a metal, metallurgists isolate the pure metal from the ore by a process called smelting. The process typically requires the use of extremely high temperatures. In order to convert copper oxide ore to pure copper metal, the ore must be heated to 1,100°C. The basic techniques used in smelting were developed thousands of years ago, when craftsmen and craftswomen first learned to use the properties of metals to create tools, weapons, and decorations.

## STEM and Oxidation-Reduction Reactions

The United States spends hundreds of billions of dollars every year to maintain or replace materials used to build roads, buildings, and pipes damaged by corrosion. Metal corrosion comprises a large portion of this cost. Metals undergo oxidation-reduction reactions upon exposure to the atmosphere.

**SMELTING**

Molten iron pours from this industrial smelter. Why is a smelter needed to produce pure iron?

San Francisco's Golden Gate Bridge is constantly being maintained to control corrosion of the steel. Steel is an alloy of iron and small amounts of other elements, such as carbon or nickel. When it corrodes, it forms rusts which weakens the bridge. Constant exposure to salty air, humidity, and condensed fog speeds up corrosion. One way to control corrosion is to protect the bridge from the elements. In the 1930s when the bridge was built, lead was used in the paint to prevent rust. The paint primer and topcoat was two-thirds (by weight) lead. Because lead is more active than iron, the lead in the paint reacted before the iron in the steel bridge reacted, thereby protecting the steel of the bridge itself. However, lead is harmful to people and the environment so it has been replaced with zinc in the primer coat of paint. Zinc protects the steel because it is more active than the iron in the bridge and corrodes more easily. Because of the constant exposure to salt-laden fog, the Golden Gate bridge is painted continuously.

**THE GOLDEN GATE BRIDGE**

The Golden Gate Bridge is subject to constant corrosion. How is the Golden Gate Bridge protected from corrosion?

A corrosion engineer is a scientist who studies the material properties of bridges, power plants, pipelines, and other types of metal structures. Corrosion engineers work to increase the structures' performance and reliability. They may design new materials that improve safety. One of the jobs of a corrosion engineer is to diagnose a problem and then to predict what will happen in the future regarding this material. By doing this, they can prevent disastrous events caused by corroded construction material like in the Golden Gate Bridge. Corrosion engineers use their knowledge of redox chemistry to investigate and resolve such problems.

# Assigning Oxidation Numbers

**Use the rules for assigning oxidation numbers to solve the problems below.**

**1.** Assign oxidation numbers to each atom in $Ga_2O_3$.

**2.** Assign oxidation numbers to each atom in $KBrO_4$.

**3.** Assign oxidation numbers to each atom in $Na_3PO_4$.

**4.** Assign oxidation numbers to each atom in $NH_2^-$.

**5.** Assign oxidation numbers to each atom in $IO_3^-$.

# Balancing Redox Reactions

**Use the half reaction method to balance the oxidation-reduction reactions below.**

**1.** $Zn(s) + NO_3^-(aq) \rightarrow Zn^{2+}(aq) + NH_4^+(aq)$

**2.** $Cu(s) + NO_3^-(aq) \rightarrow NO(g) + Cu^{2+}(aq)$

**3.** $Mn^{2+}(aq) + BiO_3^-(aq) \rightarrow MnO_4^-(aq) + Bi^{3+}(aq)$

CONCEPT

4.4

# Electrochemistry

## LESSON OVERVIEW

### Lesson Question

- What are the similarities and differences between electrolytic and voltaic cells?

### Lesson Objective

By the end of the lesson, you should be able to:

- Compare and contrast electrolytic and voltaic cells.

### Key Vocabulary

Which terms do you already know?

- ☐ anion
- ☐ cation
- ☐ cell potential
- ☐ decomposition
- ☐ double-replacement reaction
- ☐ electrochemical cell
- ☐ electrolysis
- ☐ electrolytic cell
- ☐ Faraday's law
- ☐ half-cell
- ☐ ion
- ☐ metals
- ☐ Nernst equation
- ☐ nonmetals
- ☐ oxidation
- ☐ oxidation number
- ☐ oxidation-reduction reaction
- ☐ oxidizing agent
- ☐ redox reaction
- ☐ reducing agent
- ☐ reduction
- ☐ standard reduction potential
- ☐ voltage
- ☐ voltaic cell

dlc.com/ca9085s

# Thinking about Electrochemistry

**We know that water consists of the elements hydrogen and oxygen. But how was the composition of water discovered? And why is water so important in our lives?**

dlc.com/ca9086s

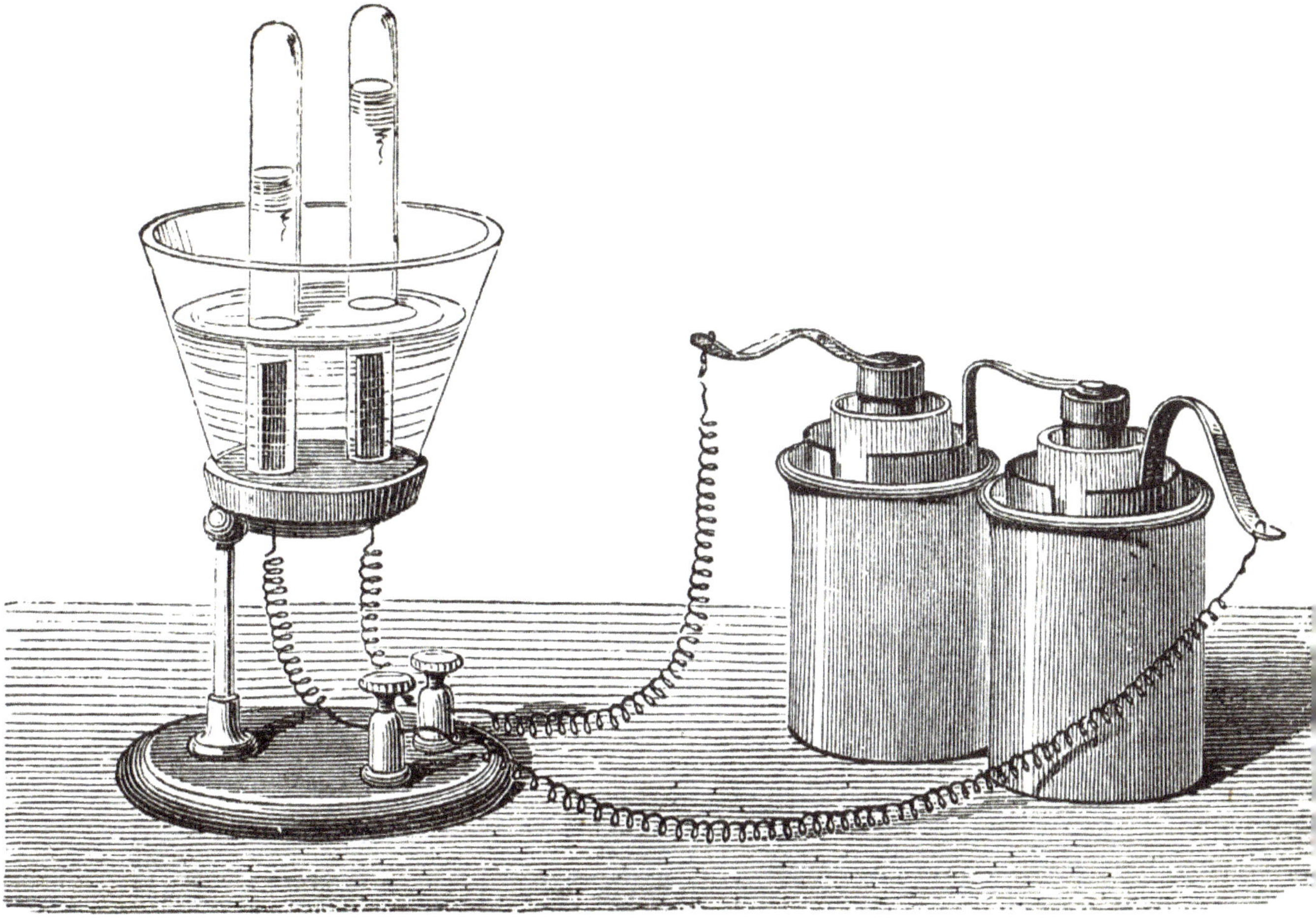

## EXPLAIN QUESTION

**What are electrolytic and voltaic cells, and what are their similarities and differences?**

**EARLY ELECTROLYSIS LABORATORY EQUIPMEN**

Electrolysis of water is the decomposition of water into oxygen and hydrogen gas due to an electric current being passed through the water. How did early chemists discover that water is not an element?

# What Are the Similarities and Differences between Electrolytic and Voltaic Cells?

## Chemical Energy and Redox Reactions

All matter contains energy, stored in the chemical bonds that join particles together to form compounds. This energy has the potential to do work. Chemical potential energy is related to the charges of the atoms involved in a chemical reaction. Electrochemistry is the study of how chemical energy is transformed into electrical energy.

During some chemical reactions, electrons are transferred from one atom to another. The atoms that lose electrons are said to undergo **oxidation**. The atoms that gain electrons are said to undergo **reduction**. A way to remember this is the mnemonic, LEO the lion goes GER. **L**oss of **E**lectron is **O**xidation and **G**ain of **E**lectrons is **R**eduction. Oxidation-reduction reactions are reactions in which electrons are exchanged, and are often referred to as redox reactions. In a **redox reaction**, the **reducing agent** is the substance that gives up electrons; it is what is being oxidized. The reducing agent causes something to gain electrons. The **oxidizing agent** is the substance that gains electrons; it is what is being reduced. The oxidizing agent causes something to lose electrons.

The **oxidation number** of a substance predicts how many electrons an atom will gain or lose. It is also related to the charge on a substance. The oxidation number generally is related to the group number. For example, sodium (Group 1) usually has an oxidation number of +1, and calcium (Group 2) usually has an oxidation number of +2. For transition elements, the oxidation number is indicated by a Roman numeral in parentheses placed next to the element name, such as iron(II). Elemental substances have an oxidation number of zero.

## Reactions in an Electrochemical Cell

To control redox reactions, harness the movement of electrons, and use electrical energy to perform work, the reactions must be performed in a closed environment called an **electrochemical cell**.

In an electrochemical cell, electrodes must be present. Oxidation occurs (electrons are released) at a location referred to as the anode. Reduction occurs (electrons are gained) at a location referred to as a cathode. The anode and cathode must be connected so that electrons and ions can move between them.

Electrons move through wires from one electrode to the other. Then ions take over and move through the solution to complete the circuit. In some cells, a salt bridge divides the two halves of the cell so that the two solutions do not mix. The salt in the bridge is composed of spectator ions, which do not interact with other ions in the solution. The spectator ions move from the salt bridge to balance the charges of the electrolytes in each **half-cell**. A salt bridge often contains a sodium chloride (NaCl) or potassium nitrate ($KNO_3$) solution, which will not interfere with the oxidation-reaction of the cell.

**ELECTROCHEMICAL CELL**

In a lemon battery, lemon juice is the electrolyte that allows ions to move between electrodes. Which type of electrochemical cell is a lemon battery?



Instead of a salt bridge, a porous barrier can be used to separate the electrolyte solutions. In this case, there are no spectator ions available to balance the charges of the ions formed during the redox reactions. Instead, some cations move through the barrier. Anions also move in the opposite direction, moving charge through the electrolytes to complete the circuit.

Remember that redox reactions involve oxidation and reduction occurring simultaneously. This statement is true of the redox reactions that occur in electrochemical cells, but the two reactions do not occur in the same location. Half-reactions take place at each electrode in the cell. For example, in a cell containing zinc and silver, zinc is oxidized and silver is reduced. The overall reaction can be described by this equation:

$$Zn + 2Ag^{+} \rightarrow 2Ag + Zn^{2+}$$

During this reaction, the following half-reactions occur:

oxidation: $Zn \rightarrow Zn^{2+} + 2e^{-}$

reduction: $2(1e^{-} + Ag^{+} \rightarrow Ag)$

There are two types of electrochemical cells. Voltaic cells use chemical reactions to generate electricity, and electrolytic cells use electricity to release chemical energy. It is important to remember that in both types of cells, the anode is the electrode at which electrons are released (oxidation) and the cathode is the electrode at which electrons are gained (reduction). The charge of the electrodes and the direction in which the current travels varies in the two types of cells, but the type of half-reaction that occurs at each type of electrode is always the same. To keep these pairings straight, it may be helpful to remember that the words *oxidation* and *anode* both begin with vowels, whereas the words *reduction* and *cathode* both begin with consonants.



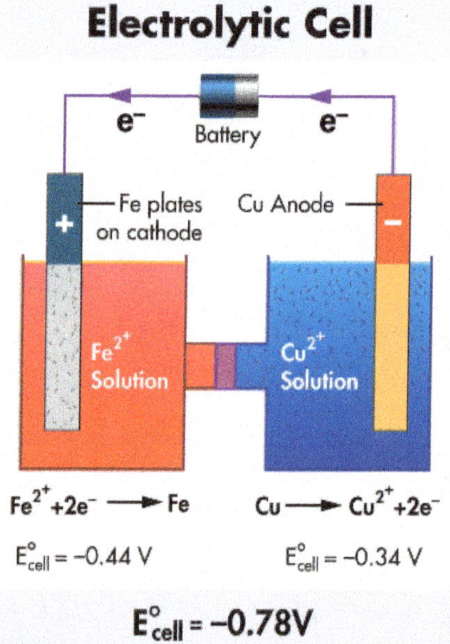


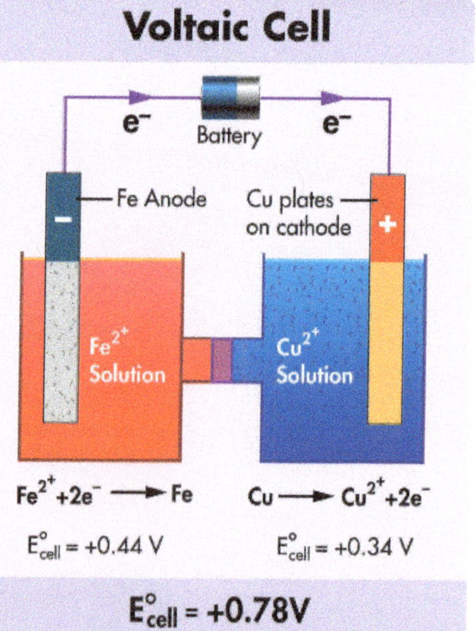


**ELECTROLYTIC AND VOLTAIC CELLS**

Electrolytic and voltaic cells both generate energy. How do these cells differ?

## Electric Potential

Potential energy refers to the amount of energy stored in an object. A rock balanced on the edge of a cliff has potential energy that will become kinetic energy if the rock falls. Even objects as small as electrons have potential energy. This potential energy varies depending on an electron's location in a circuit. Potential energy also depends on the reaction taking place at each electrode. The anode and cathode are logical locations at which to consider potential energy. **Cell potential** is the difference in electric potential energy between the two electrodes in a cell. This quantity is measured in volts (V).

Recall that the term *reduction* refers to the gaining of electrons. **Standard reduction potential** ($E^\circ$), in units of volts, measures a substance's ability to gain electrons at standard conditions of temperature and pressure. A gain in electrons is a reduction reaction, so the standard reduction potential is also a measure of a substance's ability to be reduced.

The values of standard reduction potential have been determined experimentally and are defined according to comparison with the potential of the standard hydrogen electrode, which has been assigned a value of 0.00. For a zinc-silver cell, the standard reduction potential for the reduction of silver is $E^\circ = 0.799$ V.

The standard reduction potential is also called the standard electrode potential because it can be used to describe all of the reactions that occur in a cell. The standard electrode potential of an oxidation half-reaction is equal but opposite in charge to the standard electrode potential of the reduction reaction. For the reduction of zinc, $E^\circ = -0.763$ V. For the oxidation of zinc, $E^\circ = 0.763$ V.

The standard potential for the net reaction taking place within the cell is found by adding the standard electrode potentials of the half-reactions:

| | |
|---|---|
| reduction: $2(1e^- + Ag^+ \rightarrow Ag)$ | $E^\circ = 0.799$ V |
| oxidation: $Zn \rightarrow Zn^{2+} + 2e^-$ | $E^\circ = 0.763$ V |
| cell: $Zn + 2Ag^+ \rightarrow 2Ag + Zn^{2+}$ | $E^\circ = 1.562$ V |

## Voltaic Cells

In a **voltaic cell**, sometimes referred to as a galvanic cell, metal anodes are placed in solutions of their salts and are connected by an external pathway, such as a conductive metal wire. During the chemical reaction, electrons are transferred through an external pathway, along the wire, from the anode to the cathode. Ions then are transferred through an internal pathway, through a salt bridge, from the cathode back to the anode. This movement of the ions completes a circuit. It ensures that ions are constantly available at the anode so the reaction continues and a continuous current of electrons can be created.

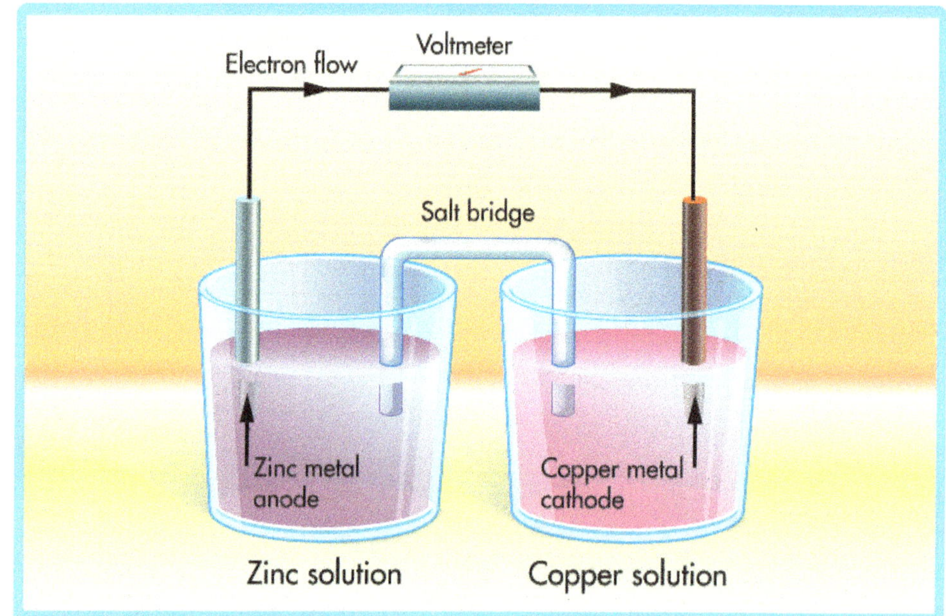


**VOLTAIC CELL**

Voltaic cells use chemical reactions to generate electricity. How does this differ from electrolytic cells? Why do electrons flow around this circuit?

The **oxidation-reduction reaction** that occurs in a voltaic cell is spontaneous. It does not require the addition of energy to proceed. In fact, a voltaic cell releases energy. As long as the necessary ions are present to keep the reaction going, the cell converts chemical energy to electric energy. At the anode, the oxidation reaction creates a source of electrons for the electric current, which gives the anode in a voltaic cell a negative charge.

Remember that the standard reduction potential (E°) of a metal compares its ability to gain electrons to that of hydrogen. This experimental value is determined using 1.00 M solutions of the metal ions, at 1 atm and 25°C. The actual potential of a cell can change, however, if the concentrations of the solutions in the cell are adjusted. When more ions are present in a solution, more electric current can be moved through a circuit. Conversely, if the solution has a lower concentration, fewer ions are available to complete the electrical circuit.

In a non-standard cell the concentrations of ions in the electrolytic solutions vary. In this case, the potential of the cell, E, can be determined using the **Nernst equation**:

$$E = E^\circ - \left(\frac{RT}{nF}\right) \times \ln\left(\frac{[ox]}{[red]}\right)$$

$E^\circ$ = standard reduction potential of the cell, in volts

$n$ = moles of electrons exchanged

$R$ = 8.314 J/mol·K (the gas constant)

$T$ = temperature in Kelvin

$F$ = 96,485 C/mol (Faraday's constant)

[ox] = concentration of oxidized substance

[red] = concentration of reduced substance

This equation can be used to find the potential of the cell or the concentrations of either electrolyte, as long as values for two of these three variables are known.

**HYDROGEN FUEL CELL**
In a hydrogen fuel cell, hydrogen and oxygen are needed to produce electrical energy. What kinds of reactions take place between the hydrogen and oxygen in this fuel cell?

## Voltaic Cells: Sample Problem

A voltaic cell is constructed using solutions of 0.6000 M zinc nitrate ($Zn(NO_3)_2$) and 0.4000 M silver nitrate ($AgNO_3$) for the electrolytes. Determine the potential of the cell. The temperature is 25°C, $R$ = 8.314 J/mol·K, and $F$ = 96,485 C/mol.

**Solution**

As in the cell discussed previously, there are two half-reactions taking place in this cell:

| reduction: $2(1e^- + Ag^+ \rightarrow Ag)$ | $E° = 0.799$ V |
|---|---|
| oxidation: $Zn \rightarrow Zn^{2+} + 2e^-$ | $E° = 0.763$ V |
| cell: $Zn + 2Ag^+ \rightarrow 2Ag + Zn^{2+}$ | $E° = 1.562$ V |

This cell is non-standard because the concentrations of the electrolytes are not 1.0 M. Therefore, the Nernst equation must be used to calculate the cell potential.

$$E = E° - \left(\frac{RT}{nF}\right) \times \ln\left(\frac{[ox]}{[red]}\right)$$

The following information is known:

$E° = 1.562$ V

$R = 8.314$ J/mol·K

$T = 25°C = 286$ K

$F = 96{,}485$ C/mol

[ox] = concentration of zinc = 0.6000 mol/L

[red] = concentration of silver = 0.4000 mol/L

The value of $n$ is not given but can be found by looking at the equations for the half-reactions. For every mole of zinc atoms that is oxidized, two moles of electrons are produced. Therefore, $n = 2$.

Next, substitute the known values into the Nernst equation and solve. [To calculate the natural log (ln), use a calculator.]

$$E = 1.562 \text{ V} - \left(\frac{8.314 \text{J/mol} \cdot 298 \text{ K}}{2 \times 96{,}485 \text{ C/mol}}\right) \times \ln\left(\frac{0.6000 \text{ mol/L}}{0.4000 \text{ mol/L}}\right)$$

$$E = 1.557 \text{ V}$$

The cell potential at the given concentrations of the electrolytes is 1.557 V.

## Electrolytic Cells

An **electrolytic cell** involves a non-spontaneous oxidation-reduction reaction. Energy must be added for this reaction to proceed. An electrolytic cell uses electrical energy, often in the form of a battery, to break down compounds and release chemical energy in a process called **electrolysis**. For example, when electricity is applied to water molecules, oxygen gas and hydrogen gas are produced from the molecules' **decomposition**.

In an electrolytic cell, electrons are transferred between two metal electrodes, just as in a voltaic cell. Also, ions are transferred through an electrolyte from one electrode to the other, just as in a voltaic cell. In an electrolytic cell, however, the electrical energy applied to the cell drives the reaction. Remember that oxidation always occurs at the anode. In this case, the anode has a positive charge. It attracts the negatively charged anions that have "extra" electrons and oxidizes them. Similarly, the negatively charged cathode attracts the cations. The cations will gain electrons and be reduced. The electrons flow from anode to cathode, completing the circuit.

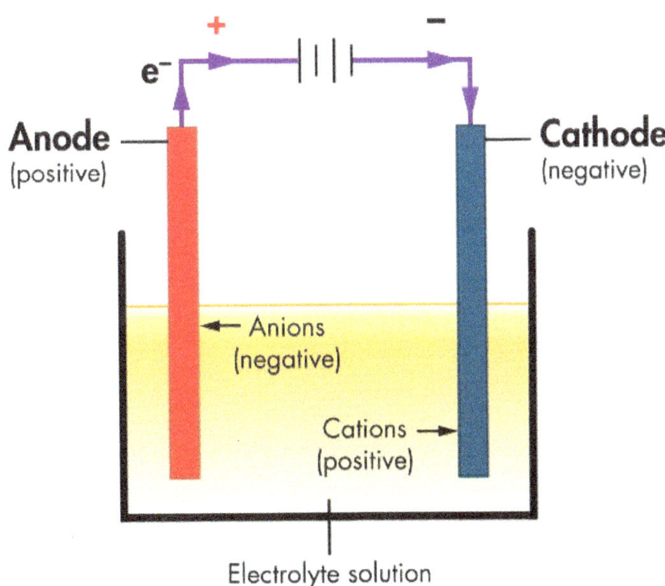


**ELECTROLYTIC CELL**
Electrolytic cells use electrical energy to decompose compounds. Why is this a non-spontaneous reaction?

Faraday's law of electrolysis can be applied to determine quantitative relationships in electrochemical cells. Faraday's law states that the amount of substance produced or consumed at an electrode in an electrolytic cell is proportional to the amount of electricity that passes through the electrolytic cell. In other words, the more electricity introduced to the cell, the more the reactant decomposes.

The amount of electricity in a current running for a specific time can be calculated using **Faraday's law**:

$$Q = I \times t$$

Where:

$Q$ = coulombs (C) of electricity
$I$ = amps (A) of current
$t$ = time in seconds (s)
$F$ = 96,485 C/mol (Faraday's constant)

A coulomb is a unit of measurement for electricity equal to 1 amp·second, so 1 C = 1 A·s.

An amp or ampere is the current that flows with an electron charge of 1 C/s.

The amount of electricity required to deposit an amount of metal also can be calculated using Faraday's constant and the moles of electrons that pass through a circuit:

$$Q = n(e^-) \times F$$

Where $n(e^-)$ = moles of electrons that are released or gained at an electrode.

## Electrolytic Cells: Sample Problem

What mass of copper could be produced from a copper(II) sulfate ($CuSO_4$) solution, using a current of 1.50 A over a period of 30.0 seconds?

**PURE COPPER**
An electrochemical process is used to produce nearly pure copper metal. What happens during this electrochemical process?

**Solution**

First, determine the amount of electricity that has been used. The given values tell us that $I = 1.50$ A and $t = 30.0$ s, so these can be substituted directly into the equation:

$$\begin{aligned} Q &= I \times t \\ &= 1.50 \text{ A} \times 30.0 \text{ s} \\ &= 45.0 \text{ A}\cdot\text{s} \\ Q &= 45.0 \text{ C} \end{aligned}$$

Next, rearrange the equation using Faraday's constant, and solve to find the number of electrons that have passed through the circuit.

$$\begin{aligned} Q &= n(e^-) \times F \\ n(e^-) &= \frac{Q}{F} \\ &= \frac{45.0 \text{ C}}{96{,}485 \text{ C/mol}} \\ n(e^-) &= 4.66 \times 10^{-4} \text{ moles of electrons} \end{aligned}$$

The chemical equation for this reaction shows that for every two moles of electrons in the cell, one mole of copper is produced:

$$Cu^{2+} + 2e^- \rightarrow Cu(s)$$

In other words, the moles of copper produced equal half the moles of electrons and can be calculated by multiplying the moles of electrons by 0.5.

$$\begin{aligned} n(\text{Cu}) &= 0.5 \times n(e^-) \\ &= 0.5 \times (4.66 \times 10^{-4}) \\ n(\text{Cu}) &= 2.33 \times 10^{-4} \text{ mol Cu} \end{aligned}$$

The periodic table shows that the molar mass of copper is 63.55 g/mol. This can be used to find the mass of the copper produced in the reaction:

$$\begin{aligned} \text{mass of Cu} &= n(\text{Cu}) \times \text{molar mass} \\ &= (2.33 \times 10^{-4} \text{ mol}) \times (63.55 \text{ g/mol}) \\ \text{mass of Cu} &= 1.48 \times 10^{-2} \text{ g} \end{aligned}$$

The mass of copper produced from a copper(II) sulfate ($CuSO_4$) solution using a current of 1.50 A over a period of 30.0 seconds is $1.48 \times 10^{-2}$ g.

### Consider the Explain Question

**What are electrolytic and voltaic cells, and what are their similarities and differences?**

Go online to complete the scientific explanation.

dlc.com/ca9087s

### Check Your Understanding

**Go online to check your understanding of this concept's key ideas.**

dlc.com/ca9088s

## Applying Electrochemistry

There are many important applications of electrochemistry in industry and nature. When **metals** are coated with other metals, this electroplating process uses electrochemistry. Electroplating is used in industry to create a barrier on the substance that protects it against atmospheric conditions such as corrosion. This allows for plated parts to last for longer periods of time, which may result in replacements becoming more infrequent. Electrical equipment can be plated with silver to make it more conductive and more effective. Another application of electroplating is coating less expensive and more durable metals with a finer metal, such as silver or gold, to create objects such as jewelry. This reduces costs. Jewelry that is pure silver or gold is much more expensive than those that are electroplated.

**ELECTROPLATING**

Electroplating covers jewelry or other objects with a thin layer of gold or silver. Would the electroplating process need an input of electrical energy?



Photosynthesis is an electrochemical process found in nature. Through a multistep process, plant cells transform energy from sunlight (light energy) into chemical energy the plant can use later on. Electrochemical processes are also used in the human body to generate electrical impulses in the nervous system. Sodium and potassium ions move in and out of cells, which creates electrical current. Some animals, such as eels, rays, and skates (similar to rays), are believed to generate varying levels of electricity within the organs of their bodies.

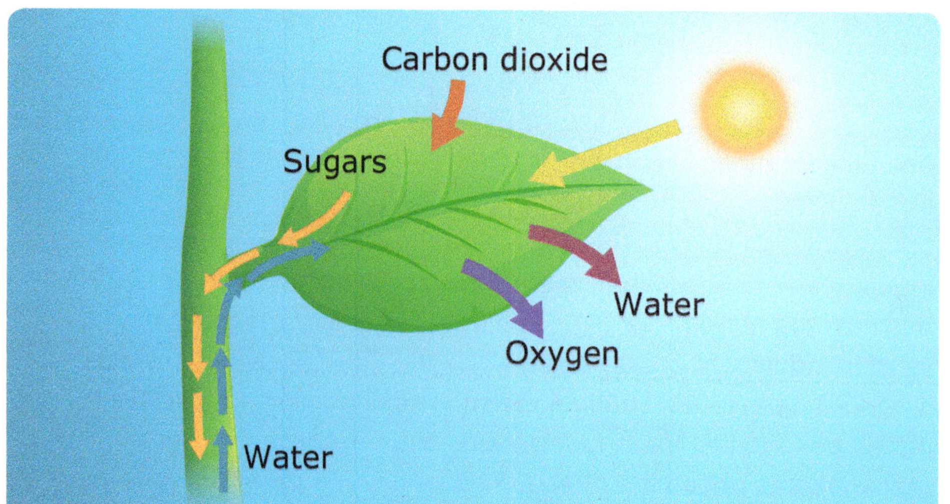


**PHOTOSYNTHESIS**

In photosynthesis, plants convert energy from the sun into chemical energy. How is photosynthesis an electrochemical process?



## STEM and Electrochemistry

Chemical engineers apply the principles of science, technology, and math to solve problems that pertain to our daily lives. Because of their multifaceted expertise, chemical engineers have been at the forefront of innovation in energy production. When fuel prices rise, many consumers begin looking for cheaper (and cleaner) forms of energy. In the car industry, many engineers are working to make cars that can run on a hybrid form of gas and electricity.

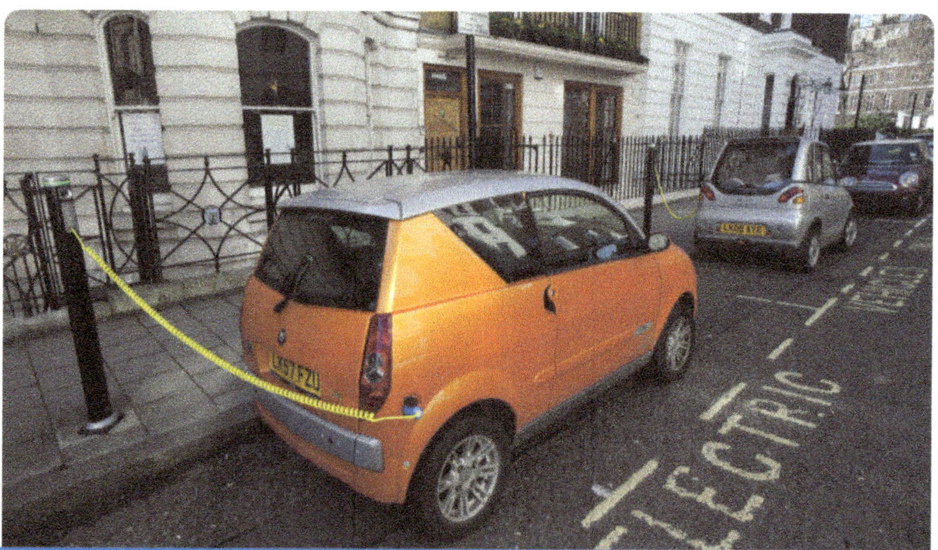


**ELECTRIC CAR**

These electric cars are plugged into special charging stations. What kind of energy conversion take place as an electric car charges?

Engineers are constantly attempting to design batteries that have a higher energy density, lower weight, better safety, higher reliability, and lower cost. One type of battery that is currently being researched is one that could reverse the process of converting chemical energy to electrical energy. It would allow electrical energy to be converted back to chemical energy for later use as electrical energy.

# Nernst Equation

**Use the Nernst equation to solve the problems below.**

1. The standard cell potential, *E*°, for a voltaic cell made using zinc and copper under standard conditions is 1.10 V. Zinc is oxidized to $Zn^{2+}$ at the anode and copper ions, $Cu^{2+}$, are reduced to copper metal at the cathode in this cell. What is the cell potential if the concentration of zinc ion, $Zn^{2+}$, is 0.100 M and the concentration of copper ion, $Cu^{2+}$, is 1.00 M in their respective cells? The temperature is 25.0°C.

2. Iron and nickel are used to make a voltaic cell at 25°C. At the anode, iron metal is oxidized to the $Fe^{2+}$ ion. At the cathode, nickel ion, $Ni^{2+}$ is reduced to nickel metal. What is the cell potential if the concentration of $Fe^{2+}$ ion in the anode compartment is $1.00 \times 10^{-5}$ M and the concentration of $Ni^{2+}$ ion in the cathode compartment is 0.200 M? *The* cell potential under standard conditions is 0.21 V.

3. The standard cell potential, *E*°, for a magnesium-silver voltaic cell is 4.36 V. In this cell, magnesium is oxidized to $Mg^{2+}$ at the anode and silver ions, $Ag^{+}$, are reduced to silver at the cathode. What is the cell potential if $[Mg^{2+}]$ is 1.00 M and $[Ag^{+}]$ is $5.00 \times 10^{-5}$ M in this cell? The temperature is 25.0°C.

CONCEPT
5.1

# The Composition and Behavior of the Atmosphere

dlc.com/ca9089s

## LESSON OVERVIEW

### Lesson Questions

- How did the atmosphere form?
- What is the composition of Earth's atmosphere?
- What affect does the sun have on Earth's atmosphere?
- How are human activities altering the atmosphere?

### Lesson Objectives

By the end of the lesson, you should be able to:

- Describe the evolution of Earth's atmosphere.
- Describe the chemical composition of the atmosphere.
- Explain the physical processes which drive the circulation of the atmosphere.
- Identify human activities that alter the composition of the atmosphere.

### Key Vocabulary

Which terms do you already know?

- [ ] absorption
- [ ] atmosphere
- [ ] carbon dioxide
- [ ] conduction
- [ ] Coriolis effect
- [ ] density
- [ ] electromagnetic wave
- [ ] gas
- [ ] giant impact hypothesis
- [ ] global warming
- [ ] global wind
- [ ] gravitational force
- [ ] greenhouse gas
- [ ] heat
- [ ] heat energy
- [ ] methane
- [ ] mixture
- [ ] outgassing
- [ ] ozone layer
- [ ] pressure
- [ ] radiation
- [ ] surface area
- [ ] thermal energy
- [ ] transpiration
- [ ] ultraviolet radiation

## Thinking about the Composition and Behavior of Atmospheres

Have you ever wondered what it would be like to inhabit another planet, for example, Venus? The planet is named for the Roman goddess of love and beauty, and it shines so brightly in the sky at sunrise and sunset that people sometimes call it the Morning Star or the Evening Star. Would it be as wonderful to live on Venus as it appears from Earth?

dlc.com/ca9090s

**EXPLAIN QUESTION**

**How do changes in the composition of the atmosphere affect its properties and behavior?**

**VENUSIAN ATMOSPHERE**

From space, Venus's atmosphere appears as swirling, yellow clouds that cover the entire planet. How does the composition of Venus' atmosphere determine conditions on the planet's surface? Would the atmosphere of Venus support human life?

# How Did the Atmosphere Form?

Many scientists propose that Earth's "first **atmosphere**" developed from a process called **outgassing**. For many Earth scientists, outgassing denotes the release of gases from volcanoes to the atmosphere and oceans. This early atmosphere would have been composed of primarily hydrogen and large amounts of water vapor, **carbon dioxide**, and hydrogen sulfide, but it also would have contained smaller amounts of nitrogen, carbon monoxide, **methane**, and other gases.

**OUTGASSING**

Large quantities of gases are released into the atmosphere during volcanic eruptions. Which gases do volcanoes emit?



Today, it is generally believed that Earth's first atmosphere was excised from the planet by a catastrophic impact with a huge colliding body called Theia. Scientists call this the **giant impact hypothesis**. It is hypothesized that Theia, a celestial body the size of Mars, displaced huge volumes of atmospheric material when it collided with primitive Earth about 4.5 billion years ago and that much of that material later condensed, forming the Moon. The intense energy released by the impact left much, if not all, of Earth in a molten and desolate state.

Air Molecules at Sea Level

Air Molecules at High Altitude

**AIR DENSITY CHANGES WITH ALTITUDE**

The density of air at sea level is higher than at high altitude. What factors explain why there fewer air molecules per unit volume of air at high altitude?



Earth's atmosphere provides the right composition of gases to support many living things, but the right temperature conditions are also required for survival.

## What Effect Does the Sun Have on Earth's Atmosphere?

The sun is the primary source of energy that heats Earth's **atmosphere**, and without solar energy Earth's atmosphere would be extremely cold and the surface would freeze. An easy way to observe the sun's influence on Earth's atmosphere is to notice the difference in air temperature between day and night. For example, it is common to experience a decrease in air temperature at sunset, and it is common for air temperature to continuously drop throughout the night until sunrise the following morning.

Air temperature doesn't always drop throughout the night; sometimes, it increases after sunset. This is because **thermal energy** is transferred through the atmosphere in three ways—**radiation**, **conduction**, and convection.

Radiation is the transfer of energy by electromagnetic waves, a form of energy released by moving charged particles. As the temperature of a substance increases, its atoms move faster and faster. Because of its high temperatures, the sun is an extremely large source of electromagnetic radiation. This electromagnetic radiation travels through space, and a miniscule amount of it reaches Earth.

Conduction is the transfer of thermal energy by collisions between particles of matter. It occurs only when substances are in contact with each other. For example, conduction occurs when the person touches the object. The object's particles collide with the particles that make up the person's hand, transferring energy that is experienced as **heat**. Convection is the transfer of thermal energy by a flowing substance, and it occurs in fluids, such as water or air. When discussing the atmosphere, convection commonly refers to the vertical circulatory movement of air. It is one of the primary mechanisms for circulating thermal energy in Earth's atmosphere. A simple way to observe convection is to watch smoke or steam rise over a campfire. Convection in Earth's atmosphere results from differences in **density** and the action of gravity, so how does this happen?

The density of a fluid changes as the volume occupied by the fluid changes, and as the volume of a fixed amount of mass increases, the density decreases. Heating a fluid, like air, causes its density to decrease.



**DENSITY**

When a fluid is heated, its volume increases while its density decreases. How is this observation related to changes in the motion of the fluid's molecules?

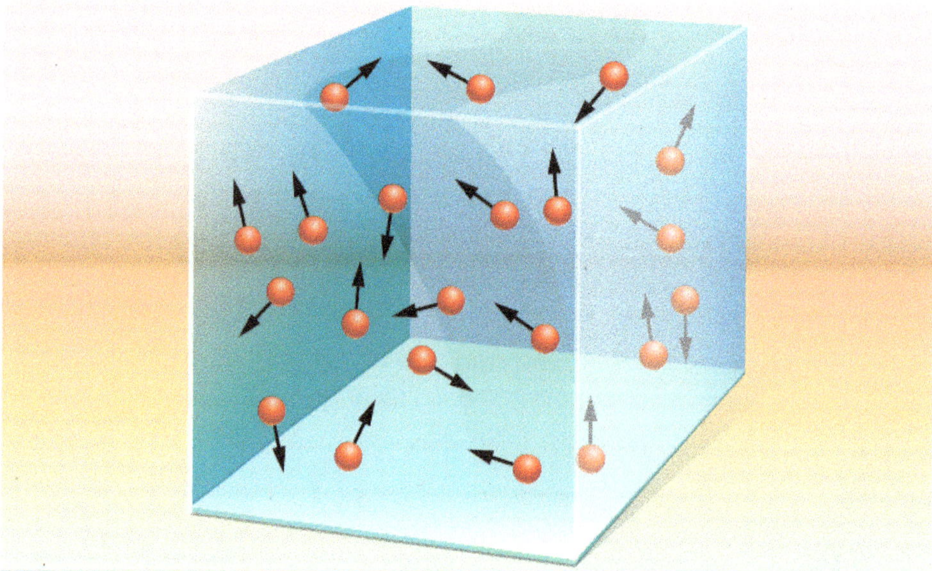

In the air, volumes of dense fluid are pulled more strongly by gravity than are equal volumes of less dense air. This is because fluids of high density have more mass in a given volume than do fluids of lower density. Gravity pulls the higher-density fluid downward beneath the fluid of lesser density. In Earth's atmosphere, dense air is pulled toward Earth's surface with more force than is less dense air, while the lower-density air is forced upward. Warm air rises because it is forced upward by the surrounding cool, dense air.

## Uneven Heating of Earth's Surface

Different characteristics of Earth's surface cause different locations to be heated unevenly, resulting in some areas being relatively warm, while other areas remain relatively cool. There are several reasons why Earth's surface is heated unevenly. Four major reasons include:

- the curvature of Earth's surface
- the distribution of land and water
- differences in ground cover
- atmospheric transparency

The curvature of Earth's surface is the most prominent reason the equator is warm and the poles are cold. A given portion of sunlight is concentrated in a small **surface area** at the equator, but that same amount of sunlight is spread out over a much larger surface area at the poles. Since the angle of incoming solar radiation, called insolation, is low to the horizon, sunlight is spread out over a larger area at the poles. At the equator, the angle of insolation is much greater. The difference in the amount of sunlight per unit of surface area makes the equator warm and the poles cold.

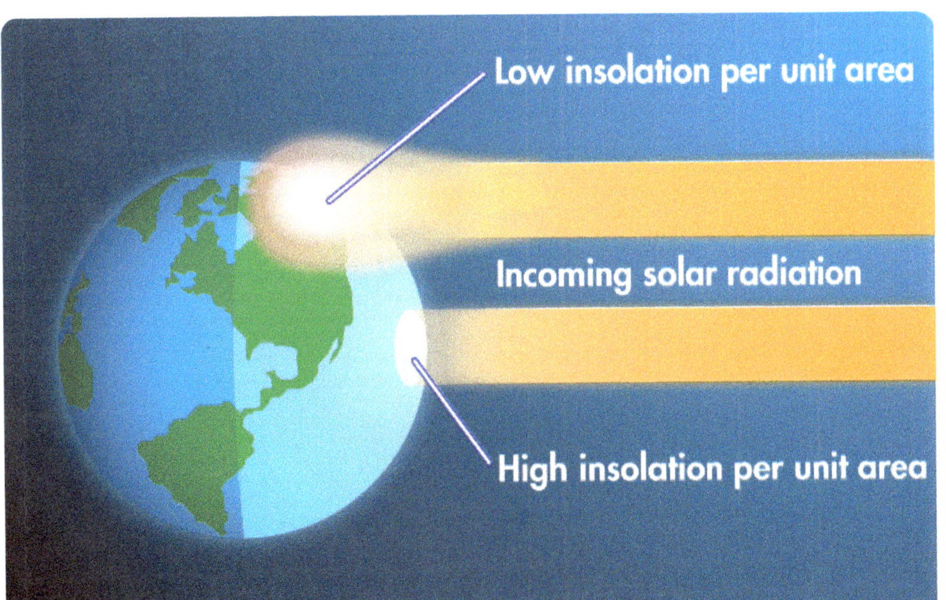


**ANGLES OF INSOLATION**

At Earth's poles, the angle of insolation is relatively low (closer to 0°), while at Earth's equator, it is relatively high (closer to 90°). How does the angle of insolation influence the amount of the sun's energy entering the atmosphere?

Earth's surface is about 30 percent land and 70 percent water. Land absorbs and releases thermal energy much more rapidly than water does, as water absorbs heat more slowly and holds onto it longer. Because land absorbs heat more quickly, surface layers can become extremely hot. Land also transfers its heat more rapidly to the atmosphere. Because the atmosphere is less transparent, areas that are covered by layers of cloud receive less solar radiation. The clouds reflect some of the solar radiation back into space. Smoke from forest or grass fires and dust from volcanic eruptions can also reduce surface temperatures.

**LAND SURFACES AND WATER SURFACES**

More than 70 percent of Earth's surface is covered by water. Why does water heat up and cool off more slowly than land?



## Forming Wind

Wind is caused by differences in air **pressure**, so why is the uneven heating of Earth's surface important? Atmospheric circulation is driven mostly by solar radiation, and winds and convection currents in the atmosphere are the result of uneven distribution of thermal energy in the atmosphere. When air is heated, the air expands and becomes less dense, forming a low-pressure zone at Earth's surface. Cooler surrounding air moves in, which forces the warmer air upward. In cool areas, the air loses thermal energy and contracts, and the cooler, denser air then sinks toward the ground, resulting in the formation of a high-pressure zone at Earth's surface.

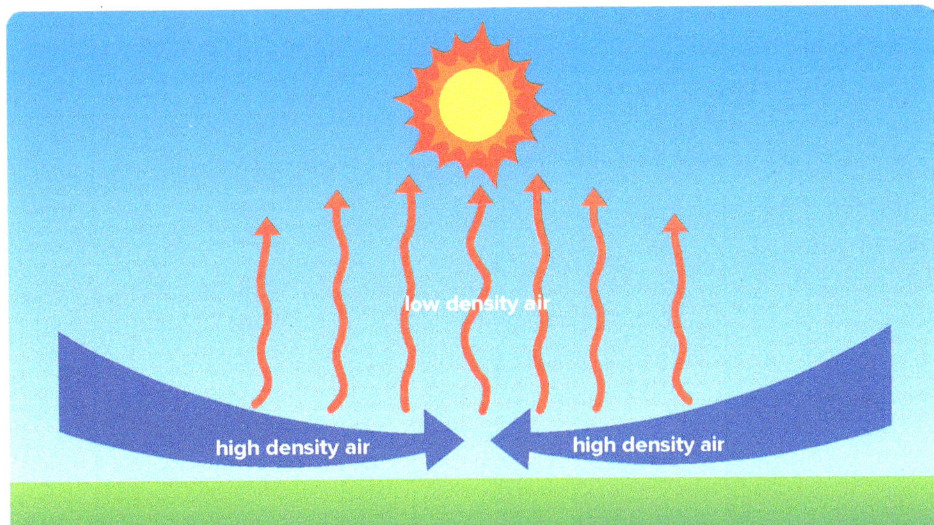


**RISING AIR**
Low-density air in a low-pressure zone is forced upward by the surrounding high-density, high-pressure air. What happens to the temperature of air as it rises?



Wind is basically the large-scale movement of air from one location to another across Earth's surface. The direction of air movement is down the pressure gradient, as air moves from areas of high pressure to areas of low pressure. A pressure gradient is the extent to which pressure changes over a given distance. Generally, cool, dense air flows from cool, high-pressure regions into warmer, low-pressure regions, resulting in the cool, dense air forcing the warm, less dense air out of the area.

Winds are commonly named for the direction from which they blow. For example, westerlies blow from the west and northerlies blow from the north. In fact, practically all winds at Earth's surface can be named in this manner, and many winds are named after the general location from which they blow. For example, ocean breezes blow from the ocean, lake breezes blow from lakes, and mountain breezes blow from mountains. Winds get their names from an extensive range of sources, such as local geography, historical events, legends, or other considerations.

## The Coriolis Effect

Air pressure is not the only thing that affects the direction of air flow, as Earth's rotation also affects the direction of air movement. Imagine being on a rotating merry-go-round and throwing a ball directly to a friend on the opposite side of the merry-go-round. The friend will not be able to catch the ball, but why not? When viewed from the merry-go-round, the ball will appear to follow a "curved" path instead of a straight path, but when viewed from above the merry-go-round, the ball will follow a straight path while the friend moves in a curved path. Scientists call this observation the **Coriolis effect**, and by substituting the rotating Earth for the merry-go-round and the wind for the ball, you will understand how the Coriolis effect manifests on Earth.

Some references describe the Coriolis effect as the apparent deflection of a moving object's straight path by Earth's rotation. The deflection is to the right in the northern hemisphere, whereas the deflection is to the left in the southern hemisphere. Because the object continues in a straight path when it is viewed from a stationary reference, others describe the Coriolis effect as the deflection of a moving object when it is viewed from a rotating frame of reference. Ultimately, the Coriolis effect is simply about inertia, or the tendency of any moving body to continue in the direction in which it is initially propelled.

While atmospheric circulation is driven by solar radiation, it is modified by the Coriolis effect. The Coriolis effect is often observed in the direction of air movement in high-pressure and low-pressure systems in the northern and southern hemispheres. In the northern hemisphere, surface air moves clockwise around a high-pressure center, and it moves counterclockwise around a low-pressure center, but the opposite is true in the southern hemisphere, as surface air in the southern hemisphere moves counterclockwise around a high-pressure center and clockwise around a low-pressure center.

**HURRICANES AND THE CORIOLIS EFFECT**

In the northern hemisphere, a low pressure system, such as a hurricane, rotates counterclockwise due to the Coriolis effect. How does the Coriolis effect determine a hurricane's direction of rotation in the southern hemisphere?

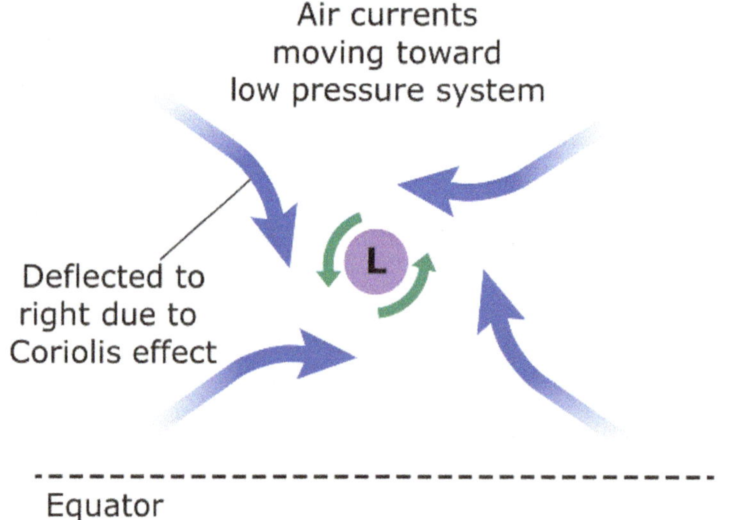

## Global Wind Systems

Global winds are large-scale movements of air that help distribute thermal energy over Earth's surface, and they often result from differences in the **absorption** of solar energy between Earth's climate zones. The two main factors relating to global winds are the uneven heating of Earth's surface between the equator and the poles and the Coriolis effect.

**Global wind** systems are often illustrated by a model showing Earth's major wind bands. In the northern hemisphere of this model, the northeasterly trade winds blow from about 30° north latitude toward the equator. The westerlies blow from about 30° north latitude toward 60° north latitude. The polar easterlies blow from the North Pole toward 60° north latitude. A mirror image of these wind bands is located in the southern hemisphere, and these six major wind bands distribute thermal energy over Earth's entire surface.

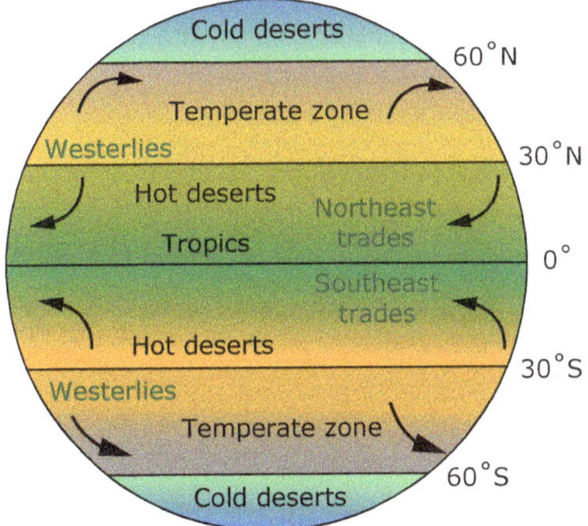


**THE GLOBAL SURFACE WINDS**

The pattern of the global winds depends on the latitude. Can you explain why the winds vary in this way?

Four major jet streams, a relatively strong wind concentrated within a narrow band in the atmosphere, are referred to as global winds because they circumnavigate the globe. Wind speeds in a jet stream often exceed 115 mph and sometimes exceed 230 mph.

Both the northern hemisphere and the southern hemisphere have a polar jet stream and a subtropical jet stream. These four high-altitude (or upper-level) jet streams are usually found at altitudes between 10 km and 15 km (6 mi–9 mi) in the atmosphere, and all four of these jet streams are westerly winds that wrap around Earth in a meandering, snake-like fashion.

**Polar and Subtropical Jet Streams**

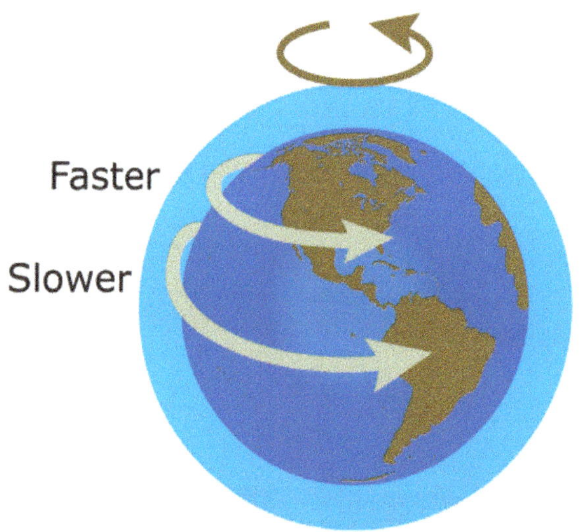


**JET STREAMS IN THE NORTHERN HEMISPHERE**

Polar jet streams have faster winds than subtropical jet streams. How does the Coriolis effect influence these speeds?



Jet streams play a major role in directing the movement of Earth's weather systems. The four high-altitude jets often appear on weather maps and in weather reports. Other jet streams, such as low-level jets, also affect weather systems, but these jets are usually classified as local winds.

## Local Winds

Local winds are winds that tend to blow over a relatively small area, and as a result, they are often caused or affected by regional conditions, such as mountains, large bodies of water, local pressure differences, and other influences. Some examples of local winds include land breezes, sea breezes, mountain breezes, valley breezes, lake breezes, and monsoons.

Sea breezes (also called ocean breezes) are winds that blow from a sea or an ocean to land. Recall that land absorbs and releases thermal energy much faster than water does. During daylight hours, land heats more rapidly than the sea, so the land then heats the air above it, causing it to expand and rise, resulting in the development of a lower-pressure zone over land, while a higher-pressure zone remains over the sea. This difference in air pressure causes a cool breeze to blow from the ocean to the land.

Land breezes are winds that blow from land to a sea or an ocean. At night, land cools more rapidly than the sea. The air above the land becomes cooler and denser relative to the air over the sea. This time, the difference in air pressures causes a breeze to blow from the land.

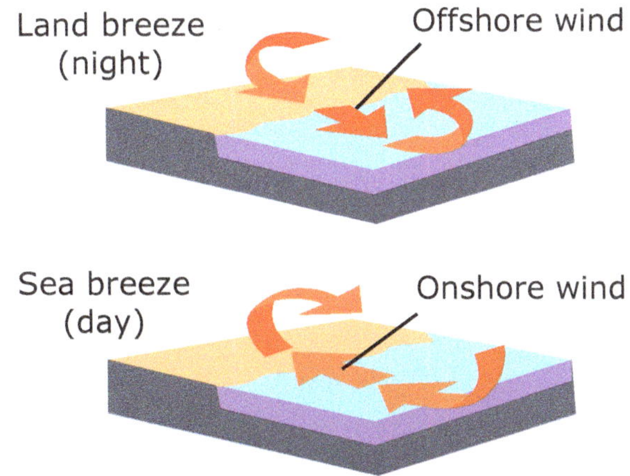


**SEA AND LAND BREEZES**

Differences in the way water and land heat up leads to high and low pressure areas. How do these differences in air pressure cause breezes on the sea shore?

Mountain breezes flow down the slopes of mountains toward valleys. At night, the exposed mountain slopes cool more rapidly than the valleys below. Air along the mountain slopes becomes cooler and denser than the air in the lower valleys. Then the cool, dense mountain air flows downslope into the valleys.

Valley breezes blow upward along mountain slopes. They usually occur during the day when the land in the valley is hotter than the mountain slopes.

## Mountain Breezes

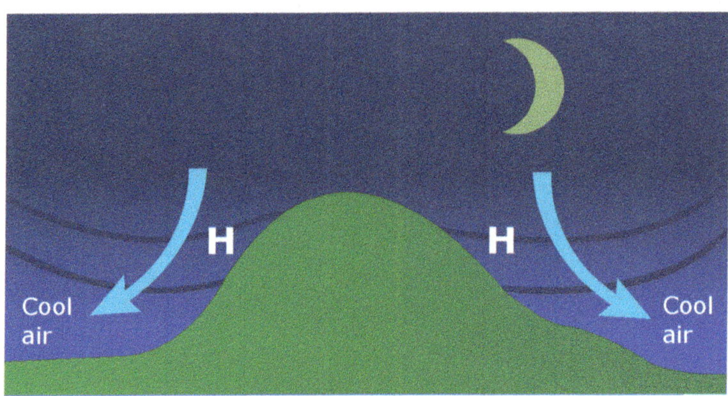


## Valley Breezes

**MOUNTAIN AND VALLEY BREEZES**

The height and shape of a mountain affects the local climate in various ways. What patterns in topography relate to patterns in a mountain climate?

Some regions have specific winds that strongly influence local weather and climate. A good example of such a wind is a "fall" wind. Meteorologists call this type of wind a katabatic wind. A katabatic wind is a wind that blows downslope. This type of wind gets its name from a Greek word that means "moving downhill." It is a technical name for a wind that carries high-density air that "falls" down a steep slope.

Katabatic winds typically start out as cold, dense air at high elevations. They flow rapidly downslope under the force of gravity. Some katabatic winds can rush down steep slopes at hurricane speeds, but most are not that fast. A good example of a katabatic wind is the Santa Ana wind, a hot, dry wind that blows from the mountain passes and narrow canyons of southern California and then across the Los Angeles basin. The wind starts out as a cool, dense, dry, high-pressure air mass in the Great Basin, and then the dense air flows across the high plateau of the Mojave Desert and spills over mountain ranges into the Los Angeles basin. The air is compressed and warmed as it is forced downslope by the high-pressure system. The wind picks up speed and its temperature increases when it pours through the narrow canyons. Hot, dry, fast-moving Santa Ana winds can quickly dry out vegetation and increase the danger of wildfires.

## How Are Human Activities Altering the Atmosphere?

The **atmosphere** of Earth has evolved over billions of years. It is the product of chemical and biological processes. Human activities alter the chemistry of the atmosphere. Burning wood or fossil fuels produces tiny particles we call smoke, resulting in smog when human activities are concentrated in cities. Smog is a major health problem in some cities because it causes or exacerbates lung disease, and clean air regulations aimed at the reduction of emissions from chimneys, vehicles, and other sources have proven effective in smog mitigation in some countries.

**FACTORIES AND AIR POLLUTION**

Emissions from factory smokestacks can release a variety of air pollutants. Which gases have the greatest effects on the atmosphere?

Less obvious than smoke are the invisible gases produced during combustion. Many fuels contain traces of sulfur and, when burned, sulfur forms oxides that enter the air. On contact with water in the lungs or in rain, these oxides form acids. Acids cause lung damage and pollute lakes and rivers. Nitrogen oxides also form by the combustion of fuel, and these oxides contribute to acid rain. **Carbon dioxide** is a naturally present **gas** in the atmosphere in very low concentrations, but human activities, through combustion and cement manufacture, have increased the total carbon dioxide level. Since carbon dioxide is important in regulating the temperature of the planet, an increase in its concentration causes a rise in temperature. For this reason, carbon dioxide is called a **greenhouse gas**. **Methane** is another greenhouse **gas** and is the major component of natural gas. Human activities expel methane into the atmosphere, contributing to global climate warming, and most fuels contain other impurities. Coal contains small quantities of heavy metals, and these are released into the air when coal is burned and later deposited from rain.

Human activities can impact the amount of water in the atmosphere, and the most significant human-induced humidity changes result from the removal of vegetation, particularly forests. Plants slowly release water into the air from their leaves in a process called **transpiration**, and removal or reduction of plant cover decreases transpiration and reduces humidity, cloud formation, and rainfall.

We are protected from the worst of the sun's ultraviolet (U.V.) rays by a layer of ozone in the upper atmosphere. Ozone is a form of oxygen that absorbs U.V. Without ozone, life on Earth would be exposed to harmful U.V. that damages DNA, causing mutations and cancers. Some chemicals produced by humans destroy ozone, and the best known of these are the chlorofluorocarbons that were used in refrigerators and air conditioners prior to the 1990s. These substances thinned the **ozone layer** and created an enlarged ozone hole over Antarctica, causing more U.V. to reach the lower atmosphere. Scientists discovered what was happening and encouraged nations to stop using these harmful chemicals, resulting in the ozone layer slowly repairing itself.

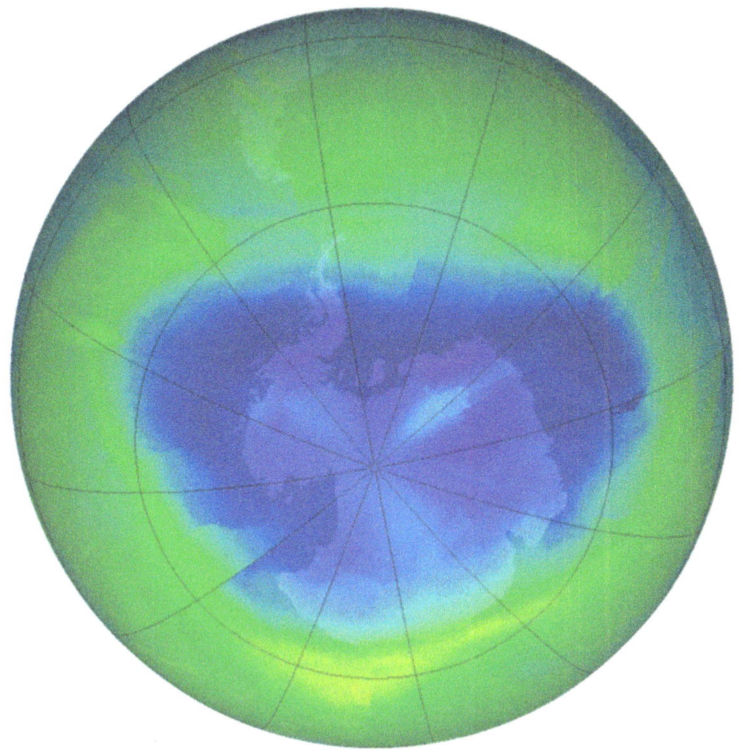

**OZONE LAYER**

Chlorofluorocarbons damaged the ozone layer. How have nations worked together to fix this problem?



### Consider the Explain Question

**How do changes in the composition of the atmosphere affect its properties and behavior?**

Go online to complete the scientific explanation.

dlc.com/ca9091s

### Consider the Explain Question

**Go online to check your understanding of this concept's key ideas.**

dlc.com/ca9092s

## in Action

# Applying the Composition and Behavior of the Atmosphere

Many places on Earth have combinations of geographical features that make them unique. These local geographies contribute to wind patterns or storm systems that are identified with the areas. For example, the residents of southern California are well aware of the hot, dry Santa Ana winds that can occur from October through March, downing trees and drying vegetation, which provides fuel for wildfires. Nor'easters are macro-scale storm systems along the northeast coast of the United States that sometimes bring blizzards and hurricane-force winds. Alberta Clippers are low-**pressure** systems that get tied up with the jet stream and barrel through central and eastern areas of North America during winter. Chinook winds are warm, dry winds that flow down the eastern slopes of the Rocky Mountains. They are popularly called "snow eaters" because they can devour snow one foot deep in less than a day.

**SANTA ANA WINDS**

The geography of Southern California leads to powerful Santa Ana winds every year. How do these winds contribute to wildfires?

Every place in the United States has some type of recurrent wind or atmospheric circulation that affects the local residents. Learning about the general mechanisms that affect atmospheric circulation can provide a basis for understanding local winds and weather patterns. Understanding the local geography can also help increase your understanding of these winds. Meteorologists at your local weather channels are good places to get more detailed information regarding local winds. Why is studying local weather patterns important for the meteorologist? Predicting weather conditions ahead of a disaster is important for meteorologists because they are able to give citizens proper warning about potentially dangerous weather conditions. When studying local winds, two of the data points that are collected include wind speed and atmospheric pressure. An anemometer is used to measure the wind speed and wind direction by using cups attached to blades, which spin from the wind on a rotor that is connected to a generator. The speed is calculated based on the amount of electricity produced. Wind direction is determined through the use of a wind vane that is attached to the anemometer. A barometer is used to gather data about atmospheric pressure, where the most common type, an aneroid, uses a small flexible metal box that measures the expansion and contraction as the atmospheric pressure changes. This information is sent as electronic signals to a computer that can then be mapped over the region to plot trends in atmospheric pressure. Meteorologists are able to analyze trends with weather conditions so that when changes occur, the public can be notified in a timely manner.

**WEATHER MAPS**

Meteorologists prepare weather maps that show current and predicted weather conditions. How could the forecast shown on this map help to save lives?

Do you know a characteristic wind system near where you live? What changes do meteorologists identify with the wind system when severe weather is heading to the area? Discuss it with someone and see if they know any other wind systems.

## STEM and the Composition and Behavior of the Atmosphere

What can we do with all this wind? Can we somehow capture it and convert it into usable renewable energy? Yes! We can use wind technologies to harness the energy in the wind and convert it to usable electrical power. How can we do this? Through wind power. Wind power is the power generated by harnessing the kinetic energy of wind. One way that scientists and engineers have been able to capture the kinetic energy from the wind and convert it into electrical energy is through the use of wind turbines. A wind turbine is engineered so that the wind turns the propellers around a rotor, which is connected to a generator which creates electricity. This wind power is considered a green energy because it is an alternative to using fossil fuels. Wind energy provides one of the least expensive ways to produce electricity. A group of wind turbines in the same location is called a wind farm. In 1980, the first large wind farm in the United States was installed to send power to a utility company, and since then, additional wind farms have been constructed to harness the kinetic energy of the wind and transform it into electrical power. In 2008, the U.S. Department of Energy projected that wind power would generate 20 percent of the electrical power in the United States by 2030. By 2015, the use of wind power to generate electricity had increased enough that this projection was updated to 20 percent by 2030 and 35 percent by 2050.

**WIND TURBINE**

Air interacts with the blades of a turbine to cause them to turn. What factors do engineers consider when designing the turbine blades to maximize efficiency?

How does the kinetic energy from the wind get converted into electrical energy? A wind turbine converts kinetic energy from wind into mechanical energy. The kinetic energy from wind turns a wind turbine, which turns an electric generator. The generator converts the mechanical energy into electricity. The electricity is then fed into a local utility grid and distributed to customers.

Wind energy engineering is an up-and-coming profession that many have become interested in. As we try to locate alternative energy resources to that of fossil fuels, many are seeking out wind energy engineering. These scientists typically study energy engineering or mechanical engineering concepts. Their work may include wind farm design, work on turbines (both manufacturing and/or construction), or they are present at the installation site to ensure that the construction is of high quality.

**WIND POWER**

Wind and solar energy are two sources of renewable energy. How does Earth's atmosphere affect the amount of energy that is generated from these power sources?

Many different engineers are utilized when constructing a wind farm. Environmental engineers are used to assess the environmental impact to the area. Civil engineers are needed for the infrastructure and support systems that make up the turbine. Mechanical engineers are utilized to test the machines that make up the turbine. Electrical engineers are needed to work on the use of electricity to operate the turbine. These are just a few of the engineers needed for a wind farm project. The U.S. government is in support of wind energy because they are looking for a way to curb the tremendous fossil fuel consumption that has taken place for several years.

Wind energy is a clean, domestic, renewable form of energy that does not emit greenhouse gases or pollutants. Wind energy has the potential to create jobs and bring billions of dollars of economic activity to states willing to have wind farms developed on their land. The increase in wind power can be accredited to tax incentives that have been given to those who were willing to have the large turbines installed on their property. With its low cost to run and the advances in technologies since its beginning in 1980, it is one of the lowest-cost competitive renewable energies. With this in mind, many believe wind energy is the way of the future for electrical energy.

**A RENEWABLE ENERGY SOURCE**

The United States is working on harnessing the renewable resource of wind power. How is wind energy a form of converted solar energy?

CONCEPT 5.2

# Solutions



## LESSON OVERVIEW

### Lesson Questions

- What are the various parts of a solution?
- How do various factors affect the solubility of a solute?
- How are concentrations of solutions using molarity, mass percent, volume percent, and parts per million (ppm) calculated?
- What are the various colligative properties of solutions?

### Key Vocabulary

Which terms do you already know?

- ☐ boiling point elevation
- ☐ colligative property
- ☐ concentration
- ☐ dilute solution
- ☐ dissolve
- ☐ electrolyte
- ☐ freezing
- ☐ freezing point depression
- ☐ heterogeneous
- ☐ homogeneous
- ☐ ideal solution
- ☐ mass percent
- ☐ molality
- ☐ molarity
- ☐ mole
- ☐ mole fraction
- ☐ nonpolar covalent bond
- ☐ osmotic pressure

dlc.com/ca9093s

## Lesson Objectives

By the end of the lesson, students should be able to:

- Distinguish between the parts of a solution.
- Summarize factors that affect the solubility of a solute.
- Calculate concentrations of solutions using molarity, mass percent, volume percent, and parts per million (ppm).
- Describe the various colligative properties of solutions.

## Key Vocabulary continued

- ☐ parts per million (ppm)
- ☐ polarity
- ☐ precipitate
- ☐ solubility
- ☐ solubility product constant (Ksp)
- ☐ solute
- ☐ solution
- ☐ solvent
- ☐ surface area
- ☐ universal solvent
- ☐ volume percent

# Thinking about Solutions

**Have you ever craved a sports drink after a long sweaty workout? Why do professional athletic teams provide sports drinks instead of water to their players on the field?**

dlc.com/ca9094s



**NURSE HANGING AN IV BAG**

After surgery or serious injury, a patient may need an intravenous (IV) feed to supplement various salts in the blood. What kinds of solutes are used in an IV feed?

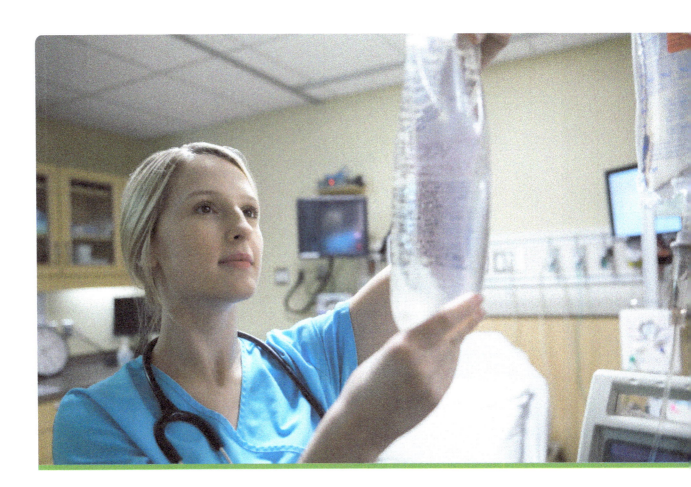

**EXPLAIN QUESTION**

**What factors influence the rate at which a solute dissolves?**

# What Are the Various Parts of a Solution?

## Solutions

Solutions are **homogeneous** mixtures in which the components are distributed evenly and will not separate. Colloids and suspensions are **heterogeneous** mixtures, not solutions. They are composed of small particles distributed throughout a medium, but the particles are large enough that they can often be filtered out. They are not truly dissolved.

Solutions can be solid, gaseous, or liquid. For example, steel is a solid **solution** in which carbon is mixed evenly with iron. (Solid solutions that contain metals are called alloys.) Air is a solution that contains primarily gases, but might also contain tiny particles in solid or liquid form. In chemistry, many of the solutions discussed are in liquid form.

There are two necessary components to every solution. A **solute** is the substance that is dissolved in a solution. The **solvent** is the substance that dissolves the solute. Usually, there is more solvent than solute in a solution. For example, if you add one spoonful of sugar to a jug of water, the water is the solvent and the sugar is the solute. When two liquids are mixed together, the solute is the substance of the lesser quantity. For example, rubbing alcohol is a liquid mixture of 70 percent isopropyl alcohol and 30 percent water. In this solution, water is considered the solute.

## Aqueous Solutions

Many aqueous solutions are formed in nature, where water commonly dissolves other substances in the environment. In fact, water is often referred to as the **universal solvent** because so many other substances will **dissolve** in it. Water is unique because it naturally exists in all three states of matter—solid, liquid, and gas. Water's unique properties are primarily caused by the **polarity** of the water molecule. The positive and negative areas of charge are attracted to many other types of particles, and can dissolve them easily.

An **electrolyte** is any substance that dissolves to form free ions, which make the solution electrically conductive. Batteries contain electrolytes because they allow negative charge (electrons) to flow from one location to another in an electric current. The human body needs electrolytes to support cellular processes, such as metabolism and muscle function. Electrolytes also help keep the body's fluid levels in balance.

A nonelectrolyte is a solute that dissolves in water to form a nonconductive solution. Sugar is a nonelectrolyte. When mixed with water, sugar creates a nonconductive solution.

## How Do Various Factors Affect the Solubility of a Solute?

### Solubility Rate

The **solubility** of a substance indicates how much of that substance will **dissolve** in a specified amount of **solvent** under certain conditions. Solubility is usually measured in terms of **concentration** at a given temperature and pressure.

There are many factors that will increase or decrease the rate of solubility. As the **surface area** of a **solute** is increased, the rate of dissolving increases. Surface area is the total area a substance has on its exterior. For example, a teaspoon of sugar contains about the same mass of sugar found in a sugar cube. However, the teaspoon of loose sugar will dissolve faster because it has more surface area available to interact with the solvent.

As the temperature of a solvent is increased, the rate of dissolving usually increases. This occurs because temperature is related to the kinetic energy of the particles involved. The more kinetic energy solvent particles have, the faster they move. The faster they move, the more they can interact with a solute. Kinetic energy can also be added to the system by stirring or shaking it.

**ADDING SUGAR TO COFFEE**

Sugar is among the most soluble everyday substances. What would you observe when you couldn't add any more sugar to a cup of coffee?

## Chemical Makeup and Solubility

If the solute is a gas, increasing the temperature has the opposite effect on the solubility. The increased kinetic energy of the **solution** allows the gas particles to move more freely, and they are more likely to escape from the solution. The solubility of the gas will decrease.

Carbonated beverages exhibit another solubility rule known as Henry's law. This law states that if temperature does not change, the amount of gas that dissolves in a liquid is proportional to the partial pressure of that gas in equilibrium with that liquid. For example, inside an unopened can of soda the pressure is much higher than that of the surrounding area. This pressure keeps carbon dioxide gas dissolved in the liquid soda solution. When the can is opened and pressure is decreased, the carbon dioxide escapes the solution.

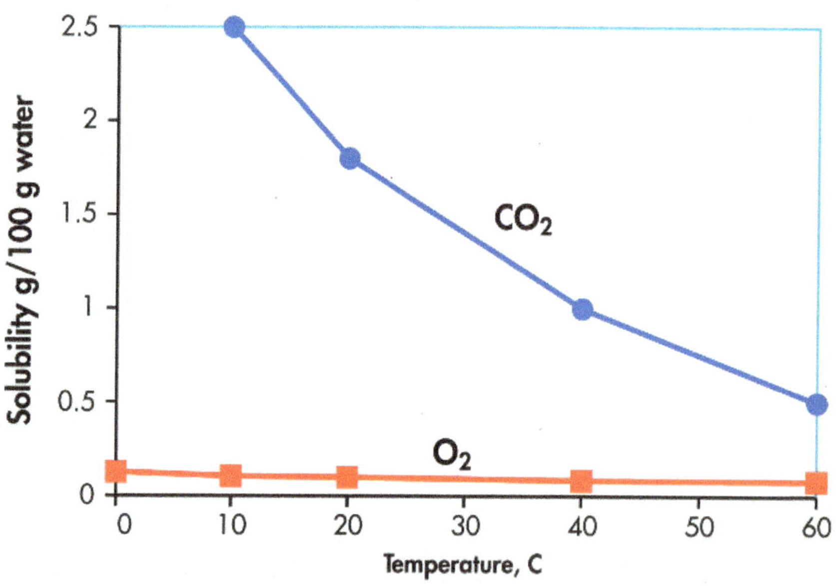


**SOLUBILITY OF GASES VERSUS TEMPERATURE**

The solubility of gas in water decreases as the temperature of the water increases. Why do $CO_2$ and $O_2$ have such different solubility?

The chemical makeup of the solvent and solute are also related to solubility. Remember that some molecules have a negative or positive charge because the electrons are not distributed evenly around them. These are referred to as polar molecules. Solutes of a specific **polarity** (negative or positive) will only dissolve in solvents with a similar polarity. The rule for this is “like dissolves like.” For example, the oil and vinegar in salad dressing do not mix because vinegar is polar and oil is nonpolar. However, water and vinegar will form a solution because they are both polar.

## Saturated, Unsaturated, and Supersaturated Solutions

When a large amount of salt is mixed into cold water, some of it does not dissolve. In fact, the salt remains solid and settles on the bottom of the container.

Because of the way in which particles react, a solvent can dissolve only a certain amount of a solute. A saturated solution is a solution in which no more solute can be dissolved. In fact, if additional solute were added it could cause some dissolved solute to become solid and settle out of the solution. The solid that forms and separates from a solution is known as a **precipitate**.

An unsaturated solution is a solution that contains less solute than can be dissolved under existing conditions. If more solute is added, it will simply dissolve into the solvent.

Under special conditions, a saturated solution can be made to dissolve more solute. Most often, this requires the addition of heat or pressure to the system. For example, water at room temperature will dissolve a certain amount of sugar. However, when the solution becomes saturated no additional sugar will dissolve. The solution becomes cloudy until the extra sugar settles to the bottom of the container. If the water is heated, more sugar can be dissolved into it. When the solution is allowed to cool back down, the extra sugar will remain dissolved. As a result, the final, cooler solution will contain more sugar than would normally dissolve in that volume of water at that temperature. This solution is supersaturated.



## Electrolytes and Nonelectrolytes

In a common science experiment, two electrodes are stuck in a lemon and attached to a conductivity meter. Amazingly, the lemon conducts electricity! Lemon juice contains citric acid which is an **electrolyte**. An electrolyte is a substance that conducts electricity if it dissolves in water. Acids and bases are electrolytes, and so are most salts. A nonelectrolyte is a substance that does not conduct electricity when dissolved in water. Sugar and ethanol are nonelectrolytes.

When an acid such as HCl dissolves in water, it produces ions. The same is true for a base such as NaOH. Similarly, when salts such as NaCl or $KNO_3$ dissolve in water, they produce ions.

$$HCl \rightarrow H^+ + Cl^-$$
$$NaOH \rightarrow Na^+ + OH^-$$
$$NaCl \rightarrow Na^+ + Cl^-$$
$$KNO_3 \rightarrow K^+ + NO_3^-$$

These ions are the reason why solutions of electrolytes can conduct electricity. Negative and positive ions move in the solution. They can carry electrical charge from one electrode to the other. Strong acids and bases are strong electrolytes because they dissociate completely into ions. Weak acids and bases are weak electrolytes because they dissociate only partly.

The conductivity of a solution is a measure of how well it conducts electricity. Conductivity depends on how many ions are present in the solution. For one electrolyte, a concentrated solution usually has a higher conductivity than a **dilute solution**.

Conductivity can be used to measure the **concentration** of a solution. Different electrolytes have different conductivities. The relationship between concentration and conductivity can be found for each electrolyte. (This data can be measured in the laboratory or obtained from reference books.) Then this relationship can be used to identify an unknown concentration by measuring its conductivity.

**IONS AND CONDUCTIVITY**

Calcium carbonate produces ions as it dissolves in water. Why would the solution conduct electricity?

# How Are Concentrations of Solutions Using Molarity, Mass Percent, Volume Percent, and Parts per Million (ppm) Calculated?

## Concentrations of Solutions

The **concentration** of a **solution** is a measure of how much **solute** is dissolved in a given amount of **solvent**. There are three ways the **concentration** of a solution can be described.

The first method for describing concentration is known as **molarity**. Molarity ($M$) gives the number of moles of solute divided by the total number of liters of the solution.

$$M = \frac{\text{moles of solute}}{\text{liters of solution}}$$

Another way to describe solution concentration is the **mole fraction**. The **mole** fraction ($X$) of a solution is the ratio of moles of the solute to the total number of moles in a solution. Since mole fraction describes a percentage, it can never be greater than one.

$$X = \frac{\text{moles of solute}}{\text{moles of solute and solvent combined}}$$

The final way to describe concentration is **molality**. Molality ($m$) is the number of moles of solute dissolved in one kilogram of the solvent.

$$m = \frac{\text{moles of solute}}{\text{1 kg of solvent}}$$

These methods of calculating concentration are most useful when solute concentrations are quite high.



## Concentrations of Solutions: Sample Problem

A solution is formed by combining 0.250 moles of sodium chloride (NaCl) in 3.61 liters of water. Assume that the addition of NaCl does not significantly change the volume of the solution. The volume of the solution, then, is 3.61 L. Compare the molarity and the molality of NaCl in the solution. What is the mole fraction of NaCl in the solution? Assume that one liter of water has a mass of one kilogram.

**Solution:**

First, to calculate the molarity of the solution, substitute the given information into the molarity ratio:

$$M = \frac{\text{moles of solute}}{\text{liters of solution}}$$

$$= \frac{0.250\text{ mol}}{3.61\text{ L}}$$

$$M = 0.0693\text{ mol/L}$$

Notice that two extra digits are included because this is an intermediate calculation. Carrying extra digits can avoid roundoff errors in the final answer. The molarity of the NaCl in the solution is 0.0693 mol/L. Notice that a volume of 3.61 L of solution is used in the calculation. The actual volume would be slightly higher due to the addition of NaCl. Notice also that the answer is expressed with the correct number of significant digits, which in this case is three.

To find molality, substitute the given information into the molality ratio:

$$m = \frac{\text{moles of solute}}{\text{kg solvent}}$$

$$= \frac{0.250\text{ mol}}{3.61\text{ kg}}$$

$$m = 0.0693\text{ mol/kg}$$

$$X = \frac{\text{moles of solute}}{\text{moles of solute and solvent combined}}$$

The value for the moles of solute NaCl is given as 0.250 mol. The total number of moles of solute and solvent combined must next be calculated.

The solution contains 3.61 kg of water, but the number of moles of water must be determined. Because the molar masses of hydrogen and oxygen are given in grams, it is convenient to convert the mass of water in the solution to grams, as well.

$$3.16\text{ }\cancel{\text{kg}} \times \frac{1{,}000\text{ g}}{1\text{ }\cancel{\text{kg}}} = 3{,}160\text{ g}$$

The mass of the water in grams is 3,610 g. This value can be used to determine the moles of water in the solution.

From the periodic table, we can find the following atomic masses:

$$\text{hydrogen (H)} = 1.008 \text{ g}$$
$$\text{oxygen (O)} = 16.00 \text{ g}$$

These masses of two moles of hydrogen and one mole of oxygen are added together to find the molar mass of water.

$$(2 \times 1.008 \text{ g/mol}) + 16.00 \text{ g/mol} = 18.016 \text{ g/mol}$$

Next, we determine the number of moles in 3,160 g of water.

$$3{,}160 \cancel{\text{g}} \times \frac{1 \text{ mol}}{18.016 \cancel{\text{g}}} = 175.39964 \text{ mol}$$

The total moles of the solute and solvent equals

$$0.250 \text{ mol NaCl} + 175.39964 \text{ mol } H_2O = 175.64964 \text{ mol total}$$

The mole fraction of NaCl in the solution is:

$$X = \frac{\text{moles of solute}}{\text{moles of solute and solution}}$$
$$= \frac{0.250 \text{ mol}}{175.64964 \text{ mol}}$$
$$X = 0.0014253$$

Note that extra digits were used during this last set of calculations so that the moles of solute and total moles in the solution could be properly compared. Notice that two extra digits are carried. Carrying extra digits can avoid roundoff errors in the final answer. The final answer, however, should again have four significant digits. The mole fraction of NaCl in the solution is about 0.00142. That is, about 0.14 percent of the solution is NaCl.

## Using Molarity to Calculate the Dilutions of Solutions

A dilution is making a solution that has smaller amounts of solute than the original solution. In other words, it has a lower concentration than that of the original. Dilution is accomplished by adding more solvent to the solution to create a smaller ratio of solute to solvent.

To find a dilution of an existing solution of known concentration, the formula, $\text{Molarity}_1 \times \text{Volume}_1 = \text{Molarity}_2 \times \text{Volume}_2$, ($M_1V_1 = M_2V_2$), can be used. $\text{Molarity}_1$ is the known concentration of the initial solution and $\text{Volume}_1$ is the known volume of the initial solution. Either the desired molar concentration must also be known or the new volume required. These would be $\text{Molarity}_2$ and $\text{Volume}_2$ values, respectively. The fourth and final value can then be calculated.

An example of how this is done is as follows:

You have five liters of a 2 M hydrochloric acid solution. What would the final concentration be if you added enough water to get a final volume of 20 liters?

$$M_1 = 2\text{ M},\ V_1 = 5\text{ liters},\ M_2 = \text{unknown, and } V_2 = 20\text{ liters}$$

Rearranging the formula to solve for the variable in question would be $M_2 = M_1V_1/V_2$. Substituting in the known values would give $M_2 = 2\text{ M} \times 5\text{ L}/20\text{ liters}$. Solving the equation would give a value of 0.5 M.

## Additional Methods for Calculating Concentrations

In the sample problem above, concentration was calculated using molarity, molality, and mole fraction. Other methods of calculating concentrations are related to the mass and volume of the solution's components instead of molar values.

**Mass percent** is the mass of a given solute divided by the mass of the solution. The mass percent of each component in a solution should add up to equal 100 percent.

$$\text{mass \%} \times \frac{\text{mass of solute}}{\text{mass of solution}} \times 100\%$$

Another method, **volume percent**, is often used to describe the concentrations of liquid solutions. For example, you may have seen 70 percent isopropyl alcohol in a first-aid kit. It is made using a ratio of 70 parts isopropyl alcohol and 30 parts water. Because volumes do not always increase when solutions are formed, the volume percent concentrations of each solute in a solution may not add up to be exactly 100 percent.

If a solution is formed from liquid or gas components, the volume of the solute is used to calculate volume percent concentration.

$$\text{volume \%} = \frac{\text{volume of solute}}{\text{volume of solution}} \times 100\%$$

Molarity is generally used when the concentration of a solute is very high. Using another method, referred to as parts per million (ppm), is better suited for solutions with low concentrations of a solute. This concentration is calculated using the ratio of the mass of solute to the mass of solution, then applying the ratio to a quantity of one million particles. In other words, the calculation shows how many parts of solute would be found in one million parts of solution.

$$\text{ppm} = \frac{\text{mass of solute}}{\text{mass of solution}} \times 10^6$$



## Calculating Concentrations: Sample Problem

A solution is formed when 5.000 g of sodium bicarbonate, $NaHCO_3$, is added to 2.000 L of water. Calculate the mass percent of sodium bicarbonate in solution. What is the concentration of sodium bicarbonate, in ppm? Assume one liter of water has a mass of one kilogram.

**Solution:**

To calculate the mass percent of sodium bicarbonate in solution, the mass of the solution must first be determined. The mass of the solute is given as 5.000 g. The mass of the solvent must be determined in kilograms, and then converted to grams.

$$\text{mass of solvent} = 2.000\ \cancel{L} \times \frac{1.0\ \text{kg}}{1\ \cancel{L}} \times 2.000\ \text{kg}$$

$$\text{mass of solvent} = 2.000\ \cancel{kg} \times \frac{1000\ \text{g}}{1\ \cancel{kg}} \times 2{,}000\ \text{g}$$

The total mass of the solution, then, combines the two values.

$$\text{mass of solute} + \text{mass solution} = \text{total mass of solution}$$

$$5.000\ \text{g} + 2.000 \times 10^3\ \text{g} = 2{,}005\ \text{g}$$

Next, we substitute the correct values into the equation for mass percent.

$$\begin{aligned}\text{mass percent} &= \frac{\text{mass of solute}}{\text{mass of solution}} \times 100\% \\ &= \frac{5.000\ \text{g}}{2{,}005\ \text{g}} \times 100\% \\ &= 0.2494\%\end{aligned}$$

The mass percent of the sodium bicarbonate in the solution is 0.2494%.

To determine the concentration of the $NaHCO_3$ solution in parts per million, the mass values used above are substituted into this equation.

$$\begin{aligned}\text{parts per million} &= \frac{\text{mass of solute}}{\text{mass of solution}} \times 10^6 \\ &= \frac{5.000\ \text{g}}{2{,}005\ \text{g}} \times 10^6 \\ &= 2{,}494\ \text{ppm}\end{aligned}$$

For every million parts of solution, there are 2,494 parts of solute.

# What Are the Various Colligative Properties of Solutions?

## Colligative Properties

Colligative properties depend on the number of particles of **solute** present in a **solution**. They are not determined by the properties of the solute or the **solvent**, and always follow some specific patterns.

The primary property which is affected by colligative factors is vapor pressure. In a liquid, vapor pressure is caused by particles escaping and changing to a gas state. If the solution is in a closed container, they cannot escape the entire system, so they occupy the space above the solution and below the top of the container. As temperature increases, more particles evaporate and vapor pressure increases.

When a solute is added to a solvent, it becomes more difficult for the particles of the solvent to evaporate because there is less **surface area** available. In other words, some of the surface is blocked by the solute particles. The more solute particles there are, the more vapor pressure decreases.

The change in vapor pressure affects two specific colligative properties. First, when a solid solute is added to a liquid to create a solution, the boiling point of that solvent will increase. This **colligative property** is described as **boiling point elevation**. Because particles of solute on the surface of the solution make it more difficult for evaporation to occur, boiling requires extra energy.

**BOILING POINT AND SOLUTES**

Bubbles form at the top of this solution as it boils. Why would you expect the boiling point of this solution to be higher than the boiling point of pure water?

Second, adding solid solute to a solvent will decrease the **freezing** point of the solvent used in the solution. This colligative property is known as **freezing point depression**. Freezing point depression is also proportional to the number of solute particles in the solution and is independent of the type of particle. The solute particles interfere with the solvent's formation of crystals. The freezing process does not occur until the temperature is lowered further.

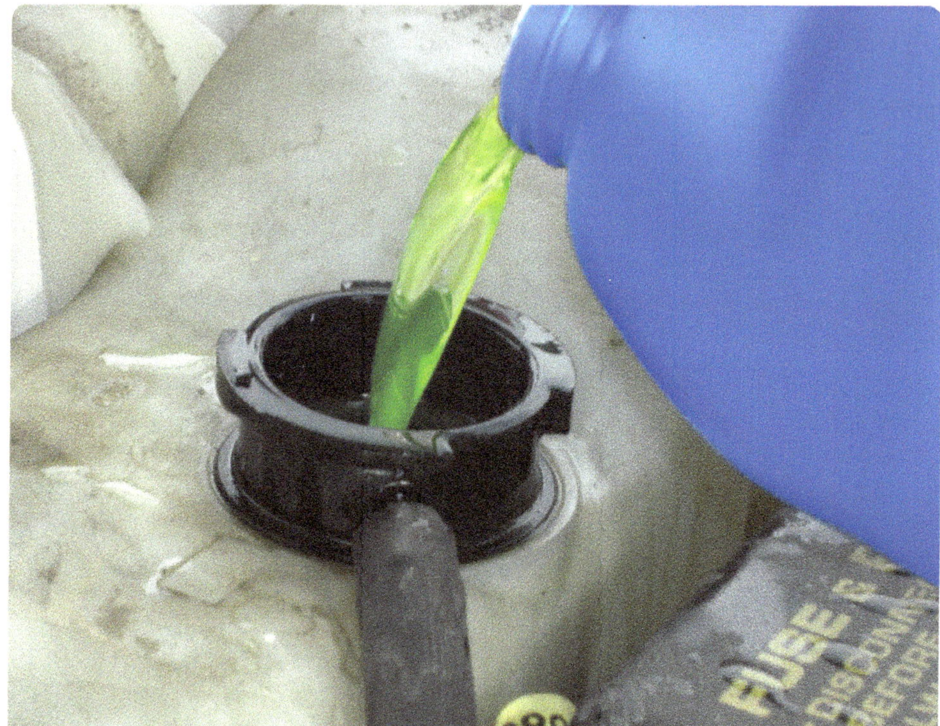


**ANTIFREEZE**
Colligative properties are solution properties that depend on the concentration of solutes in a solution. How does antifreeze keep the coolant in your car from freezing?



To calculate changes in boiling or freezing point, the **concentration** of the solution must be known. **Molality** is proportional to the change in boiling and freezing points, but the proportional relationship depends on the specific pure solvent or solution that is present. Constants for boiling point elevation ($K_b$) and freezing point depression ($K_f$) have been determined through research.

Based on this proportional relationship, the change in temperature required for boiling can be expressed by the equation:

$$\Delta T_b = iK_b m$$

Similarly, the change in temperature required for freezing is calculated using the equation:

$$\Delta T_f = iK_f m$$

The Greek symbol delta, $\Delta$, represents "change." The $m$ in each equation is the molality. The van't Hoff factor, $i$, is determined by the solute's behavior in the particular solvent. For example, solutes that do not dissociate (or break apart into two components) in water, $i = 1$. For solutes that completely dissociate into two ions, $i = 2$. Sodium chloride, NaCl, completely dissociates into two ions, $Na^+$ and $Cl^-$. Some solutes completely dissociate into three ions, and for those $i = 3$. For example, CaCl completely dissociates into three ions, one $Ca^+$, and two $Cl^-$. The values of $i$ continue to increase as the number of ions created during dissociation increases.

## Calculating Freezing Point: Sample Problem

A solution is made with 12.55 g of sodium chloride, NaCl, and 180.4 mL of water. The water is 35.0°C. What is the freezing point of this solution? Assume the NaCl dissolves completely in the water. The density of water at 35.0°C is 0.9941 g/mL, and the $K_f$ of $H_2O$ is 1.858°C kg/mol.

**ICY ROAD**

Salt is added to roads when temperatures start to fall. How does salt improve driving safety in winter?

**Solution:**

To find the freezing point of the solution, use the formula:

$$\Delta T_f = iK_f m$$

First, it is known that $i = 2$ because two ions ($Na^+$ and $Cl^-$) form when NaCl dissociates.

The value $K_f$ of water is given as 1.858°C kg/mol.

The value of $m$ must be calculated. This is the molal **concentration** of the solution, or:

$$m = \frac{\text{moles of solute}}{\text{1 kg of solvent}}$$

To find the moles of NaCl, first use atomic masses from the periodic table to find the molar mass of NaCl.

Sodium (Na) = 22.99 g
Chlorine (Cl) = 35.45 g

$$\text{molar mass} = 22.99\text{ g/mol} + 35.45\text{ g/mol}$$
$$= 58.44\text{ g/mol}$$

One **mole** of NaCl has a mass of 58.44 g.

The solution contains 12.55 g of NaCl. This is divided by the molar mass to find the number of moles present.

$$\text{number of moles} = 12.55\ \cancel{g} \times \frac{1\text{ mol}}{58.44\ \cancel{g}} = 0.214750\text{ mol}$$

There are 0.214750 mol of solute present. Now the total moles of solute and solvent must be found. The volume of the water in the solution is 180.4 mL, and the density of the water is given as 0.9941 g/mL.

$$\text{mass of solvent} = 0.9941\ \frac{g}{\cancel{mL}} \times 180.4\ \cancel{mL} = 179.336\text{ g}$$

Again, two extra digits are carried. Because the molality ratio uses the unit of kilograms, the mass of the water must be converted from grams.

$$\text{mass of solvent} = 179.3\ \cancel{g} \times \frac{1\text{ kg}}{1{,}000\ \cancel{g}} = 0.179336\text{ kg}$$

The values can then be substituted into the molality ratio to find $m$.

$$m = \text{moles of solute/kg of solvent} = \frac{0.214750\text{ mol NaCl}}{0.179336\text{ kg } H_2O} = 1.19747\text{ mol/kg}$$

Now all of the values can be substituted into the formula to find the freezing point depression.

$$\Delta T_f = iK_f m$$

$$\Delta T_f = 2 \times 1.858°\text{C}\,\frac{\cancel{\text{kg}}}{\cancel{\text{mol}}} \times 1.19747\,\frac{\cancel{\text{mol}}}{\cancel{\text{kg}}}$$

$$= 4.44980°\text{C}$$

Rounding to four significant digits, $\Delta T_f = 4.450°\text{C}$. The freezing point of pure water is 0°C. Apply the freezing point depression as calculated above to determine the new freezing point:

$$0°\text{C} - 4.450°\text{C} = -4.450°\text{C}$$

Adding 12.55 g of sodium chloride to 180.4 mL of water will decrease the freezing point to more than 4.4 degrees lower than normal.

## Osmotic Pressure

The transport of solutions through structures is part of many natural processes. Plant and animal cells, for example, rely on the movement of solutions through their bodies to transport nutrients, waste, and other biological chemicals. These cells are surrounded by semipermeable membranes that control substances that enter and exit cells. Only solute particles of a certain size can pass through the membrane. Typically, this means that solutes are filtered out by the membrane, while the pure solvent passes through it. This movement is called osmosis.



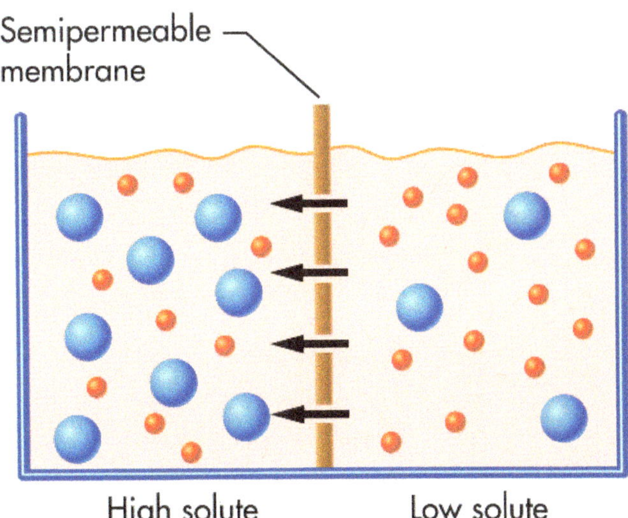


**OSMOSIS**

A semipermeable membrane will allow only particles of certain sizes to pass through. What moves through the membrane to balance the concentration of the solutes?

Think about a solution of sugar and water that is passed through a semipermeable membrane. The sugar molecules will only be found on one side of the membrane (the solution side), and pure water will be on the other side (the solvent side).

The water molecules move back and forth through the membrane, but more of them tend to move from the solvent side to the solution side. This creates **osmotic pressure** on the solution side of the membrane, which could eventually increase so that osmosis stops completely. Factors, such as temperature, also affect osmotic pressure, since the energy of the particles will affect how they move through the semipermeable membrane.

### Consider the Explain Question

**What factors influence the rate at which a solute dissolves?**

Go online to complete the scientific explanation.

dlc.com/ca9095s

### Check Your Understanding

**What are three factors that impact the rate at which a solution can be made?**

dlc.com/ca9096s

# STEM in Action

## Applying Solutions

Ice cream starts to melt when left out of the freezer for too long. When making homemade ice cream in an old-fashioned churn, a **solution** of water, ice, and salt is used to lower the temperature of the milk, sugar, and cream. This allows the ingredients to freeze and form a creamy mixture.

The **freezing** point of pure water at sea level is 0°C. The freezing point of a 20 percent salt solution is approximately −16°C. Adding salt to ice in a homemade ice cream maker lowers the temperature of the ice/salt slurry, and the ice cream inside the container freezes. Without the addition of salt, the ice-cream would not become cold enough for the ingredients to freeze.

Rock salt is used on roads to prevent or reduce ice in the winter, making it safer to drive. There are many different types of salts that are used. Primarily, sodium chloride and calcium chloride are the salts of choice. They work by lowering the freezing point of water, hence allowing the water that is on the roads to stay in the liquid form. It also causes any ice to melt as it decreases the freezing and melting point.



**Salt and Freezing a Solution**
*Effect of Salt on Freezing Point of a Solution*

| Trial | Solutions | Time (totally frozen) |
|---|---|---|
| 1 | 5 ml salt in 100 ml water | 1 hr 5 min |
| 2 | 10 ml salt in 100 ml water | 1 hr 30 min |
| 3 | 15 ml salt in 100 ml water | 1 hr 45 min |
| 4 | 100 ml water in no salt | 48 min |

**EFFECT OF SALT**

The freezing point of water changes when salt is added to water. What effect does the addition of salt have on the time it takes for a solution to freeze?

Antifreeze is a substance that is added to any water-based liquid to help prevent freezing while also preventing boiling. It is commonly added to radiators of automobiles for these purposes. It works based on the colligative properties, such as **freezing point depression** and **boiling point elevation**. In the wintertime, it prevents the water-based liquids from freezing and in the summertime, from boiling.

There are different substances that act as antifreezes. The most common ones that we add to our cars are ethylene glycol and propylene glycol. Ethylene glycol is a colorless, odorless, syrupy, sweet-tasting liquid that is toxic. Because of its sweet taste, children and animals drink more of it than other poisons. Bitter flavoring is commonly added to antifreeze solutions which contain ethylene glycol. Because of their high toxicity, ethylene glycol antifreezes are being phased out.

## STEM and Solutions

Certain fuels, such as jet fuel, diesel fuel, and gasoline, are a common source of groundwater contamination. When a spill occurs, some of the organic components leak into the groundwater. The most common contaminants are benzene, toluene, ethylbenzene, and xylene. Even though these compounds are slightly soluble in water, this process can be hazardous. Over time, these compounds will degrade into other compounds, one of which is hydrogen gas. The **concentration** of hydrogen gas in a groundwater sample can be easily calculated using Henry's law and will further help a chemist determine what treatments are appropriate and the fate of contaminants at that site.

Environmental chemists use a method known as the bubble stripping method to determine the dissolved hydrogen concentrations in well water. The method is based on the principle that gases will undergo a partitioning between a vapor phase and a liquid phase that are in contact with each other. At equilibrium, the partitioning of a gas between a vapor and liquid phase can be quantified by applying Henry's law.

**CONTAMINATED WATER**

Water pollution is a serious problem for humans and wildlife. How can chemists determine the extent of contaminated water?

# Calorimetry Calculation

**Use calorimetry to solve the problems below.**

1. A calorimeter holds 105 g water at 21.0°C. A sample of hot iron is added to the water, and the final temperature of the water and iron is 28.0°C. What is the change in enthalpy associated with the change in the water's temperature? The specific heat of water is 4.18 J/(g·°C).

2. A 27.6 g sample of aluminum is warmed from 12.0°C to 45.4°C. If the specific heat of aluminum is 0.899 J/(g·°C), what is the enthalpy change of the metal?

3. While fishing, a student leaves a 7.09 g lead sinker in the sun. The sinker's temperature increases from 17.0°C to 30.0°C. If the specific heat of lead is 0.159 J/(g·°C), what is the enthalpy change of the metal?

4. A student places 995 g water in a copper pot at 21.2°C. The pot has a mass of 635 g. The student then places the pot on a stove and begins heating it. If the final temperature of the water and copper is 95.0°C, what is the total enthalpy change of the system? The specific heat of water is 4.18 J/(g·°C), and the specific heat of copper is 0.385 J/(g·°C).

## Calculating Dilutions

**Use the equation Molarity$_1$ × Volume$_1$ = Molarity$_2$ × Volume$_2$ to solve the problems below.**

1. What would the final volume be of a solution that was initially 3 liters of a 5 molar solution and ended up a dilution of 1.5 molar?

2. A chemist has 0.5 liters of 12 molar sodium hydroxide solution. How much water does she need to add to dilute the solution down to a 5 molar concentration?

3. If a chemist diluted a solution to produce 1.5 liters from an initial volume of 400 mL of a 6 molar solution, what will the resulting concentration be?

**4.** 700 mL of solution has a 2 molar concentration. How much water would it take to make this solution 0.75 molar?

**5.** A chemist mixed 2.5 liters of a dilution of 0.33 molar. If she initially started with 800 mL, what was the concentration of the original solution?

# Calculating Freezing and Boiling Point

**Use equations for freezing point depression and boiling point elevation to solve the problems below.**

**1.** A sample of glucose ($C_6H_{12}O_6$) of mass 8.44 grams is dissolved in 2.11 kg water. What is the freezing point of this solution? The freezing point depression constant, $K_f$, for water is 1.86 °C/mol.

**2.** What is the boiling point of a solution made by dissolving 10.0 g sucrose ($C_{12}H_{22}O_{11}$) in 500.0 g carbon tetrachloride ($CCl_4$)? The boiling point of carbon tetrachloride is 76.5°C and the boiling point elevation constant, $K_b$, for carbon tetrachloride is 5.03°C/mol.

**3.** Calculate and compare the freezing points of the following aqueous solutions: 0.0500 m $CaCl_2$ and 0.0500 m KCl. The freezing point depression constant, $K_f$, for water is 1.86 °C/mol.

**4.** Calculate the boiling point of a solution made by dissolving 0.01700 mol $Na_2SO_4$ in 100.0 g water. The boiling point elevation constant, $K_b$, for water is 0.5100°C/mol.

## Calculating Concentrations

**Use the definitions of mass percent, volume percent, and parts per million to solve the problems below.**

1. A solution was prepared by mixing 50.0 g $MgSO_4$ and 1000.0 mL water at 25.0°C. The density of the water at this temperature is 0.997 g/mL. What is the percent mass of $MgSO_4$ in this solution?

2. A student mixes 60.0 mL ethanol ($C_2H_6O$) with 240.0 mL water. Determine the concentration of this solution in volume percent.

3. What is the concentration of a solution (in parts per million) if 15 mg $KNO_3$ are dissolved in water to make a total solution mass of 5.0 kg?

CONCEPT 5.3

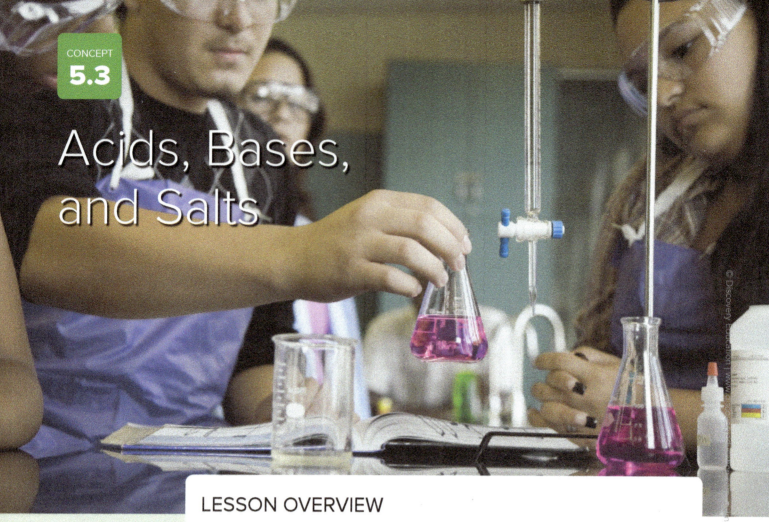

# Acids, Bases, and Salts

## LESSON OVERVIEW

dlc.com/ca9097s

### Lesson Questions

- What are the similarities and differences among the Arrhenius, Brønsted-Lowry, and Lewis acid-base theories?
- What determines if a precipitate will form when solutions are combined, and how are net ionic equations written?
- What are the similarities and differences between acid and base strength and concentration?
- How are hydronium ions formed?
- What products are formed in acid-base reactions?
- What occurs during the neutralization process?
- What are pH and pOH?

### Key Vocabulary

Which terms do you already know?

- ☐ acid dissociation constant (Ka)
- ☐ amphoterism
- ☐ aqueous solution
- ☐ Arrhenius acids and bases
- ☐ Brønsted-Lowry acids and bases
- ☐ buffer
- ☐ buffer solution
- ☐ concentration
- ☐ conjugate acid
- ☐ conjugate base
- ☐ dissociation
- ☐ electron pair acceptor
- ☐ electron pair donor
- ☐ equilibrium
- ☐ hydrogen ion

### Lesson Questions continued

- What is the correlation between logarithmic pH scale changes and corresponding changes in concentration?
- How is $K_w$ (water dissociation constant) related to pH, pOH, concentration of hydrogen ions, and concentration of hydroxide ions?

## Lesson Objectives

By the end of the lesson, students should be able to:

- Compare and contrast the Arrhenius, Brønsted-Lowry, and Lewis acid-base theories.
- Predict whether a precipitate will form when solutions are combined and write a net ionic equation for a precipitation reaction.
- Compare and contrast acid or base strength and concentration.
- Explain how hydronium ions are formed.
- Identify products formed in acid-base reactions.
- Describe the neutralization process.
- Explain pH and pOH.
- Correlate between logarithmic pH scale changes and corresponding changes in concentration.
- Relate Kw (water dissociation constant) to pH, pOH, concentration of hydrogen ions, and concentration of hydroxide ions.

### Key Vocabulary continued

- ☐ hydronium ion
- ☐ hydroxide ion
- ☐ indicator
- ☐ ionization
- ☐ ion product constant of water
- ☐ $K_w$
- ☐ Lewis acids and bases
- ☐ logarithmic
- ☐ molarity
- ☐ neutralization
- ☐ pH
- ☐ pOH
- ☐ protic acid
- ☐ salt
- ☐ strong acid
- ☐ strong base
- ☐ titration
- ☐ weak acid
- ☐ weak base

# Identifying Acids, Bases, and Salts

dlc.com/ca9098s

**If you look in your fridge or cupboard, you will see dozens or even hundreds of acids and bases. Oranges, vinegar, nuts, and dairy tend to be acidic. Leafy green vegetables, soy products, and spices tend to be basic. Acids and bases may be familiar, but what exactly are they?**

**EATING PIZZA**

Pizza tastes good, but unfortunately, eating a highly acidic food such as pizza sometimes leads to heartburn. How would you use your knowledge of chemistry to help relieve heartburn?

## EXPLAIN QUESTION

**Explain why acetic acid, $HC_2H_3O_2$, is defined as an acid and describe both conceptually and using a chemical equation what happens when $HC_2H_3O_2$, a weak acid, reacts with sodium hydroxide, NaOH, a strong base.**

# What Are Similarities and Differences among the Arrhenius, Brønsted-Lowry, and Lewis Acid-Base Theories?

## Acid and Base Theories

There are many theories describing the properties of acids and bases. Each theory has a particular application. The original theory of acids and bases was proposed by Svante August Arrhenius around 1884. His definition was specific to hydrogen and **hydroxide ion** concentrations in solution.

An Arrhenius acid is any substance that increases the number of hydrogen ions ($H^+$) in an **aqueous solution**. In the example below, hydrochloric acid is an Arrhenius acid because hydrogen ions are formed when it is added to water.

$$HCl\ (aq) \rightarrow H^+\ (aq) + Cl^-\ (aq)$$

| Anion | Corresponding Acid |
|---|---|
| fluoride ($F^-$) | hydrofluoric acid (HF) |
| chloride ($Cl^-$) | hydrochloric acid (HCl) |
| bromide ($Br^-$) | hydrobromic acid (HBr) |
| iodide ($I^-$) | hydroiodic acid (HI) |
| cyanide ($CN^-$) | hydrocyanic acid (HCN) |
| sulfide ($s^{2-}$) | hydrosulfuric acid ($H_2S$) |

**COMMON ACIDS**

Acids are formed from common anions. Why are acids formed from anions and not cations?

An Arrhenius base is any substance that increases the number of hydroxide ions ($OH^-$) in an aqueous solution. For example, sodium hydroxide is an Arrhenius base in an aqueous solution. Sodium hydroxide produces hydroxide ions when added to water.

$$NaOH\ (aq) \rightarrow Na^+\ (aq) + OH^-\ (aq)$$

| Cation | Corresponding Base |
|---|---|
| lithium ($Li^+$) | $LiOH$ |
| potassium ($K^+$) | $KOH$ |
| sodium ($Na^+$) | $NaOH$ |
| rubidium ($Rb^+$) | $RbOH$ |
| $NH_2^+$ | $NH_3$ |
| calcium ($Ca^{2+}$) | $Ca(OH)_2$ |

**COMMON BASES**

Bases are formed from common cations. Why are bases formed from cations and not anions?



Not all acids and bases produce hydrogen or hydroxide ions in solutions. The Brønsted-Lowry definition of acids and bases was created to solve this issue. It specifically describes the gain or loss of hydrogen ions (protons). Johannes Nicolaus Brønsted and Thomas Martin Lowry proposed their version of acid-base theory around 1923.

A Brønsted-Lowry acid is any substance that can donate a **hydrogen ion** (proton). For example, hydrochloric acid was described above as an Arrhenius acid. Hydrochloric acid is also a Brønsted-Lowry acid in an aqueous solution because it donates a hydrogen ion.

$$HCl\ (aq) \rightarrow H^+\ (aq) + Cl^-\ (aq)$$

A Brønsted-Lowry base is any substance that can accept a hydrogen ion (proton). For example, ammonia is a Brønsted-Lowry base in an aqueous solution. Ammonia can accept a hydrogen ion.

$$NH_3\ (aq) + H^+\ (aq) \rightarrow NH_4^+\ (aq)$$

Gilbert Lewis proposed a different mechanism for acid-base theory, also around 1923. The Lewis theory is based on the transfer of electrons.

A Lewis acid is an **electron pair acceptor**. A Lewis base is an **electron pair donor**. Water is both a Lewis acid and a Lewis base. For example, boron trifluoride can accept electron pairs, whereas fluorine can donate electron pairs.

$$BF_3\ (aq) + F^-\ (aq) \rightarrow BF_4^-\ (aq)$$



## What Determines If a Precipitate Will Form When Solutions Are Combined, and How Are Net Ionic Equations Written?

### Precipitation Reactions

Acid-base reactions are not the only reactions that occur in aqueous solutions. A precipitation reaction is a reaction in an **aqueous solution** that results in an insoluble product. In other words, the product of the reaction cannot dissolve in water. Instead, it precipitates out of the solution in solid form.

A precipitation reaction can sometimes occur when two salts are mixed together in an aqueous solution. For example, the two salts, barium chloride and sodium sulfate, undergo a precipitation reaction when mixed together in water:

$$BaCl_2\ (aq) + Na_2SO_4\ (aq) \rightarrow BaSO_4\ (s) + 2NaCl\ (aq)$$

The term *(aq)* shows that most of the compounds are dissolved in water. The term *(s)* shows that $BaSO_4$ is not dissolved but is in a solid state. As the reaction proceeds, barium sulfate precipitates out of the solution as a white powder.

Writing the equation in terms of the ions involved gives a clearer picture of what is happening in this precipitation reaction. This is called the ionic equation:

$$Ba^{2+}(aq) + 2Cl^{-}(aq) + 2Na^{+}(aq) + SO_4^{2-}(aq) \rightarrow BaSO_4(s) + 2Na^{+}(aq) + 2Cl^{-}(aq)$$

Because the sodium and chloride ions are on both sides of the equation, they can be removed from the equation. Ions that are present on both sides of an equation are called spectator ions. Once the spectator ions have been removed from the equation, what remains is called a net ionic equation:

$$Ba^{2+}(aq) + SO_4^{2-}(aq) \rightarrow BaSO_4(s)$$

How can you tell if a reaction involving ions is a precipitation reaction or not? Solubility rules are guidelines that help to identify insoluble compounds. A list of solubility rules is given on this page.

| Soluble compounds have these ions: | Insoluble compounds have these ions: |
|---|---|
| • $Cl^-$, $Br^-$, $I^-$, $NO_3^-$, $HCO_3^-$, $ClO_3^-$, $SO_4^{2-}$<br>• $Li^+$, $Na^+$, $K^+$, $Rb^+$, $Cs^+$, $NH_4^+$ | • $CO_3^{2-}$, $PO_4^{3-}$, $CrO_4^{2-}$, $S^{2-}$, $OH^-$ |
| **Except for these combinations of ions:** | **Except for these combinations of ions:** |
| Compounds containing $Ag^+$, $Hg_2^{2+}$, or $Pb^{2+}$ **AND** $Cl^-$, $Br^-$, or $I^-$ are **insoluble**<br><br>($AgCl$ is insoluble)<br><br>Compounds containing $Ag^+$, $Ca^{2+}$, $Sr^{2+}$, $Ba^{2+}$, or $Pb^{2+}$ **AND** $SO_4^{2-}$ are **insoluble**<br><br>($CaSO_4$ is insoluble) | Compounds containing $Li^+$, $Na^+$, $K^+$, $Rb^+$, $NH_4^+$ **AND** $CO_3^{2-}$, $PO_4^{3-}$, $CrO_4^{2-}$, $S^{2-}$ are **soluble**<br><br>($Na_2CO_3$ is soluble)<br><br>Compounds containing $Na^+$, $K^+$, $Rb^+$, $Ba^{2+}$ **AND** $OH^-$ are **soluble**<br><br>($NaOH$ is soluble) |

**SOLUBILITY RULES**

Solubility rules show at a glance whether or not a specific compound is soluble. How can the rules be applied to determine the outcome of a reaction?

## Sample Problem: Using the Solubility Rules

When a solution of $Pb(NO_3)_2$ is added to a solution of NaI, a reaction occurs to produce $PbI_2$ and $NaNO_3$.

(a) Write the chemical equation and identify any insoluble products. Is this a precipitation reaction or not?

(b) Write the net ionic equation.

**Solution:**

(a) Write the chemical equation and balance it.

$$Pb(NO_3)_2\ (aq) + 2NaI\ (aq) \rightarrow PbI_2\ (?) + 2NaNO_2\ (?)$$

The solubility rules state that compounds with Cl are usually soluble except when combined with $Pb^{2+}$ ions, so $PbI_2$ is insoluble.

The solubility rules state that products containing $NO_3^-$ ions and $Na^+$ ions are soluble, so $NaNO_3$ is soluble.

This is a precipitation reaction because the two reactants are dissolved in water, but one of the products is insoluble.

Chemical equation with states:

$$Pb(NO_3)_2\ (aq) + 2NaI\ (aq) \rightarrow PbI_2\ (s) + 2NaNO_3\ (aq)$$

(b) ionic equation:

$$Pb^{2+}\ (aq) + 2NO_3^-\ (aq) + 2Na^+\ (aq) + 2I^-\ (aq)$$
$$\rightarrow PbI_2\ (s) + 2Na^+ + 2NO_3^-\ (aq)$$

Removing the spectator ions that are present on both sides leaves the net ionic equation:

$$Pb^{2+}\ (aq) + 2I^-\ (aq) \rightarrow PbI_2\ (s)$$

# What Are the Similarities and Differences between Acid and Base Strength and Concentration?

## Acid or Base Strength

According to the Arrhenius definition of acids and bases, the strength of an acid or base is defined as the degree of **ionization** or **dissociation** of the acid or base in **aqueous solution**. Ionization is the process by which an atom becomes an ion through the addition or loss of an electron. Dissociation is the process by which a substance breaks up into smaller particles. All acids dissociate, but only some will ionize.

The strengths of acids and bases can be measured using the **equilibrium** constant expressions $K_a$ (for acid strength) and $K_b$ (for base strength). Strong acids and bases dissociate almost completely in water. This means that almost all of the particles in the acid or base break up into smaller particles. Weak acids and bases dissociate slightly in water. Most of the particles that make up a **weak acid** or **weak base** do not split into smaller particles.

During the process of dissociation, if the acid dissociates completely, it is considered a **strong acid**. If it does not completely dissociate, then the strength can be determined by finding the acid dissociation constant ($K_a$). To begin, it is helpful to study the acid-base reaction in the form of an equation:

$$HA = H^+ + A^-$$

In this equation, HA represents the acid, $H^+$ represents protons that are produced, and $A^-$ represents the base.

Using the concentrations of protons $[H^+]$, the **conjugate base** $[A^-]$, and the acid [HA], the acid dissociation constant can be calculated:

$$K_a = \frac{[H^+][A^-]}{[HA]}$$

The strength of an acid is measured using the acid dissociation constant ($K_a$). The higher the $K_a$ value, the stronger the acid.

Similarly, if the base dissociates completely, it is considered a **strong base**. If it does not completely dissociate, then the strength can be determined by finding the base dissociation constant ($K_b$).

In this equation, B represents the base, $OH^-$ represents the hydroxide ions that are produced, and BH represents the **conjugate acid**.

$$B = OH^- + BH^+$$

Using the concentrations of the base [B], hydroxide ions [$OH^-$], and conjugate acid [B], the dissociation constant for the Arrhenius base can be calculated:

$$K_b = \frac{[HB^+][OH^-]}{[B]}$$

The strength of a base is measured using the base dissociation constant ($K_b$). The higher the $K_b$ value, the stronger the base.

According to the Brønsted-Lowry definition of acids and bases, an acid that loses a proton during dissociation is called a conjugate base. A base that gains a proton during dissociation is called a conjugate acid. In the Brønsted-Lowry model, strength of the acid is determined by how readily the acid gives up a proton. Conversely, the strength of a base is determined by how readily it accepts the proton. In the Brønsted-Lowry model, an obvious inverse relationship can be seen between the acid and base in the acid-base pair. The stronger an acid is, the weaker its conjugate base is. While the stronger a base is, the weaker its conjugate acid.

It should be noted that some substances may act as either an acid or a base; it depends upon what other substances are around them. These substances are said to be amphoteric. **Amphoterism** occurs when a substance can act as either an acid or a base. Whether an amphoteric substance acts as an acid or base depends on the solution it is in. Water is the most common amphoteric molecule.

In the ionization of hydrochloric acid, water acts as a base.

$$\underset{\text{Acid}}{HCl\ (aq)} + \underset{\text{Base}}{H_2O\ (l)} \rightarrow \underset{\text{Acid}}{H_3O^+\ (aq)} + \underset{\text{Base}}{HSO_4^-\ (aq)}$$

However, water acts as an acid in the following reaction:

$$\underset{\text{Base}}{NH_3\ (aq)} + \underset{\text{Acid}}{H_2O\ (l)} \rightarrow \underset{\text{Acid}}{NH_4^+\ (aq)} + \underset{\text{Base}}{OH^-\ (aq)}$$

If water reacts with a compound that is a stronger acid than itself, then water acts as a base. If water reacts with a compound that is a stronger base than itself, then water acts as an acid.

# How Are Hydronium Ions Formed?

## The Formation of Hydronium Ions

A protic chemical is one that acts as a proton donor. When a **protic acid** is added to water and releases a **hydrogen ion**, the hydrogen ion immediately bonds with a nearby water atom. This forms $H_3O^+$, an ion known as hydronium.

Hydronium is a water molecule with one extra hydrogen ion attached. Hydronium is sometimes referred to as a *proton* because it can act as a hydrogen or proton donor in solution. Hydronium carries a positive charge and is usually an intermediate product of reactions.

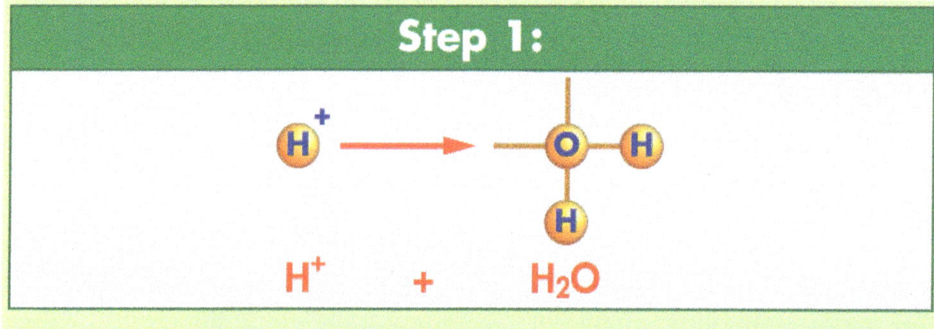


**HYDRONIUM**

When a water molecule accepts a proton, hydronium is formed. Where does this extra proton come from?

Water, naturally and continuously, dissociates into hydroxide and hydronium ions. The ions then reassociate and the reactions are repeated continually.

$$2H_2O(aq) \rightleftarrows OH^-(aq) + H_2O^+(aq)$$

# What Products Are Formed in Acid-Base Reactions?

## Products of Acid-Base Reactions

In a **strong acid-strong base** reaction, a **salt** and water will be formed as the products. Because both reactants are strong, they will dissociate completely. For example, hydrochloric acid and sodium hydroxide will form sodium chloride (a salt):

$$HCl(aq) + NaOH(aq) \rightarrow NaCl(aq) + H_2O(aq)$$

In a **weak acid**–strong base reaction, the resulting solution will be basic, and there will likely be a soluble salt. The strong base will dissociate completely. The weak acid will only dissociate a little. For example, acetic acid and potassium hydroxide will produce potassium acetate (a salt):

$$HC_2H_3O_2(aq) + KOH(aq) \rightarrow H_2O(aq) + KC_2H_3O(aq)$$

In a strong acid–**weak base** reaction, the resulting solution will be acidic, and there will likely be a soluble salt. The strong acid will dissociate completely. The weak base will only dissociate a little. For example, hydrochloric acid and ammonia form ammonium chloride (a salt):

$$HCl(aq) + NH_3(aq) \rightleftarrows NH_4Cl(aq)$$

In a weak acid– weak base reaction, the resulting solution will be neutral. The reactants are weak, so they will only dissociate a little. For example, ammonia is a weak base, and water is a weak acid.

$$NH_3(aq) + H_2O(aq) \rightleftarrows NH_4^+(aq) + OH^-(aq)$$

**NEUTRALIZATION**
This student is neutralizing strong acid with a strong base. What pH would you expect the final solution to have?



## Acid-Base Reactions: Sample Problem

Ammonium is a weak base that dissolves in water. What are the products formed in this reaction?

$$NH_4^+ + OH^-$$

**Solution:**

Ammonium is a cation and has an extra hydrogen atom. Water will take that hydrogen and form hydronium. This reaction will also form ammonia.

$$NH_4^+ + OH^- \rightleftarrows NH_3 + H_2O$$

# What Occurs during the Neutralization Process?

## Buffers and Neutralization

One way that chemists can maintain a constant **concentration** of the **pH** of a solution is through the use of **buffer** solutions. Buffer solutions are solutions that can resist a change in **pH** when small amounts of acid or base are added to it. Acidic buffer solutions have a pH of less than seven, while basic buffer solutions have a pH of over seven.

Acidic buffer solutions are made from a **weak acid** and one of its salts. This will allow the **buffer solution** to resist against a higher **concentration** of hydroxide ions ($OH^-$) being produced in the solution.

If a small amount of base is added to an acetic acid–sodium acetate solution, the $OH^-$ ions of the base react with and remove hydronium ions to form nonionized water molecules. The acetic acid molecules then ionize and restore the **equilibrium** concentration of hydronium ions.

$$CH_3COOH(aq) + H_2O(l) \rightleftarrows CH_3COO^-(aq) + H_3O^+(aq)$$

If a small amount of acid is added to an acetic acid–sodium acetate solution, the acetate ion will react with most of the added hydronium ions to form the non-ionized acetic acid molecules.

$$CH_3COO^-(aq) + H_3O^+(aq) \rightleftarrows CH_3COOH(aq) + H_2O(aq)$$

Conversely, basic or alkali buffer solutions are made from a weak acid and one if its salts. This enables the buffer to resist the higher concentration of hydrogen ions ($H^+$) being produced in the solution.

**Neutralization** is a chemical reaction. During neutralization, an acid and a base react to form water and a **salt**. The resulting solution is neutral, having the properties of neither an acid nor a base.

The concentration can be expressed in normality (N). Normality has the unit of moles per liter. Remember that **molarity** describes the moles of solute per liter of solution. Normality is very similar, but instead it describes the moles of each ion in the solution. For example, consider a solution that has a concentration (molarity) of 1 mol/L, or 1 M NaOH. Each NaOH dissociates to form one $Na^+$ ion and one $OH^-$ ion. The normality of $Na^+$ ions in the solution would be 1 mol/L, or 1 N. The normality of the $OH^-$ ions would also be 1 N.

In contrast, a solution with a concentration of 1 M $H_2SO_4$ would produce $H^+$ ions and $SO_4^{2-}$ ions. The normality of the $H^+$ would be 2 mol/L, or 2 N, and the normality of the $SO_4^{2-}$ would be 1 mol/L, or 1 N. Notice that the normality is always a whole number because ionic compounds dissociate into a whole number of ions.

An acid and a base neutralize when they are combined because all of the ions that are present reorganize themselves to form neutral components. The number of hydrogen ions must be made to equal the number of hydroxide ions in the solution so that they form water molecules. The remaining ions bond to form salts.

Neutralization calculations are done using this formula:

$$N_aV_a = N_bV_b$$

Where

$N_a$ is the normality of the acid

$V_a$ is the volume of the acid

$N_b$ is the normality of the base

$V_b$ is the volume of the base

Given the volume and concentration of an acid (or a base), you can calculate how much base (or acid) is needed to neutralize the solution.

## Neutralization: Sample Problem

What volume of 2.0 M $H_2SO_4$ (**strong acid**) is required to neutralize 250 mL of 0.5 M NaOH (**strong base**)?

**Solution:**

The equation to determine the amount needed for neutralization can be rearranged to solve for the value needed:

$$N_aV_a = N_bV_b$$

$$V_a = \frac{N_bV_b}{N_a}$$

In the acid, there are two hydrogen ions produced for every molecule of $H_2SO_4$. There are two moles of acid in one liter of the solution, which means there are four moles of hydrogen ions in one liter of solution.

$$N_a = 2.0 \text{ mol/L} \times 2 = 4 \text{ mol/L } H^+$$

In the base, there is one **hydroxide ion** produced for every molecule of NaHO. There are 0.5 moles of base in one liter of solution, which means there are 0.5 moles of hydroxide ions in the solution.

$$N_b = 0.50 \text{ mol/L} \times 1 = 0.50 \text{ mol/L } OH^-$$

The volume of the base is given as 250 mL but must be converted to liters to be consistent with the normality values.

$$V_b = 250 \cancel{\text{mL}} \times \frac{1 \text{ L}}{1{,}000 \cancel{\text{mL}}}$$

$$V_b = 0.250 \text{ L NaOH}$$

Next, the values can be substituted into the equation.

$$V_a = \frac{N_b V_b}{N_a}$$

$$V_a = \frac{0.50 \text{ N} \times 0.250 \text{ L}}{4.0 \text{ N}}$$

$$V_a = 0.03125 \text{ L}$$

Finally, the volume should be converted back to milliliters.

$$0.03125 \text{ L} \times \frac{1{,}000 \text{ mL}}{1 \text{ L}} = 31 \text{ mL}$$

Therefore, 31 mL of 2.0 M $H_2SO_4$ is required to neutralize 250 mL of 0.5 M NaOH.

# What Are pH and pOH?

## pH and pOH

It is often helpful to know about the strength of an acid or base before you use a chemical or combine it with another substance. Also, knowing whether an unknown substance is an acid or a base can be helpful in identifying it. The most common method for describing the strength of an acid or base is known as the **pH** scale.

The term **pH** is a mathematical way of comparing the strength of acids and bases. The difference of one on the pH scale indicates a tenfold difference in the strength of the compounds. An acid with the pH of one is 10 times as strong as an acid with the pH eliminate wrap of two.

To calculate pH, the strength or **concentration**, often known as **molarity** (M) of the acid must be known. By taking the negative logarithm of the molarity of the acid, the pH can be found.

$$pH = -\log[\text{molarity}]$$

For example, the pH of an acid solution with the **concentration** of 0.01 M would be two.

$$pH = -\log[\text{molarity}]$$
$$pH = -\log[0.01]$$
$$pH = 2$$

The pH scale ranges in values from one to 14. One is the most acidic, seven is neutral, and 14 is the most basic. These values represent the quantity of hydrogen ions in a solution.

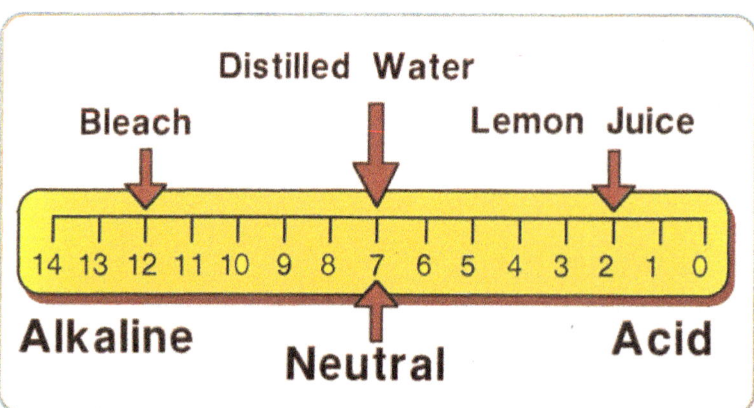


**pH SCALE**
The pH scale covers both acids and bases. How is pH related to the molarity of a solution?

Another method for describing the strength of acids and bases is the pOH scale, which describes the quantity of hydroxide ions in the solutions. It runs in the opposite direction as the pH scale. On the pOH scale, one is the most basic, and 14 is the most acidic.

Because the pH and pOH of a solution always total 14, one can be calculated from the other.

$$pH + pOH = 14$$

Indicators are useful in determining the pH of a solution. They will change color when exposed to an acid or a base of a certain strength. Indicators are often used during acid-base titrations. Some common indicators are phenolphthalein, thymol blue, and methyl orange.

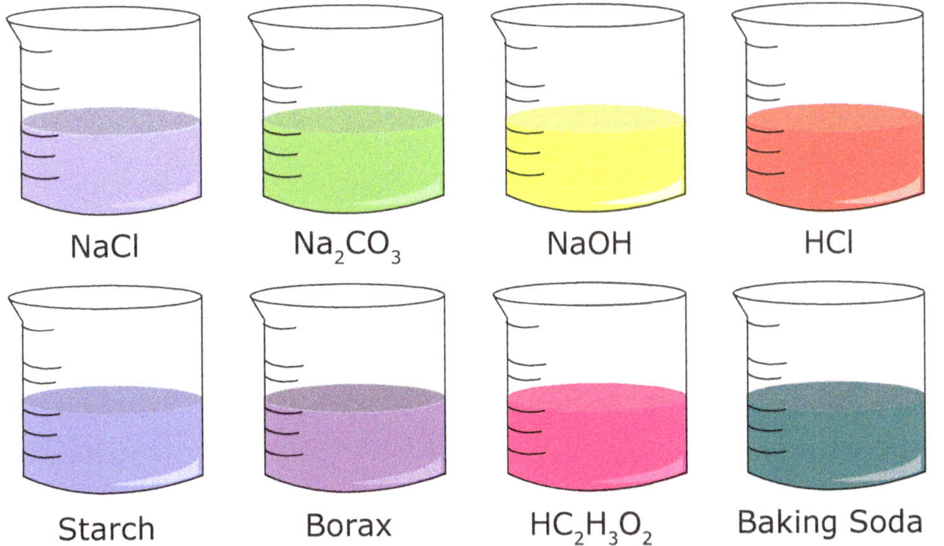




**RED CABBAGE INDICATOR**

Red cabbage juice contains a natural pH indicator that changes color with the acidity of the solution. Which colors indicate acids? Which colors indicate bases?

## Titration

Acids and bases react with each other to form a salt and water by a process called neutralization. Chemists use acid-base titrations to study neutralization reactions.

Titration is a method to calculate the concentration of a sample by analyzing neutralization during the reaction. A solution of known concentration (the titrant, or standard solution) is added in incremental measured amounts to a second solution of known volume but unknown concentration (the titer). The titrant is added until the reactants are neutralized. This is called the equivalence point. Indicators allow an observer to detect changes in the solution's pH during titration by changing color. Eventually a color change occurs with the addition of a single drop, indicating the end point of the reaction.

The volume of the unknown solution necessary to complete the reaction allows its concentration to be calculated. The end point in a titration indicates when the equivalence point has been reached and can be detected by some form of **indicator** which changes color or a pH meter.

An appropriate acid-base indicator is used to show when neutralization has occurred. In the laboratory, bromthymol blue is often used for **strong acid**–**strong base** neutralizations with endpoints around pH 7.

Indicators that change color at pH lower than seven are useful in determining the equivalence point in strong acid–**weak base** titrations. The equivalence point of a strong acid–weak base titration is acidic because the salt formed is itself a **weak acid**. Thus, the salt solution has a pH lower than seven. Methyl orange can be used as an indicator for these reactions.

Indicators that change color at pH higher than seven are useful in determining the equivalence point in weak acid–strong base titrations. The equivalence point of a weak acid–strong base titration is basic because the salt formed is itself a weak base. Thus, the salt solution has a pH higher than seven. Solutions that contain phenolphthalein turn from colorless to deep pink as the solution changes from acidic to basic at a pH higher than seven.

When a strong acid such as 1.00 M HCl is titrated with a strong base such as 1.00 M NaOH, the equivalence point occurs about pH 7. If a pH meter measures the pH of the solution during the titration, a titration curve can be produced. As shown in the figure Strong Acid–Strong Base, the pH of the solution changes during the titration. The pH of the initial base solution is high. As the acid is added, the pH increases because some of the base is neutralized. As the titration approaches neutralization, a pH 7, the pH decreases dramatically as the OH ions are used up. Once past the point of neutralization, additional acid produces a further decrease in pH. The equivalence point is the midpoint of the vertical portion of the pH titration curve and in this instance is pH 7.

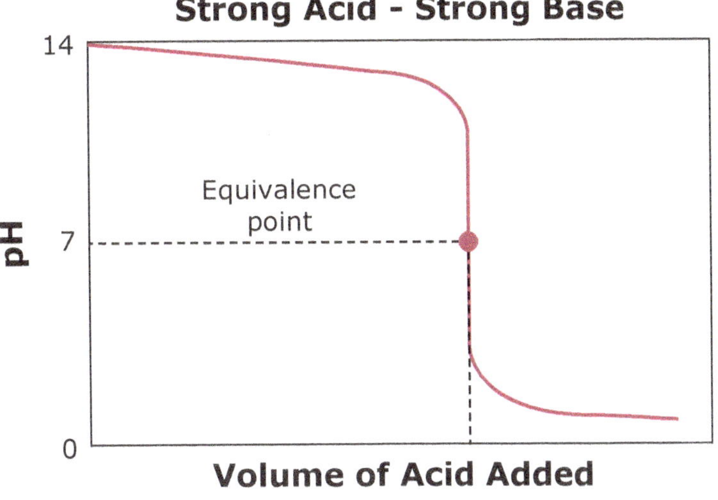


**STRONG ACID–STRONG BASE**

This graph shows the equivalence point. How is titration used to find the equivalence point?

When a strong acid is titrated with a weak base, the resulting solution will be acidic. The strong acid dissociates completely while the weak base will only dissociate a little.

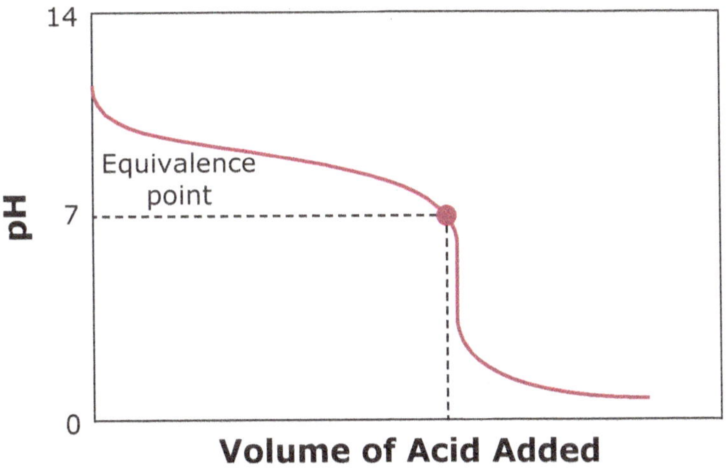


**STRONG ACID–WEAK BASE**

Find the equivalence line on this graph. Does this scenario result in a basic or acidic solution?

**Buffer** solutions contain a weak acid and its **conjugate base** (or weak base and its **conjugate acid**). A buffer can accept the addition of a strong acid or base without a significant change in pH. Buffers are common in many products where pH must remain stable over time. Blood also contains buffers that resist change in pH to keep the body working properly.

### pH and pOH: Sample Problem

Using titration, a 75 mL solution of 0.15 M NaOH is found to neutralize a 25 mL sample of HCl. What is the concentration of the HCl?

**Solution:**

Every mole of NaOH will produce 1 M of $OH^-$. Because there is 0.15 M NaOH, there will be $[OH^-] = 0.15$ M:

$$\text{molarity} = \text{moles/volume}$$
$$\text{moles} = \text{molarity} \times \text{volume}$$
$$\text{moles } [OH^-] = 0.15 \text{ M} \times 75 \text{ mL} \times \frac{1 \text{ L}}{1{,}000 \text{ mL}} = 0.01125 \text{ mol}$$

When a base neutralizes an acid, the number of moles of $OH^-$ is equal to the number of moles of $H^+$. Therefore, the number of moles of $H^+ = 0.01125$ M.

Every mole of HCl will produce one mole of $H^+$, so the number of moles of HCl equals the number of moles of $H^+$.

$$\text{Molarity of HCl} = \frac{0.01125 \text{ mol}}{25 \text{ mL}} \times \frac{1{,}000 \text{ mL}}{1 \text{ L}} = 0.45 \text{ mol/L}$$

For a 25 mL sample of HCl to neutralize a 75 mL solution of 0.15 M NaOH, a molarity of 0.45 M is required.

## What Is the Correlation between Logarithmic pH Scale Changes and Corresponding Changes in Concentration?

### Logarithmic pH Scale and Concentration

The **pH** scale is **logarithmic**. A logarithmic scale is based on the power of 10.

Therefore, a solution with a **pH** of one is 10 times more concentrated than a solution with a pH of two.

If you know the pH of a strong acid or base solution, you can calculate the concentration of the acid or the base:

$$M = \frac{1}{10^{pH}}$$

where M is the molarity.

There is a correlation between logarithmic pH scale changes and corresponding changes in concentration. Concentration refers to the amount of hydronium ions in a solution. A low pH indicates a solution has a high concentration of hydronium ions. A high pH indicates that a solution has a low concentration of hydronium ions.

The pH can be calculated when the hydronium ion concentration is known.

$$pH = -\log[H^+]$$

This is also true for pOH, where $[OH^-]$ is the hydroxide concentration.

$$pOH = -\log[OH^-]$$

For example, a glass of orange juice has a hydronium ion concentration of $[H^+] = 0.0003$.

$$pH = -\log[H^+] = -\log[0.0003] = 3.522$$

This pH is below seven; therefore, the orange juice is acidic.

| pH | Example |
|---|---|
| 1 | hydrochloric acid in stomach, battery acid (strong acid) |
| 2 | lemon juice, vinegar |
| 3 | orange juice, sodas |
| 4 | tomato juice |
| 5 | acid rain |
| 6 | milk |
| 7 | pure water (neutral) |
| 8 | eggs, baking soda, seawater |
| 9 | toothpaste |
| 10 | milk of magnesium |
| 11 | cleaning ammonia |
| 12 | soapy water |
| 13 | oven cleaner |
| 14 | drain cleaner (strong base) |

**EVERYDAY ITEMS ON THE pH SCALE**

Substances we encounter every day include strong acids, strong bases, and others that lie between these on the pH scale. How does the concentration of hydronium ions differ between each incremental change in pH?

# How Is $K_w$ (Water Dissociation Constant) Related to pH, pOH, Concentration of Hydrogen Ions, and Concentration of Hydroxide Ions?

## The Water Dissociation Constant

The water **dissociation** constant ($K_w$) describes the degree to which water dissociates into hydrogen and hydroxide ions at various temperatures. $K_w$ is related to **pH**, **pOH**, **concentration** of hydrogen ions and **concentration** of hydroxide ions.

At standard conditions, $K_w$ is always calculated as follows:

$$K_w = [H^+][OH^-] = 1.0 \times 10^{-14}$$

The acid and base dissociation constants can be used to determine $K_w$ as well:

$$K_w = K_a \times K_b$$

Salts also dissociate into ions when placed in water. The $K_a$ or $K_b$ value of the **salt** when dissolved in water determines if the salt is acidic or basic.

**BATH SALTS**

When bath salts dissolve, they add minerals and a pleasing scent to bath water. How is the water affected if bath salts have a high $K_b$ value?

## The Water Dissociation Constant: Sample Problem

Complete the following table:

| pH | $[H^+]$ | $[OH^-]$ | pOH | $K_w$ |
|---|---|---|---|---|
| 2.5 | | | | |

**Solution:**

$$K_w = [H^+][OH^-] = 1.0 \times 10^{-14}$$

$$pOH = -\log[OH^-]$$

$$pH = -\log[H^+]$$

$$pH + pOH = 14$$

$$pOH = 14 - pH = 14 - 2.5 = 11.5$$

From $pH = -\log[H^+]$, we know that $\frac{1}{10^{pH}}$ = M of hydronium ions.

From $pOH = -\log[OH^-]$, we know that $\frac{1}{10^{pH}}$ = M of hydronium ions.

$$\frac{1}{10^{11.5}} = 3.16227766 \times 10^{-12}\,M\ OH^-$$

$$K_w = [H^+][OH^-] = 3.16 \times 10^{-3} \times 3.16 \times 10^{-12} = 1.00 \times 10^{-14}$$

$K_w$ checks out, so we know our calculations are correct.

| pH | $[H^+]$ | $[OH^-]$ | pOH | $K_w$ |
|---|---|---|---|---|
| 2.5 | $3.16 \times 10^{-3}$ | $3.16 \times 10^{-12}$ | 11.5 | $1.00 \times 10^{-14}$ |

### Consider the Explain Question

dlc.com/ca9099s

**Explain why acetic acid, $HC_2H_3O_2$, is defined as an acid and describe both conceptually and using a chemical equation what happens when $HC_2H_3O_2$, a weak acid, reacts with sodium hydroxide, NaOH, a strong base.**

Go online to complete the scientific explanation.

### Check Your Understanding

dlc.com/ca9100s

**What are the quantitative values that describe acids and bases?**

# in Action

## Acids, Bases, and Salts

One of the most common uses of acids, bases, and salts is in food preparation and packaging for storage. Food that is canned must be stored at the proper **pH** to prevent spoilage. But storage at the wrong **pH** would result in food that is inedible. Canned food that is improperly prepared can be contaminated by germs, become too mushy, or lose its flavor.

By balancing the acidity of canned food, its flavor, texture, and nutrients can be preserved during storage. Some fruits and vegetables will even remain at freshness levels similar to those before canning.

Buffers are what allow this balance of acidity to happen. If you read the ingredient label of packaged foods in your refrigerator or cupboard, you may see some familiar names repeated. Some of the more common buffers are acetic acid, citric acid, and phosphoric acid. These buffers will maintain the pH of the food within a very specific range.

Each type of **buffer** has a specific range of pH at which it works. Each type of buffer also has a specific application that it is best suited for. Some buffers work best for meat and dairy while others work best for fruits, vegetables, and beans.

**CANNED FOOD**
Canning preserves food long after it would otherwise spoil. How do the chemical properties of buffers ensure that the pH of food is optimal for preservation?

## STEM and Acids, Bases, and Salts

Imagine a farmer sprinkling **salt** on farmland to enhance the growth of crops. Do you think it would work? It may sound strange, but that's exactly what many farmers do. The salt they use isn't the table salt, sodium chloride, of course. The sodium ions released by table salt would be harmful to crops. Instead, farmers use other types of salts that provide the nutrients nitrogen, phosphorus, and potassium that crops need to grow. One example of a salt in commercially available fertilizers for crops is ammonium nitrate. Ammonium nitrate encourages plant growth by slightly acidifying the soil to a pH of around six or seven.

Although salt-based fertilizers allow plants to grow faster and larger, they also have disadvantages. Because these salts are fairly soluble in water, runoff containing high levels of fertilizer can flow into nearby waterways. In high enough concentrations, the runoff causes algae to grow wild. This large amount of algae reduces the **concentration** of oxygen in the water available to plants and fish. It also feeds some plants and weeds so well that they grow and overtake native species. This results in a population decline of native species.

**FARMER SPRAYING FIELD**

Salt-based fertilizers sprayed may cause pollution of nearby waterways. What are some methods that this farmer can take to avoid polluting nearby waterways with runoff from the fertilizers?

Environmental engineers have developed a variety of technologies and procedures to combat nutrient pollution of waterways by salt-based fertilizers. One method is to build a vegetative swale alongside farmland. These U-shaped channels, filled with vegetation, are designed to catch and filter the runoff of the fertilizers. Another method is to build a detention basin to hold and filter runoff. Agricultural researchers have also developed methods that farmers can implement to reduce the detrimental effects of salt-based fertilizers. For instance, farmers are encouraged to plant cover crops in winter to reduce runoff from fields, and vegetation planted alongside streams near farmland can hinder the flow of salt-based fertilizer runoff into the water. Careful soil testing is required to ensure correct application rates for fertilizers. If soil testing reveals high acidity, lime (a base) can be added to neutralize the soil.

**SOIL TESTING**

A soil sample can be tested for pH, nitrogen, phosphorus, and other properties. What actions might a farmer take if the soil sample revealed very acidic soil?

# Using pH and pOH

**Use molarity and pH to solve the problems below.**

**1.** A student prepares a solution by placing 1.68 g $Na_2SO_4$ in a flask, adding water to bring the total volume to 100.0 mL, and mixing. What is the molarity of the solution?

**2.** It requires 25.4 mL of 0.102 M KOH to neutralize 20.0 mL of $H_3PO_4$. What is the molarity of the acid solution?

**3.** If 20.0 mL of 1.00 M $H_2SO_4$ was required to completely neutralize 35.0 mL of an aqueous ammonia solution, what is the molarity of the ammonia solution?

## Amphoterism

**Use the self-dissociation constant of water, $K_w$, and equations for pH and pOH to solve the problems below.**

1. A solution is found to have a pH of 5.40. What is the concentration of hydrogen ion in moles per liter, the concentration of hydroxide ion in moles per liter, and pOH of the solution?

2. If an HCl solution contains hydrogen ions at a concentration of $2.64 \times 10^{-2}$ M, what is the pOH of the solution?

3. Determine the hydrogen ion concentration and pH of a solution that has a hydroxide ion concentration of $5.00 \times 10^{-4}$ M.

4. Determine the hydroxide ion concentration, hydrogen ion concentration, and pH of a solution that has a pOH of 4.20.

# Neutralization

**Use the stochiometries found from neutralization reactions to solve the problems below.**

1. What volume, in milliliters, of a 0.997 M KOH solution is needed to neutralize 30.0 mL of 0.0400 M HCl?

2. A student pipets 10.0 mL 0.0500 M $H_3PO_4$ into a flask. What volume of 0.0200 M $Mg(OH)_2$ could be added that will completely neutralize the acid?

3. An unknown volume of 0.200 M acetic acid ($HC_2H_3O_2$) was completely neutralized by 22.0 mL 0.140 M ammonia ($NH_3$) in water. What was the original volume of acetic acid before neutralization?

4. Two solutions of $H_2SO_4$ and NaOH have the same molarity. What volume of the NaOH solution would be need to be added to completely neutralize 20.0 mL of the $H_2SO_4$?

## Acid-Base Reactions

**For each reaction, predict the product(s). Then apply concepts of acid-base chemistry to complete and balance each equation to see if your prediction was correct.**

**1.** $HClO_4 + KOH \rightarrow$

**2.** $NH_3 + H_2O \rightarrow$

**3.** $Mg(OH)_2 + H_2S \rightarrow$

**4.** $H_3PO_4 + LiOH \rightarrow$

**5.** $H_2O + CN^- \rightarrow$

CONCEPT
**5.4**

# Chemical Equilibrium

dlc.com/ca9101s

## LESSON OVERVIEW

### Lesson Questions

- How is an equilibrium constant ($K_{eq}$) calculated, and what does the value for $K_{eq}$ mean?
- How is Le Chatelier's principle used to predict the effect of stress applied to a system at equilibrium?
- How does the value of $K_{sp}$ relate to the solubility of a compound?

### Key Vocabulary

Which terms do you already know?

- [ ] activation energy
- [ ] calorimetry
- [ ] catalyst
- [ ] chemical potential energy
- [ ] chemical reaction
- [ ] condense
- [ ] dissolve
- [ ] endothermic
- [ ] energy
- [ ] enthalpy
- [ ] entropy
- [ ] equilibrium
- [ ] equilibrium constant ($K_{eq}$)
- [ ] exothermic
- [ ] first law of thermodynamics
- [ ] freezing
- [ ] gas
- [ ] Gibbs free energy

## Lesson Objectives

By the end of the lesson, you should be able to:

- Calculate the equilibrium constant ($K_{eq}$) for chemical reactions and interpret its meaning.
- Apply Le Chatelier's principle to predict the effect of stress applied to a system at equilibrium.
- Explain how the value of the solubility product constant ($K_{sp}$) relates to solubility of a compound.

## Key Vocabulary continued

- [ ] heat
- [ ] heat of reaction
- [ ] Hess's law
- [ ] inhibitor
- [ ] Key Vocabulary continued
- [ ] Le Chatelier's principle
- [ ] liquid
- [ ] melting
- [ ] noble gas
- [ ] non-spontaneous
- [ ] order of a chemical reaction
- [ ] precipitate
- [ ] product
- [ ] reactant
- [ ] reaction rate
- [ ] saturated
- [ ] second law of thermodynamics
- [ ] solid
- [ ] solubility
- [ ] solubility product
- [ ] solublity product constant ($K_{sp}$)
- [ ] solute
- [ ] spontaneous
- [ ] standard heat of formation
- [ ] surroundings
- [ ] system
- [ ] vaporize
- [ ] work

## Life Is a Balancing Act

dlc.com/ca9102s

**Balance is a word you have often heard in conversation, and yet it is used in multiple ways. You may have been asked to balance on your toes, your parents may have taught you how to balance your bank account, and you have heard many people say it is important to eat a balanced diet. What other things have people asked you to balance?**

**PAN SCALES**

Pan scales show the relative weight between two objects, and the pans hang level when the weights are balanced. What other objects or processes relate to the concept of balance?

**EXPLAIN QUESTION**

**What factors are necessary to consider when determining a chemical reaction at equilibrium, and why are they important?**

# How Is an Equilibrium Constant ($K_{eq}$) Calculated, and What does the Value for $K_{eq}$ Mean?

## Reversible Reactions

Many chemical reactions proceed to completion. In these complete reactions, all of the **reactant** is changed to **product**. However, not all chemical reactions use up all of the reactants. When a reaction is at **equilibrium**, quantities of both reactants and products remain present. In an equilibrium reaction, the amount of formed product and the amount of remaining reactant are balanced.

On Earth and in the atmosphere, the movement of water through the water cycle is an example of a physical change at equilibrium. During this cycle, water exists as a **solid**, **liquid**, or a **gas**. Liquid water evaporates to water vapor or freezes as ice or snow. Water vapor in the atmosphere condenses back into liquid as rain. Each of these water conversions is in equilibrium.



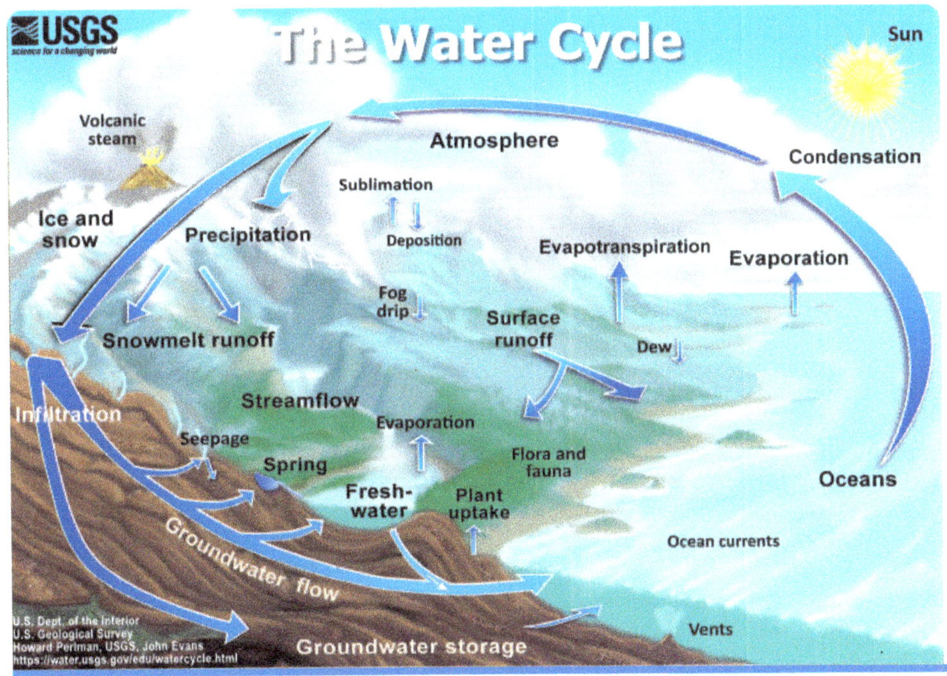


**THE WATER CYCLE**

The water cycle is an example of an equilibrium process. How is it reversible?

A **chemical reaction** at equilibrium can occur in either the forward or reverse direction, depending on the conditions of the reaction. This is a reversible chemical reaction. Typically, a one-way chemical reaction is written with an arrow pointing from the reactants to products. This indicates that the reaction converts all reactants fully into products. An example of this is the following reaction, where reactants A and B form the products C and D:

$$aA + bB \rightarrow cC + dD$$

However, a reversible reaction also proceeds in the reverse direction, from products to reactants:

$$cC + dD \rightarrow aA + bB$$

Because both the forward and reverse reaction occur in a reversible reaction, a way to write both reactions is to draw a double arrow:

$$aA + bB \leftrightharpoons cC + dD$$

**CHEMICAL EQUATIONS**

Chemical equations are used to evaluate chemical reactions. How do chemical equations represent reactions at equilibrium?

## The Equilibrium Constant

When a chemical reaction is at equilibrium, the concentration of reactants and products does not change. A constant value that describes this reaction is the equilibrium constant, or $K_{eq}$.

For the reaction, $aA + bB \leftrightharpoons cC + dD$, the equilibrium constant, $K_{eq}$, is described as:

$$K_{eq} = \frac{[C]^c[D]^d}{[A]^a[B]^b}$$

where [C] and [D] are the concentration of each product at equilibrium and [A] and [B] are the concentration of each reactant at equilibrium. The coefficients in the equation for each chemical determine the power to which each chemical is raised. For example, reactant A is raised to the power a.

Solid substances and liquid water are not included in equilibrium calculations. This is due to the structure of the particles in both liquids and solids. Both solids and liquids have essentially a constant concentration. This is because of the structure they take according to the kinetic molecular theory. In both solids and liquids, the particles are densely packed (as compared with gases) and thus it would take enormous pressure to cause even a fraction of a reduction in volume. As an example, consider the chemical reaction below:

$$C(s) + 2H_2(g) \leftrightharpoons CH_4(g)$$

The equilibrium constant is expressed as:

$$K_{eq} = \frac{[CH_4(g)]}{[H_2(g)]^2}$$

Note that C(s) is not included in the expression because it is a solid, as indicated by the letter "s." $H_2$ is raised to the second power because of the coefficient 2 in front of hydrogen gas in the balanced equation.

At equilibrium, both the forward reaction and the reverse reaction occur, but the amount of each reactant and product formed remain constant. Thus, at equilibrium the reaction rates of the forward and reverse reaction are equal, or:

$$\text{rate}_f = \text{rate}_r$$

In this equation, $\text{rate}_f$ is the **reaction rate** of the forward reaction, and $\text{rate}_r$ is the reaction rate of the reverse reaction.

In a reversible reaction, such as the formation of dinitrogen tetroxide ($N_2O_4$) gas from nitrogen dioxide ($NO_2$) gas, the initial forward reaction is rapid because only the reactant is present:

$$2NO_2(g) \leftrightharpoons N_2O_4(g)$$

As the number of product molecules increases, the reverse reaction begins, and the reverse reaction rate increases. Both the forward and reverse reaction take place as the **system** reaches equilibrium. Once the reaction reaches equilibrium, both forward and reverse reactions still occur. When a reaction is at equilibrium, the concentrations of reactants and products are not necessarily equal, but the rates of the forward and reverse reactions are always equal.

**CHANGE IN CONCENTRATION AT EQUILIBRIUM**

In the reversible reaction $2NO_2(g) \leftrightharpoons N_2O_4(g)$, initially only $NO_2$ is present. As equilibrium is established, both $NO_2$ and $N_2O_4$ exist. What do you think would change the ratio of final products?

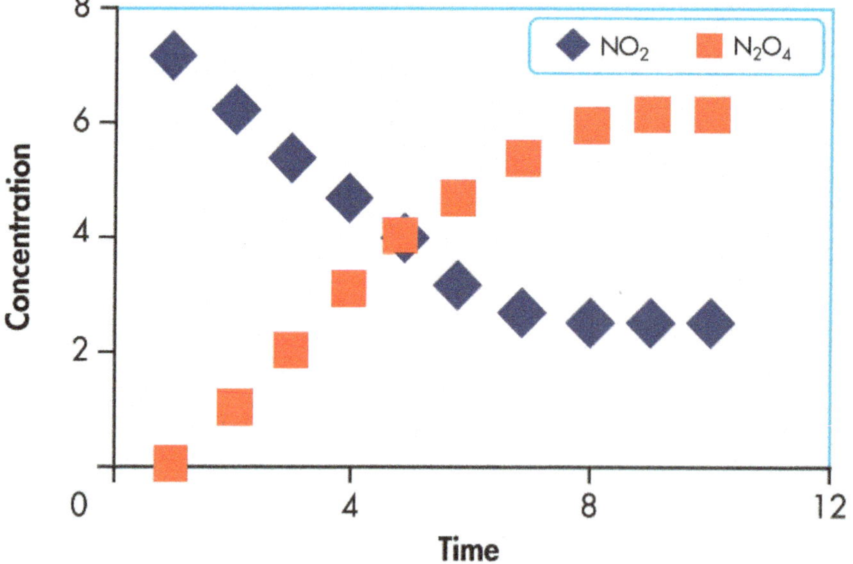

## Equilibrium Constant: Sample Problem

Find the equilibrium constant for the reaction between $NO_2$(g) and $N_2O_4$ (g) given the equilibrium concentration: $NO_2 = 0.150$ M and $N_2O_4 = 0.127$ M.

$$K_{eq} = \frac{[N_2O_4]}{[NO_2]^2}$$

$$= \frac{0.127}{(0.150)^2}$$

$$K_{eq} = 5.64$$

## ICE Method

The equilibrium concentrations of reactants and products can be calculated if the initial concentrations of reactants and the $K_{eq}$ value of the reaction are known. The equilibrium constant equation is used to determine these values. An ICE (Initial Change Equilibrium) table is used to track the initial concentrations, the change in concentrations, and the equilibrium concentrations of all chemicals in an equilibrium reaction.

Before a reaction begins, the concentration of products is zero, and the concentration of reactants will likely be known. The change in concentration is dependent on the coefficient of each substance.

Consider the following chemical reaction:

$$H_2(g) + I_2(g) \leftrightharpoons 2HI(g)$$

Considering this reaction mathematically, each mole of either reactant or product is assigned a whole number value based on the number of moles present. Initially the concentration of the product, HI(g), is zero because no reaction has yet occurred. One mole of each of these reactants combines to form two moles of product. The change in concentration of HI(g) is 12x, where x represents the number of moles.

The equilibrium concentration of the reactants is equal to the initial concentration minus the change in concentration. The change in concentration for both $H_2$(g) and $I_2$(g) is $-x$. Alternatively, the equilibrium concentration of the products is the sum of the initial concentration (or zero, in this case) and the change in concentration.

### ICE Method: Sample Problem

The $K_{eq}$ of the reaction of $N_2O_4(g)$ forming $NO_2(g)$ is $4.4 \times 10^{-3}$ M. If 0.026 M $N_2O_4$ reacts, what is the equilibrium concentration of $NO_2$?

$$N_2O_4(g) \leftrightharpoons 2NO_2(g)$$

**Solution:**

Write the initial equilibrium equation:

$$K_{eq} = \frac{[NO_2]^2}{[N_2O_4]}$$

Set up an ICE table for the reaction:

| | $[N_2O_4]$ | $[NO_2]$ |
|---|---|---|
| [Initial] | 0.026 M | 0 M |
| [Change] | $x$ | $2x$ |
| [Equilibrium] | $0.026 - x$ | $0 + 2x$ |

Substitute the equilibrium concentrations into the equilibrium equation.

$$K_{eq} = 4.4 \times 10^{-3} = \frac{[2x]^2}{[0.026 - x]}$$

Get all terms on the same side of the equation. Then rearrange them in the following format: $ax^2 + bx + c = 0$.

$$4.4 \times 10^{-3}(0.026 - x) = (2x)^2$$

$$1.14 \times 10^{-4} - 4.4 \times 10^{-3}x = 4x^2$$

$$4x^2 + 4.4 \times 10^{-3}x - 1.14 \times 10^{-4} = 0$$

Use the quadratic equation ($x = \frac{-b \pm \sqrt{b^2 - 4ac}}{2a}$) to solve for $x$.

Because a concentration cannot be a negative value, the only possible value for $x$ is 0.0048. Insert this value for $x$ in the equilibrium concentration for $NO_2(2x)$. Then solve for the equilibrium concentration of $NO_2$.

$$[NO_2]_{equilibrium} = 2x$$

$$[NO_2]_{equilibrium} = 2(0.0048 \text{ M})$$

$$[NO_2]_{equilibrium} = 0.0096 \text{ M}$$

The concentration of $NO_2$ at equilibrium is 0.0096 M.

# How Is Le Chatelier's Principle Used to Predict the Effect of Stress Applied to a System at Equilibrium?

Just as a strong wind can disrupt a hummingbird trying to maintain its position near a flower, an outside force can disrupt a chemical **system**. The system changes to offset the stress introduced into the system.

Several factors can disrupt a system at **equilibrium**. **Le Chatelier's principle** describes how the system works. Le Chatelier's principle states that when a system at equilibrium undergoes a change, the equilibrium will shift to oppose this change and a new equilibrium will be established. Equilibrium shifts occur when there are changes to concentration, temperature, volume, or partial pressure in the system where the reaction takes place.

## Changing the Reactant or Product Concentration of a System at Equilibrium

Changing the concentration of either a **reactant** or **product** affects an equilibrium system. If reactant or product is added to a system, the equilibrium shifts in the direction opposite to the substance added. For example, consider the reaction:

$$N_2O_4(g) \leftrightharpoons 2NO_2(g)$$

If $N_2O_4$ is added after equilibrium is already established, the shift will be in the direction of the products ($NO_2$). Alternatively, if $N_2O_4$ is removed from the system, the equilibrium shifts in the direction of the reactants.

When a reactant or product is added or removed, the equilibrium shifts according to the quantity of substance added or removed. An ICE (Initial Change Equilibrium) table can be used to quantify the effect of adding or removing a substance to a system at equilibrium.

**REVERSING CHEMICAL REACTIONS**

The equilibrium of a system cannot be altered by adding more precipitate to the reaction. Are precipitates on the reactant or the product side of a chemical reaction?

## Changing the Reactant or Product Concentration of a System at Equilibrium: Sample Problem

The equilibrium concentrations for the reactants and products were measured for the following reaction:

$$N_2O_4(g) \leftrightharpoons 2NO_2(g)$$

At equilibrium, the concentration of $N_2O_4$ was 0.090 M and $NO_2$ was 0.020 M. The $K_{eq}$ of the reaction is $4.4 \times 10^{-3}$ M. If 0.090 M $N_2O_4$ were added to the reaction after equilibrium, what is the new equilibrium concentration of $NO_2$?

**Solution:**

Write the initial equilibrium equation:

$$K_{eq} = \frac{[NO_2]^2}{[N_2O_4]}$$

Set up an ICE table for the reaction:

| | $[N_2O_4]$ | $[NO_2]$ |
|---|---|---|
| [Initial] | 0.090 M + 0.090 M | 0.020 M |
| [Change] | $x$ | $2x$ |
| [Equilibrium] | 0.18 M $- x$ | $0.020 + 2x$ |

Substitute the new equilibrium concentrations into the equilibrium equation.

$$K_{eq} = 4.4 \times 10^{-3} = \frac{[0.020 + 2x]^2}{[0.18 - x]}$$

Get all terms on the same side of the equation and rearrange in the following format: $ax^2 + bx + c = 0$.

$$4.4 \times 10^{-3}(0.18 - x) = (0.020 + 2x)^2$$

$$7.92 \times 10^{-4} - 4.4 \times 10^{-3}x = 4x^2 + 0.080x + 4.0 \times 10^{-4}$$

$$4x^2 + 0.0844x - 3.92 \times 10^{-4} = 0$$

Use the quadratic equation ($x = \frac{-b \pm \sqrt{b^2 - 4ac}}{2a}$) to solve for $x$.

$$x = \frac{-0.0844 \pm \sqrt{(0.0844)^2 - 4(4)(-3.92 \times 10^{-4})}}{2(4)}$$

$$x = \frac{-0.0844 \pm 0.1157}{8}$$

$$x = 0.0039, -0.025$$

Because a concentration cannot be a negative value, the only possible value for $x$ is 0.0039. Insert this value for $x$ in the equilibrium concentration for $NO_2(0.020\ M + 2x)$. Then solve for the new equilibrium concentration of $NO_2$.

$$[NO_2]_{equilibrium} = 0.020\ M + 2x$$

$$[NO_2]_{equilibrium} = 0.020\ M + 2(0.0039\ M)$$

$$[NO_2]_{equilibrium} = 0.020\ M + 0.0078\ M$$

$$[NO_2]_{equilibrium} = 0.028\ M$$

The concentration of $NO_2$ at the new equilibrium is 0.028 M.

## Temperature, Pressure, and Volume Changes to Equilibrium Systems

Changing the temperature of a chemical reaction affects an equilibrium reaction. Chemical reactions can be exothermic or endothermic. At equilibrium, a reaction will be exothermic if energy is given off, while the opposite endothermic reaction requires energy input.

For the reaction of $N_2O_4$ forming $NO_2$, the forward reaction is endothermic:

$$N_2O_4(g) + \text{energy} \rightarrow 2NO_2(g)$$

Increasing the temperature adds energy to the system, shifting the equilibrium of this reaction in the endothermic direction to create more product, or $NO_2(g)$. Alternatively, decreasing the temperature shifts the equilibrium of a reaction in the exothermic direction. In this reaction, the exothermic direction is toward the reactants on the left-hand side of this equation.

Pressure can also disrupt the equilibrium of a chemical system where gases are involved. If a reaction at equilibrium experiences an increase in pressure or a decrease in volume, it shifts in the direction that contains fewer moles of gas. Alternatively, a reaction at equilibrium that experiences a decrease in pressure shifts in the direction with more moles of gas. Consider the chemical reaction below:

$$N_2O_4(g) + \text{energy} \leftrightharpoons 2NO_2(g)$$

In this equation, there are 1 mole of reactants and 2 moles of product. Increasing the pressure, or decreasing the volume, shifts the equilibrium in the direction of the fewest number of moles. In the given example, the reaction proceeds toward the reactant. Alternatively, lower pressure or greater volume pushes the reaction to the right, in the direction of the product.

Some conditions do not affect the equilibrium of a chemical system. Because a **noble gas** is inert, adding noble gases to an equilibrium system at constant volume does not affect equilibrium. Additionally, adding a **catalyst** increases the rate of a chemical reaction but does not affect the direction of the equilibrium of a chemical reaction.

## How Does the Value of $K_{sp}$ Relate to the Solubility of a Compound?

### Solubility

Some ionic solids, when mixed with water, **dissolve** easily. These solutes, such as NaCl(s), have a high **solubility** in water. They readily form ions in water. Other ionic solids, such as AgCl(s), **precipitate** out of solution when just a small amount is mixed with water. These substances have a low solubility in water. The equation to describe the reaction of AgCl(s) in water is:

$$AgCl(s) \leftrightharpoons Ag^{+}(aq) + Cl^{-}(aq)$$

The **equilibrium** constant for this reaction would not include AgCl(s) because it is a **solid**. Therefore, the equilibrium constant would be written as:

$$K_{eq} = [Ag^{+}(aq)][Cl^{-}(aq)]$$

Because this reaction describes the solubility of the ionic solid, this equilibrium constant is called the solubility **product** constant, $K_{sp}$. So, $K_{eq}$ can be replaced with $K_{sp}$, in the following way:

$$K_{sp} = [Ag^{+}(aq)][Cl^{-}(aq)]$$

A table is used to find the solubility product constants of different ionic substances with low solubility. In this table, a salt that has a smaller $K_{sp}$ is less soluble than a salt with a higher $K_{sp}$.

**Solubility Product Constants (at 25°C)**

| Substance | $K_{sp}$ | Substance | $K_{sp}$ | Substance | $K_{sp}$ |
|---|---|---|---|---|---|
| $AgBr$ | $7.70 \times 10^{-13}$ | $BaSO_4$ | $1.08 \times 10^{-10}$ | $MnCO_3$ | $1.82 \times 10^{-11}$ |
| $AgBrO_3$ | $5.77 \times 10^{-5}$ | $CaCo_3$ | $8.70 \times 10^{-9}$ | $NiCo_3$ | $6.61 \times 10^{-9}$ |
| $Ag_2CrO_3$ | $6.15 \times 10^{-12}$ | $CdS$ | $3.60 \times 10^{-29}$ | $PbCl_2$ | $1.62 \times 10^{-5}$ |
| $AgCl$ | $1.56 \times 10^{-10}$ | $Cu(IO_3)_2$ | $1.40 \times 10^{-7}$ | $PbI_2$ | $1.39 \times 10^{-8}$ |
| $Ag_2CrO_4$ | $9.00 \times 10^{-12}$ | $CuC_2O_4$ | $2.87 \times 10^{-8}$ | $Pb(IO_3)_2$ | $2.60 \times 10^{-13}$ |
| $Ag_2Cr_2O_7$ | $2.00 \times 10^{-7}$ | $FeC_2O_4$ | $2.10 \times 10^{-7}$ | $SrCO_3$ | $1.60 \times 10^{-9}$ |
| $AgI$ | $1.50 \times 10^{-16}$ | $FeS$ | $3.70 \times 10^{-19}$ | $TlBr$ | $3.39 \times 10^{-6}$ |
| $AgSCN$ | $1.16 \times 10^{-12}$ | $Hg_2SO_4$ | $7.41 \times 10^{-7}$ | $ZnCO_3$ | $1.45 \times 10^{-11}$ |
| $Al(OH)_3$ | $1.26 \times 10^{-33}$ | $Li_2CO_3$ | $1.70 \times 10^{-2}$ | $ZnS$ | $1.20 \times 10^{-23}$ |
| $BaCO_3$ | $8.10 \times 10^{-8}$ | $MgCO_3$ | $2.60 \times 10^{-5}$ | | |

When an ionic solid dissolves in water, the ion product can be used to determine if a precipitate forms. The ion product is the amount of each ion at any given time point. If the solution becomes overly "full" of ionic solid, it is said to be supersaturated, and a precipitate forms.

The amount of ion in solution relative to the value of $K_{sp}$ determines whether a solution is **saturated**. An unsaturated solution has an ion product value less than the value of $K_{sp}$, while a saturated solution has an ion product value equal to $K_{sp}$. In both cases, a precipitate will not form. However, when the ion product amount exceeds the $K_{sp}$, the solution becomes supersaturated and a precipitate forms.

**SOLUBILITY PRODUCT CONSTANT**

The solubility product constant for different salts can indicate if a precipitate will form when substances react. Which compounds in the table have the lowest and highest solubility?

## Solubility: Sample Problem

The solubility constant of AgBr is $7.7 \times 10^{-13}$. What is the concentration of silver ions in solution at equilibrium?

$$AgBr(s) \leftrightharpoons Ag^+ (aq) + Br^- (aq)$$

**Solution:**

Write the equation for the solubility product constant of this reaction.

$$K_{sp} = [Ag^{+}(aq)][Br^{-}(aq)]$$

Because the molar ratio of $Ag^{+}$ and $Br^{-}$ is 1:1, the concentrations of $Ag^{+}$ and $Br^{-}$ at equilibrium are equal. Therefore, the value of each can be replaced with *x*.

$$7.7 \times 10^{-13} = [x][x] = x^2$$

Take the square root of each side to solve for *x*.

$$\sqrt{7.7 \times 10^{-13}} = \sqrt{x^2}$$

$$8.77 \times 10^{-7} = x$$

At equilibrium, the concentration of silver ions is $8.77 \times 10^{-7}$ M.

### Consider the Explain Question

**What factors are necessary to consider when determining a chemical reaction at equilibrium, and why are they important?**

dlc.com/ca9103s

Go online to complete the scientific explanation.

### Check Your Understanding

**Go online to check your understanding of this concept's key ideas.**

dlc.com/ca9104s

## STEM in Action

# Applying Chemical Equilibrium

Can we do something to a **chemical reaction** to encourage it to produce more **product**?

Many chemical reactions are at **equilibrium**. Equilibrium occurs when the rate of forward reaction equals the rate of reverse reaction. The reaction is no longer able to make more product than it already has. We use chemical compounds in our homes, gardens, and in factories, and we need to produce these compounds in large quantities. One of the most widely used chemical reactions in industrial processes is the Haber-Bosch process. In this reaction, nitrogen **gas** and hydrogen **gas** combine to form ammonia. This occurs in the following reaction:

$$N_2(g) + 3H_2(g) \leftrightharpoons 2NH_3(g)$$

The Haber-Bosch process is the primary method by which ammonia is made commercially. Ammonia is the main ingredient in nitrogenous fertilizers. Understanding the equilibrium reaction in the production of ammonia was so important that in the early 1900s two scientists, Fritz Haber and Carl Bosch, won the Nobel Prize for developing the process.

**AMMONIA PLANT**
Industrial processes produce millions of tons of ammonia annually. How are conditions in the plant optimized to maximize the yield of ammonia?

Haber and Bosch found that **heat**, high pressure, and a **catalyst** could increase the quantity of ammonia. Yet, when the temperature was too high, the reverse reaction occurred. At high heat, ammonia broke down into nitrogen and hydrogen gas. Haber and Bosch determined the optimum conditions of temperature and pressure to achieve the highest yield of ammonia. These conditions are used today to produce cheap fertilizer for agricultural use.

The concepts behind this process are stated in Le Chatelier's principle. It states that if a **system** at equilibrium is subjected to factors of temperature change, pressure change, volume change, or a change in concentration, the system will readjust itself to counteract the change and achieve a new equilibrium.

Haber and Bosch determined that a certain temperature increase—but not one that was too high—and an increase in pressure were needed for the highest output of ammonia.

### STEM and Forensic Chemists

Working through the math of a chemical reaction seems to have only theoretical importance. But the basic laws of chemistry have proven to hold vital keys to advancing the analysis **work** of forensic chemists.

**FORENSIC EVIDENCE**

Forensic chemists collect samples. What do you think they could be investigating?

Forensic chemists use their knowledge of chemical reactions to reveal an unknown factor in a reaction. The career field for forensic chemists is vast, as there are detectives that solve mysteries across many different situations. There are forensic chemists at work with local, state, and federal police departments. They work for fire departments, police departments, water quality boards, and for the FBI. They solve crimes, prove environmental damage, determine cause of death, identify bomb signatures, reveal fires set by arsonists, and even prove murder. The field of forensic chemistry continues to expand as knowledge of chemical reactions is further combined with advancing technology and design.

Forensic chemists may find themselves working as fire investigators. Starting with the charred remains at a scene, investigators collect chemical samples and work backward to identify the original reactants found at the scene. They must calculate the rates of reactions at different temperatures to determine the quantity of the original reactants. Identifying the reactants and in what amounts they occur can prove whether the fire was accidental or arson—a fire set on purpose. Certain gases, while in equilibrium at normal temperatures, may combust at high temperatures and high pressure. Leaking containers of compressed oxygen used in welding, under the right conditions, may cause **spontaneous** fires. Fire investigators must understand the chemistry to find the initial cause.

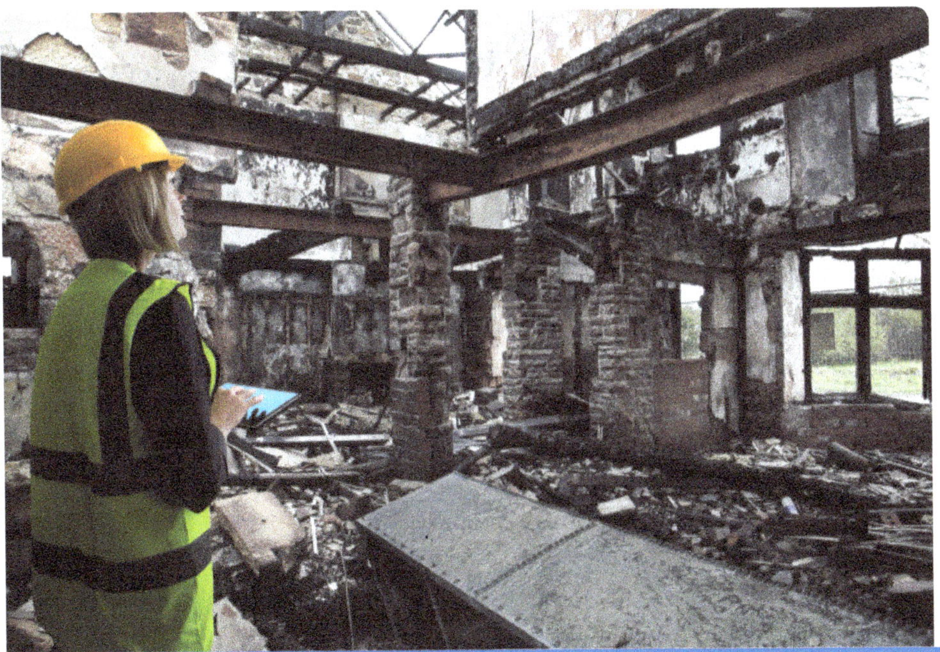

**ARSON INVESTIGATOR**
Arson investigators begin by collecting samples. They apply their knowledge of chemistry to work out the variables of the reactions. What kinds of reactions do you think they would need to know?

Forensic chemists may also work for the FBI or the ATF (Department of Alcohol, Tobacco, Firearms, and Explosives) identifying the blend of explosive chemicals used in a device. Tracing the signature of the materials may help lead investigators to the criminal.

Forensic toxicologists investigate nonbiological substances, chemicals, or poisonous products that may invade our environment, our water supply, or our bodies. They study the way chemical substances are absorbed and react in water or the air. They trace how they are distributed and metabolized in the body. For example, in the body the system of oxygen absorption in hemoglobin can be affected by pressure or concentration, since it is a system in equilibrium. Here a combination of math, biology, and chemistry are necessary to determine whether physical factors affected a person, such as a lack of oxygen present in the environment (concentration), or if the addition of a potentially poisonous substance, such as carbon monoxide, was introduced. Understanding the law of conservation of mass means their calculations will accurately determine the quantities of a chemical. This allows them to determine the intent behind the addition of a substance to a system. Were the amounts found in keeping with an accidental poisoning, or was the death deliberate? Forensic toxicologists offer scientific evidence, too, and are often called upon to testify at a trial.

Forensic teams comb crime scenes for forensic evidence. A forensic chemist may spend weeks sifting through this evidence in a lab. What sort of techniques could he use?

## Changing Concentration

**Set up an ICE table to solve each of the problems below.**

**1.** Hydrofluoric acid, HF, dissociates according to the following equation.

$$HF(aq) \leftrightharpoons H^{+}\ (aq) + F^{-}\ (aq)$$

The equilibrium constant, $K_{eq}$, for this system is $7.21 \times 10^{-4}$ at 25°C, and the equilibrium concentrations are: [HF] = 0.974 M, [H ] = 0.0265 M, and [F ] = 0.0265 M. If the concentration of HF is increased by 0.50 M, what will be the new concentrations of $H^{+}$ and $F^{-}$?



**2.** The equilibrium constant for the following reaction is $1.00 \times 10^{-2}$ at the relevant temperature.

$$2HI(g) \leftrightharpoons H_2(g) + I_2(g)$$

Data for concentrations at equilibrium are given in the table below.

| [HI] (mol/L) | $[H_2]$ (mol/L) | $[I_2]$ (mol/L) |
|---|---|---|
| 0.0830 | 0.0083 | 0.0083 |

What will the equilibrium concentrations of the hydrogen and iodine ions be when the system returns to equilibrium?

**3.** Consider the following system:

$$2HOCl(g) \leftrightharpoons H_2O(g) + ClO_2(g)$$

At equilibrium, the concentration of each component is: [HOCl] = 0.040 M, [$H_2O$] = 0.230 M, and [$ClO_2$] = 0.230 M. If the concentration of HOCl is increased by 0.050 M from the equilibrium conditions, what will be the new concentrations of $H_2O$ and $ClO_2$ once the system returns to equilibrium?

**4.** Household ammonia is a solution of ammonia in water. In this solution, ammonia reacts with water to form ammonium hydroxide, which ionizes according to the equation below. At 298K, $K_{eq} = 1.8 \times 10^{-5}$ for this reaction.

$$NH_3(aq) + H_2O(l) \leftrightharpoons NH_4^+(aq) + OH^-(aq)$$

The effectiveness of the cleaner decreases as the amount of ammonia in the solution decreases. However, the cost to produce the solution increases when more ammonia is added. Solution A was made using $[NH_3] = 7.2 \times 10^{-3}$ M. In Solution B, $[NH_4^+] = 5.2 \times 10^{-4}$ M. Which solution cleans better? Which solution is less costly to produce? Explain your answers in terms of the amount of ammonium in the solutions.

**5.** Carbonated beverages contain bubbles of carbon dioxide that make the beverage fizzy. However, these beverages also contain carbonic acid, which can slowly wear away the enamel of the teeth of people who drink them. When carbon dioxide reacts with water, it reacts to form carbonic acid. Carbonic acid then dissociates, as shown by the set of equations below.

$$CO_2(aq) + H_2O(l) \leftrightharpoons H_2CO_3(aq)$$

$$H_2CO_3(aq) + H_2O(l) \leftrightharpoons H_3O^+(aq) + HCO_3^-(aq)$$

The equations can be combined.

$$CO_2(aq) + 2H_2O(l) \leftrightharpoons H_3O^+(aq) + HCO_3^-(aq)\ K_{eq} = 4.6 \times 10^{-7}$$

If $[CO_2] = 0.15$ M in a sealed carbonated beverage, what is the concentration of hydronium ions in the soda? Assume no other acids are added. As bubbles escape from the beverage after it is opened, the $[CO_2]$ decreases to $1.5 \times 10^{-5}$. What happens to the concentration of the hydronium ions in the beverage? Although “flat” beverages might not taste as good, would they be less damaging to teeth?

## ICE Method

**Set up an ICE table to solve each of the problems below.**

**1.** Nitric acid dissociates in water according to the following equation:

$$HNO_2(aq) \leftrightharpoons H^+(aq) + NO_2^-(aq)$$

The equilibrium constant, $K_{eq}$, at 25°C, for this equilibrium is $4.0 \times 10^{-4}$. If 1.0 mol $HNO_2$ is dissolved in water to make 0.50 liters of solution, calculate the concentrations of $H^+$ and $NO_2^-$ ions that will be present at equilibrium.

**2.** A solution of $H_2$ and $I_2$ reacts and reaches equilibrium at 721 K. The equilibrium concentrations are $[H_2] = 0.460$ M, $[I_2] = 0.390$ M, $[HI] = 3.00$ M. Calculate $K_{eq}$ at 721 K.

**3.** At 25°C, the equilibrium constant, $K_{eq}$, for the following reaction is $4.6 \times 10^{-2}$. If you start with 0.100 M reactant A, what will be the equilibrium concentrations of products B and C?

$$2A(g) \leftrightharpoons B(g) + C(g)$$

# Solubility

**Use the definition of the solubility constant to solve each of the problems below.**

1. The solubility constant of MnS is $2.3 \times 10^{-13}$ at 25°C. What is the concentration of sulfide ions in a saturated solution of MnS at equilibrium?

2. Iron carbonate, $FeCO_3$, is added to water until the saturation point is reached. What concentration of carbonate ion, $CO_3^{2-}$, is present in this solution? The solubility constant for $FeCO_3$ is $2.1 \times 10^{-11}$ at 25°C.

3. A saturated solution is made using lead iodide in water. What is the concentration of lead ions in this solution at equilibrium if the solubility constant for $PbI_2$ is $1.4 \times 10^{-8}$ at 25°C?

4. The solubility constant of $Ba(OH)_2$ is $5.0 \times 10^{-3}$ at 25°C. Calculate the equilibrium concentration of hydroxide ions in a saturated solution.

## ICE Method

**Set up an ICE table to solve each of the problems below.**

**1.** At 25°C, the equilibrium constant, $K_{eq}$, for the following reaction is $2.7 \times 10^{-10}$. If you start with 1.0 M $COCl_2$, what will be the equilibrium concentrations of CO and $Cl_2$?

$$COCl_2(g) \leftrightharpoons CO(g) + Cl_2(g)$$

**2.** HCN dissociates in water according to the following equation:

$$HCN(aq) \leftrightharpoons H^+(aq) + CN^-(aq)$$

The equilibrium constant, $K_{eq}$, at 25°C, for this equilibrium is $6.2 \times 10^{-10}$. If HCN is dissolved in water to make a 2.00 M solution, calculate the concentrations of $H^+$ and $CN^-$ ions that will be present at equilibrium.

**3.** Suppose the following system establishes equilibrium in water:

$$2A(aq) \leftrightharpoons B(aq) + C(aq)$$

where A, B, and C represent the reactants and products in a chemical reaction. The equilibrium constant, $K_{eq}$, at 25°C, for this system is $9.1 \times 10^{-8}$. If reactant A is added to water to give a final concentration of 1.0 M, what concentrations of products B and C will be present after the system has reached equilibrium?

CONCEPT 5.5

# Reaction Rate



## LESSON OVERVIEW

dlc.com/ca9105s

### Lesson Questions

- What factors influence reaction rate, and what is their impact?
- How do rate laws describe chemical reactions?

### Lesson Objectives

By the end of the lesson, you should be able to:

- Identify factors that influence reaction rate and explain their impact.
- Apply rate law equations to predict reaction rates.
- Determine the overall order of a chemical reaction.

### Key Vocabulary

Which terms do you already know?

- [ ] activation energy
- [ ] catalyst
- [ ] chemical reaction
- [ ] collision theory
- [ ] enzyme
- [ ] inhibitor
- [ ] order of a chemical reaction
- [ ] oxidation
- [ ] reactant
- [ ] reaction rate
- [ ] surface area

## Thinking about Reaction Rate

It is time for a picnic. Imagine you are planning a cookout for your friends, and one of your tasks is to help prepare the food. As you are busy preparing the potato salad, hamburgers, and cut-up apples, you consider putting the food on the table now, instead of waiting until your friends arrive. Why might this be a problem?

dlc.com/ca9106s

**EXPLAIN QUESTION**

What factors influence a reaction rate, and what is their impact on reaction rate?

**REFRIGERATED FOODS**

A refrigerator keeps food cold. Why do cold temperatures help to preserve food?

# What Factors Influence Reaction Rate, and What Are Their Impact?

## Factors That Influence Reaction Rate

The physical and chemical properties of elements determine how atoms interact when they come into contact with one another. Some elements, such as fluorine, are quite reactive, and they form compounds with many other elements. Others, such as the noble gases, do not interact with other elements at all.

Similarly, the specific substances that interact in a **chemical reaction** may interact quickly, slowly, or not at all. The chemical properties of the reactants play a key role in determining the rate of a chemical reaction. For example, a reaction between potassium and water will progress at a faster rate than a reaction between iron and water.

Another important factor in **reaction rate** is the amount of each **reactant** present. For example, in order for rust to form, iron and water must both be present, or **oxidation** will not occur. If oxygen is not present in a system, the combustion reaction that burns gasoline cannot occur. Often, the amount of a reactant present is described using concentration, particularly when working with solutions.

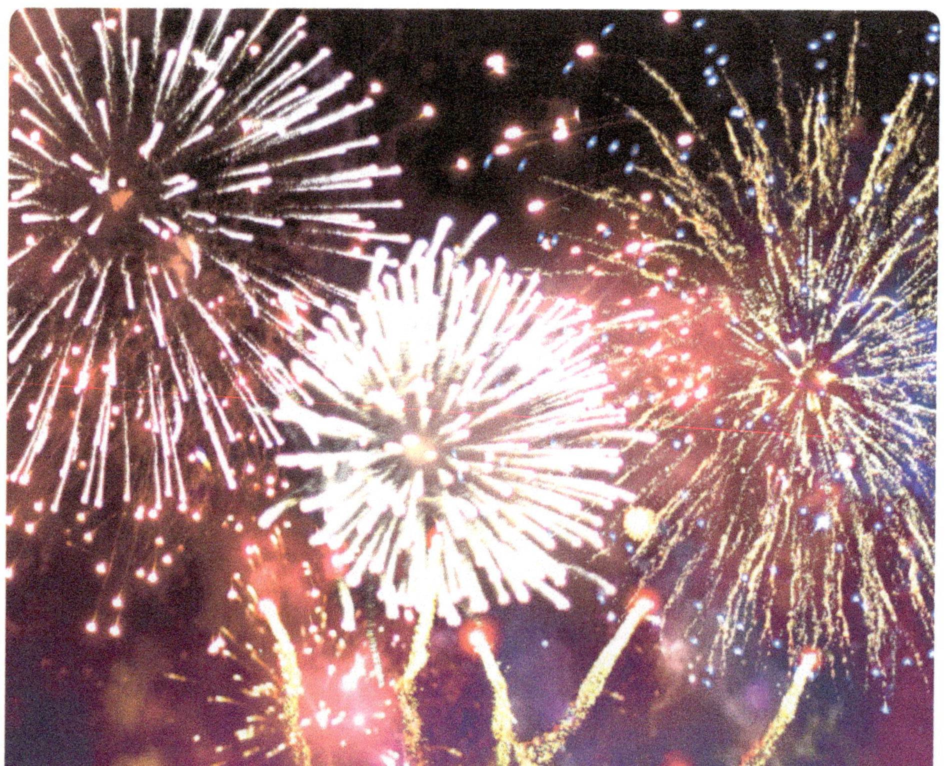

**FIREWORKS**

The chemical reactions in fireworks are very rapid. How could a scientist study these reactions?

The effect of concentration on reaction rate is partly explained by **collision theory**. Remember that, according to kinetic molecular theory, atoms and ions are constantly in motion. Even in a solid, the particles have energy and regularly collide with one another.

During a chemical reaction, chemical bonds break and atoms are rearranged into new compounds. In order for this to happen, the atoms involved must collide with one another with a certain amount of energy. Usually, the more collisions that occur between atoms of the reactants, the faster the reaction proceeds. Logic tells us that the more atoms there are present (the higher the concentration), the more collisions will occur.

The size of the reactant's particles is another factor. Differently sized particles have different surface areas. For example, in a cube of sugar, there are six large, exposed sides that might react with another substance. The sugar that lies in the center of the cube is kept out of the reaction, at least for a little while. If the cube is broken apart, however, the **surface area** of each smaller crystal is exposed. Though each particle is smaller, there are many more of them and the exposed surface area increases. The atoms that make up the smaller crystals of sugar are free to collide with other atoms. This allows the entire reaction to proceed more rapidly.

The conditions of the environment in which a reaction takes place strongly affect the rate of the reaction. These include temperature and pressure. For example, when the temperature of a system is increased, the kinetic energy of all the particles in the system increases as well. The number of collisions between atoms will also increase, causing the reaction to proceed more rapidly. When there is more energy in a system, some of it is absorbed by the molecules present, which makes it easier for molecular bonds to be broken. The atoms can then be rearranged to form new substances more quickly.

If at least one of the reactants is a gas, the pressure of the system may also affect the reaction rate. Remember that when held under high pressure, gas particles cannot move as freely. If a gas is held in a system by pressure, its particles cannot escape. They remain in the system and participate in the reaction. There are more of them present, so the reaction proceeds at a faster rate than it would if some of the gas particles spread out into the space surrounding the system.

There are times when extra substances present in the system might affect the rate of a reaction. Often, this involves the **activation energy**, or the amount of energy required for a reaction to proceed. Catalysts are substances that speed up reactions, but they remain unchanged when the reaction is complete. A **catalyst** will reduce the amount of activation energy needed for a reaction to occur. Metals and special proteins, called enzymes, often act as catalysts. Nickel, for example, acts as a catalyst in the reaction used to make margarine. Cars are equipped with catalytic converters, which use metals to remove dangerous substances from engine exhaust. Food digestion, respiration, and most other biological processes rely on catalysts.

**CATALYTIC CONVERTER**

The presence of platinum causes reactions in a catalytic converter that break down air pollutants such as carbon monoxide and nitrogen oxides into carbon dioxide, nitrogen gas, and water. Why is platinum not broken down during these reactions?

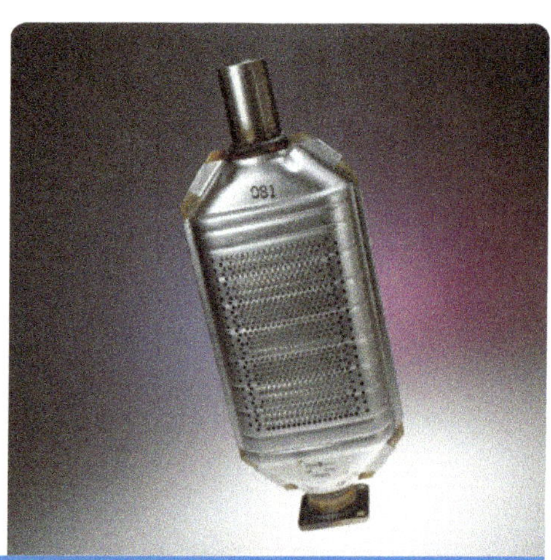

The use of an **inhibitor**, on the other hand, will decrease reaction rate. An inhibitor acts in the opposite way that a catalyst does, increasing the energy required for the reaction to proceed. Some inhibitors, in fact, work by limiting the action of the catalysts. Many enzymes can act as inhibitors when they are present during a chemical reaction.



## How Do Rate Laws Describe Chemical Reactions?

### Rate Laws

Whenever specific chemical compounds come into contact with one another, they will react in a predictable way. Experiments can be completed and repeated to determine the behavior of reactants in a system, including the rate at which the substances react.

As a reaction occurs, the reactants go through a chemical change. Molecular bonds are broken, and atoms are rearranged. The concentration of the remaining reactants can be measured. Monitoring the changes in concentrations provides information that can be used to calculate the **reaction rate**.

A rate law is a mathematical equation that describes the reaction rate of a specific reaction. The rate law relates the molar concentrations of the reactants to $k$, the rate constant, which remains the same for a reaction as long as temperature does not change.

A basic **chemical reaction** can be described using a balanced chemical equation.

$$aA + bB \rightarrow C$$

The rate law equation uses information about the reaction.

$$r = k[A]^x[B]^y$$

Here, the variable $k$ is the rate constant. The values in square brackets represent the concentration levels of the reactants A and B. Notice that the coefficients $a$ and $b$ from the initial reaction have no effect on the rate calculation and can be ignored. The variables $x$ and $y$ describe the order of the chemical reactions. These values must be determined experimentally and will be discussed in the next section.

Reaction rates for all reactants are determined experimentally. It is not possible to look at a chemical equation and know which **reactant** will drive the rate of reaction.

**RUST ON IRON**

The oxidation of iron is a slow reaction that causes corrosion to the metal as iron oxide forms. What can be added to the metal that would prevent or slow the formation of rust?

### Rate Laws: Sample Problem

A solution of 0.1 M acetone ($C_3H_6O$) reacts with a solution of 0.1 M bromide ($Br_2$). Assume that the value of *x* is 0, the value of *y* is 1, and *k* is $1.64 \times 10^{-4}$/s. Find the reaction rate.

**Solution:**

Substitute the given values into the rate law equation.

$$r = k[A]^x[B]^y$$

$$r = 1.64 \times 10^{-4}/s \times (0.1\ mol/L)^{0x}\ (0.1\ mol/L)^1$$

$$r = 1.64 \times 10^{-4}/s \times 1 \times (0.1\ mol/L)$$

$$r = 1.64 \times 10^{-5}\ mol/L \cdot s$$

The reaction will use $1.64 \times 10^{-5}$ mol/L of reactants per second.



### The Order of Reactions

The order of reaction can be determined with respect to each of the reactants as well as for the overall reaction. Recall that the concentration of a reactant can affect the rate of a reaction. The order of reaction describes how changing the concentration of a reactant would affect the reaction rate.

The order is described in the rate law equation using the variables *x* and *y*. When finding the order of reaction with respect to each reactant, however, only one concentration is used in the rate law equation.

$$r = k[A]^x$$

A zero-order reaction has only one reactant. In addition, the reaction rate is independent of the concentration of the reactant. Even if the concentration of the reactant is doubled, the rate of the reaction does not change. This is usually seen in decomposition reactions, such as the decay of a nitrous oxide gas molecule into nitrogen gas and oxygen gas. Because any number raised to the power of 0 is equal to 1, in a zero-order reaction $r = k$.

$$r = k[A]^x$$

$$r = k[A]^0$$

$$r = k \times 1$$

$$r = k$$

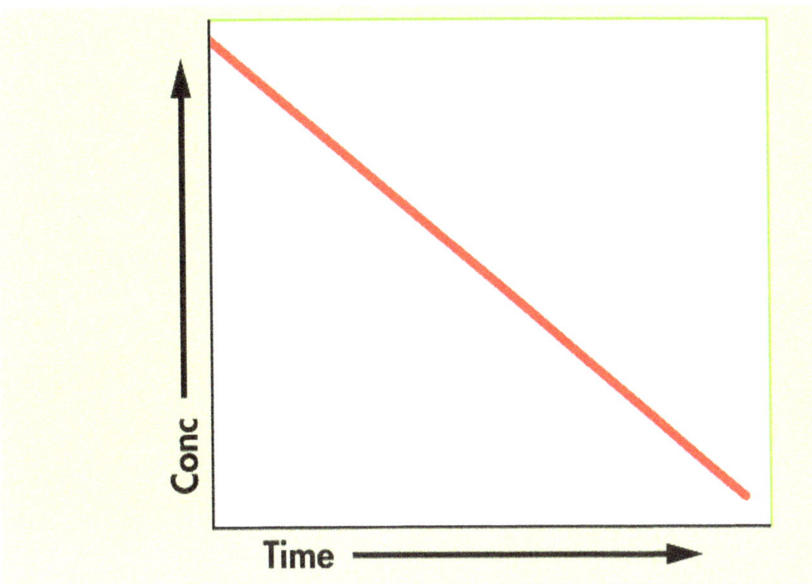


**ZERO-ORDER REACTION**

In a zero-order reaction, concentration decreases linearly over time. How many reactants are involved in these reactions?

A first-order rate of reaction is directly proportional to the concentration of the reactant. In other words, as the concentration of the reactant increases, the rate of the reaction increases. Because any number raised to the power of 1 is equal to that number, for a first-order reaction, $r = k[A]$.

$$r = k[A]^x$$

$$r = k[A]^1$$

$$r = k[A]$$

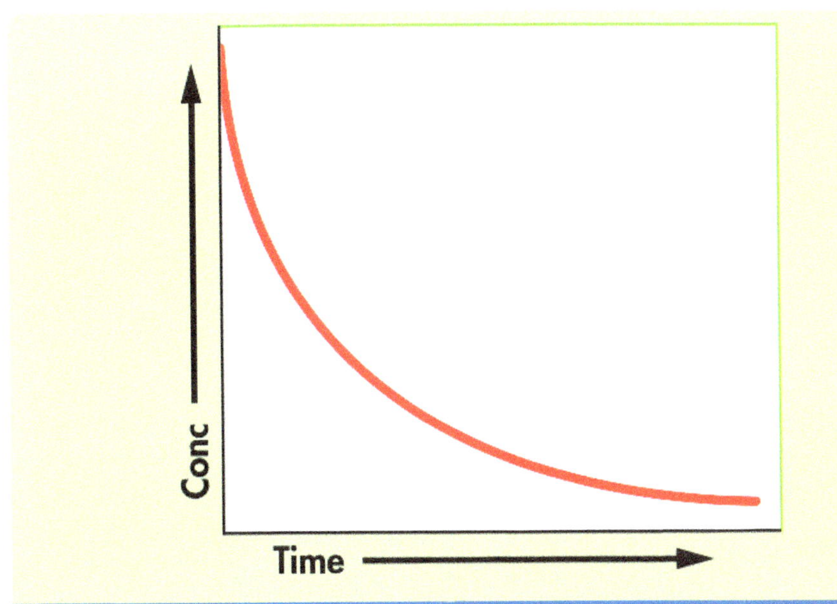


**FIRST-ORDER REACTION**

In a first-order reaction, concentration decreases exponentially over time. How does the concentration of the reactant affect the reaction rate?

The rate of a second-order reaction is more complicated. This rate can be proportional to the square of the concentration of one of the reactants. This means that if the concentration is doubled, the reaction rate increases by a factor of four. It might also be proportional to the product of the concentrations of two reactants. For a second-order reaction, the exponent $x$ is equal to 2, and $r = k[\mathrm{A}]^2$ or $r = k[\mathrm{A}][\mathrm{B}]$.

**SECOND-ORDER REACTION**

In a second-order reaction, the rate is proportional to the square of one of the reactant's concentration. How do changes in the reactant concentration affect the reaction rate?

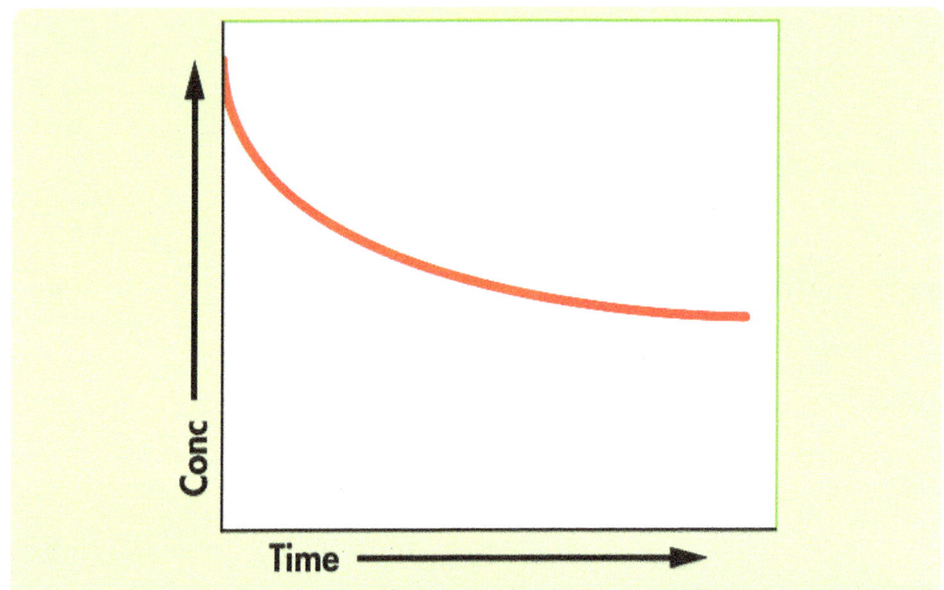


The overall **order of a chemical reaction** is found by adding the orders of each reactant. For example, if the order of the reaction with respect to the first reactant is 0 and the order of the reaction with respect to the second reactions is 2, the overall order of the reaction is 2.

## The Order of Reactions: Sample Problem

Write the rate law equation and determine the overall order of the chemical reaction between acetone ($C_3H_6O$) and iodide ($I_2$), given the experimental data in the equation and in the table below.

$$C_3H_6O + I_2 \rightarrow C_3H_5I + HI$$

**SAMPLE DATA TABLE**

Review the sample data table. What can you observe when the concentrations of the different reactants are changed?

| Trial | $[C_3H_6O]$ | $[I_2]$ | Initial rate of reaction (mol/L·s) |
|---|---|---|---|
| 1 | 0.1 M | 0.1 M | $2.52 \times 10^{-2}$ |
| 2 | 0.1 M | 0.2 M | $2.50 \times 10^{-2}$ |
| 3 | 0.2 M | 0.1 M | $5.14 \times 10^{-2}$ |

**Solution:**

Substitute the chemical formulas of the reactants into the rate law equation.

$$r = k[A]^x[B]^y$$

$$r = k[C_3H_6O]^x[I_2]^y$$

Determine $x$ and $y$ from the experimental data in the table. Comparing the data for trials 1 and 2, note that as the concentration of the acetone is unchanged and the concentration of the iodine is doubled, the rate did not change significantly. This means the reaction is zero-order with respect to iodine. Therefore:

$$r = k[C_3H_6O]^x [I_2]^0$$

Because the reaction rate more than doubled when the concentration of acetone was doubled in trial 3, the reaction is first order with respect to acetone. Therefore

$$r = k[C_3H_6O]^1 [I_2]^0$$

Any number to the zero power equals one. Therefore:

$$r = k[C_3H_6O]^1 = k[C_3H_6O]$$

The orders with respect to $[C_3H_6O]$ and $[I_2]$, 1 and 0, respectively, have a total value of 1. Therefore, the overall order of this reaction is 1.

### Consider the Explain Question

**What factors influence a reaction rate, and what is their impact on reaction rate?**

Go online to complete the scientific explanation.

dlc.com/ca9107s

### Check Your Understanding

**What are the factors that influence reaction rate?**

dlc.com/ca9108s

# STEM in Action

## Applying Reaction Rate

The rate of reaction is proportional to **surface area**. The larger the surface area that is exposed, the faster the **reaction rate**. Some substances that have high surface areas need to have special handling due to their highly combustible nature when in the presence of oxygen.

A dust explosion is the rapid combustion of fine particles suspended in the air, sometimes in an enclosed location. These explosions can occur where any dispersed powdered combustible material is present in high enough concentrations in the atmosphere or other oxidizing gaseous medium such as oxygen. Grain dust explosions are one type of dust explosion that are often severe and cause loss of life and major property damage. In the certain conditions, grain dust is more explosive than dynamite. Over the last 35 years, there have been over 500 explosions in grain handling facilities across the United States which have killed and injured many people. Because of the highly reactive nature of grain dust, special grain dust elevators have been designed to help minimize problems.

**GRAIN DUST**

Grain dust, a combination of cornstarch and corn oil, is extremely flammable when it comes in contact with air. How do corn storage facilities take precautions against grain dust?

OSHA (Occupational Safety and Health Administration) standards require that both grain dust and ignition sources must be controlled in grain elevators to prevent these often deadly explosions. A grain elevator is a tower containing a bucket elevator or a conveyor, which scoops up grain from a lower level and deposits it in a silo or other storage facility. Although the grain elevator is not a new idea, current elevators are constructed much differently than they were 100 years ago. Modern day elevators are made of steel and/or concrete and have proper ventilation and dust-control measures in order to minimize explosions. Although the grain elevators have not stopped explosions from happening, they have certainly helped in reducing the number of casualties.

## STEM and Reaction Rate

An industrial chemist is a chemist who works to develop and manufacture products and processes for the better good of society and the environment. One important development by industrial chemists was the use of catalysts. World history would have been different without catalysts, because wars would have been less destructive, transportation would be much slower, and Earth's population would be much smaller. These surprising claims are true because most industrially produced chemicals depend on catalytic action, thanks to the work of industrial chemists.

Some of the more important industrial chemicals produced by catalyzed reactions are explosives, fertilizers, gasoline, and polymers. Ammonia ($NH_3$) is manufactured by the Haber process, which would not be possible without an iron **catalyst**. Nitric acid ($HNO_3$) is produced by a different reaction, which is catalyzed by platinum (Pt). Together these chemicals are used to manufacture the powerful explosive ammonium nitrate ($NH_4NO_3$). This one compound is said to have changed the course of World War I. The historical importance of ammonium nitrate as a fertilizer is even greater. Without this and other fertilizers, the world's population would be much smaller due to agricultural limitations. Sulfates, another important class of fertilizers, are produced from sulfuric acid ($H_2SO_4$), which results from a reaction catalyzed by vanadium (V) } oxide ($V_2O_5$).

NASA engineers have also been working hard to develop low-temperature **oxidation** catalysts throughout the past two decades. They were initially intended to support space-based $CO_2$ lasers but are now being used in several different facets of everyday life. These catalysts are used to sense and remove two deadly gases, carbon monoxide and formaldehyde, from houses and other buildings. They can also be used in automotive gas and diesel catalytic converters to improve their efficiency. Engineers continue their work in these low-temperature catalysts for use in commercial and industrial applications.

CONCEPT 5.6

# Ocean Water



## LESSON OVERVIEW

### Lesson Questions

- What are the characteristics of ocean water?
- What Role Does Carbon Dioxide Play in Ocean Acidification?
- How does ocean water move?

### Key Vocabulary

Which terms do you already know?

- [ ] biomass
- [ ] carbonate
- [ ] carbon sink
- [ ] climate
- [ ] continental rise
- [ ] continental shelf
- [ ] continental slope
- [ ] Coriolis effect
- [ ] deep ocean current
- [ ] density
- [ ] evaporation
- [ ] fresh water
- [ ] greenhouse gas
- [ ] liquid
- [ ] neap tide
- [ ] ocean current
- [ ] oceanography
- [ ] ocean wave
- [ ] photosynthesis

dlc.com/ca9109s

## Lesson Objectives

By the end of the lesson, you should be able to:

- Understand the characteristics of ocean water.
- Understand the movement of ocean water.

## Key Vocabulary continued

- ☐ plankton
- ☐ rip current
- ☐ salinity
- ☐ salt water
- ☐ solar energy
- ☐ spring tide
- ☐ surface current
- ☐ thermocline
- ☐ tide
- ☐ tsunami
- ☐ turbidity current
- ☐ undertow
- ☐ upwelling
- ☐ wave
- ☐ wave height
- ☐ wavelength
- ☐ wave speed
- ☐ wind energy

## Thinking about Ocean Water

Imagine spending a hot summer day at the beach with your friends—running across the warm sand and splashing in the cool, salty water. It seems strange that the water can feel so cool when the air feels so warm. Waves are crashing onto the shore, carrying sand and pebbles back and forth along with them. What forms the waves? Why is understanding the behavior of waves and ocean water important?

dlc.com/ca9110s

**A MID-OCEAN GOLF COURSE IN BERMUDA**

Can you describe how ocean water interacts with the air, land, and living things in this image?

**EXPLAIN QUESTION**

**How do the characteristics and properties of ocean water affect weather and climate?**

# What Are the Characteristics of Ocean Water?

## Gases and Other Materials in Ocean Water

Ocean water contains a variety of dissolved gases. The most abundant gases are nitrogen, oxygen, and carbon dioxide. All of these gases can be exchanged with the atmosphere above Earth's oceans. A proportion of the carbon dioxide enters oceans through respiration by marine organisms or via underwater volcanic vents. Marine algae and other plants add oxygen through the process of **photosynthesis**. Water temperature determines how much gas can be stored in the water.

**DISSOLVED GASES IN OCEAN WATER**

A variety of gases are dissolved in ocean water. What is the biological role of these gases?

Ocean water contains many other materials, besides gas, including dissolved minerals, such as salt and an abundance of suspended microorganisms. Many of these microorganisms are types of **plankton**.

Plankton is a word that comes from the Greek meaning "to drift." It describes any organism that drifts with ocean currents. Phytoplankton are drifting plants that use photosynthesis for nutrient production. These tiny organisms are the basis of food chains in most parts of oceans. Zooplankton are animals that feed on phytoplankton.

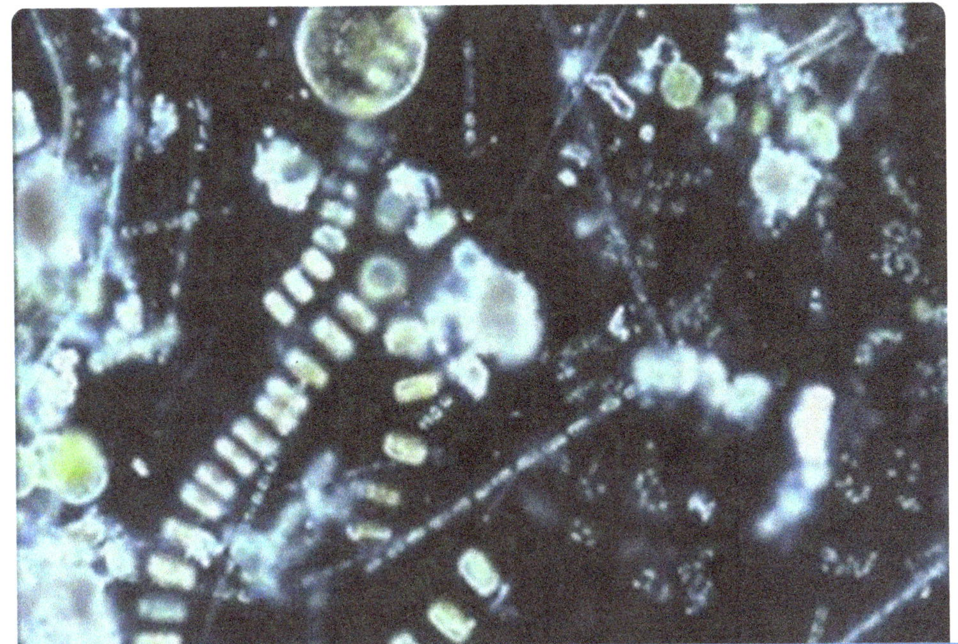

**PHYTOPLANKTON**
Phytoplankton are marine microorganisms that convert sunlight into energy through photosynthesis. How are these organisms similar to plants on land?



## Distribution of Marine Organisms

Marine organisms are plants, animals, and microorganisms that live in **saltwater** environments. Many factors influence the distribution of organisms throughout the world's oceans. These factors include temperature, **salinity**, depth, and the direction and speed of ocean currents.

Organisms have different tolerances to salinity and temperature, which limit their range—no marine organism could survive in all parts of Earth's oceans. Water depth is a limiting factor because below a certain depth there is increasing water pressure, and there is very little light, which plants need to grow. Many of the most abundant marine organisms, such as phytoplankton, are photosynthetic and require sunlight to produce energy. Because plankton drift with ocean currents, their range is also determined by where ocean currents carry them.

Plankton abundance depends on other factors as well. Because phytoplankton are plants, they require nutrients, such as phosphorus, which is more abundant in colder waters. The distribution of phytoplankton often changes with the seasons, especially in temperate climates. Nutrient-rich areas where **upwelling** occurs typically have an abundance of phytoplankton. Zooplankton are found in abundance around phytoplankton—their principle food source.

**ZOOPLANKTON**

Zooplankton are marine microorganisms that feed on phytoplankton. How are zooplankton affected by variation in water chemistry?



## Salinity

Salinity is a measure of the dissolved solids contained in ocean water. When describing oceans, salinity usually refers to salt content. The salts in ocean water are composed of a variety of elements, including sodium, potassium, chlorine, and iron. Because salinity is a measure of concentration, it is calculated by dividing the amount of solids by the total volume of water.

The salinity of water is determined by a variety of factors, including temperature, **evaporation**, precipitation, runoff from the continents, and water depth. Temperature changes affect the amount of material that is able to be dissolved in water. Warmer water is able to dissolve solids faster than cold water.

Evaporation, precipitation, and runoff all affect salinity in the same way. The controlling factor is the total volume of water compared to the amount of dissolved solids in the water. As water evaporates, the total volume of water decreases and the minerals in the water become more concentrated. Precipitation and runoff from the continents increase the amount of water in the oceans, increasing the volume without increasing the amount of dissolved solids to any great extent. Therefore, increased precipitation and runoff decrease salinity.

Because deep ocean areas have a much larger volume of water than shallow areas, deep ocean water is less affected by evaporation and precipitation, and runoff only happens near the shore. This is one reason why deep ocean water generally has a higher salinity than shallow water.

## Ocean Water Density

The **density** of ocean water can vary with temperature and salinity changes. Density is a property of matter that describes its mass per unit of volume. Water with a higher salinity is denser because it contains more dissolved material. If equal volumes of fresh and salt water were put on a scale, the salt water would weigh more. People float easier in salt water because it is denser than **fresh water**.

Temperature also affects the density of water in Earth's oceans. Cold water is slightly denser than warm water. Temperature is a measure of average molecular motion in a substance. At lower temperatures, water molecules are moving slower; the molecules are packed more tightly together, so the water is denser. This is true until the water freezes. Ice is one of the few solids with a lower density than its **liquid** form. That is why ice floats in water.



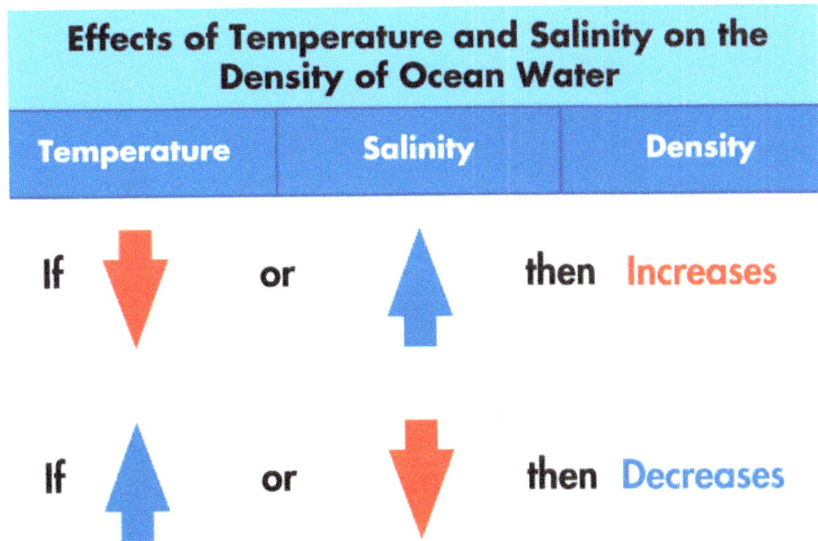


**EFFECTS ON DENSITY**
Changes in salinity and temperature affect the density of ocean water. What type of water would be denser: warm fresh water or cold salt water?

### Water Depth and Temperature

The temperature of ocean water usually decreases as depth increases. Most of the heat in oceans comes from **solar energy**. When solar radiation enters the water it transfers energy to water molecules, increasing the water's temperature. Because shallow water can absorb the solar energy, little of this warming energy is able to reach the deepest parts of the oceans. In general, the farther from the surface the water is, the colder it will be.

Oceans can be divided into different zones by temperature. The mixed layer at the surface is the area most affected by solar radiation. The temperature of this layer may change with the seasons. Below this, the water receives very little of the sun's energy, and the amount absorbed decreases rapidly with depth. This is known as the **thermocline**. The temperature drops off steadily in this zone. Sunlight does not reach the deep water zone, so the temperature doesn't change much with increased depth.

There is also a relationship between the temperature and composition of ocean water. Liquids that are colder are capable of holding more dissolved gases. For example, cold soda will stay carbonated longer than warm soda. Cold ocean water contains more dissolved oxygen and carbon dioxide than warm ocean water.

## What Role Does Carbon Dioxide Play in Ocean Acidification?

The oceans play a critical role in Earth's biogeochemical cycles. Many of the substances that enter the ocean and become dissolved in water are either precipitated out or, more commonly, incorporated into phytoplankton and passed along the ocean food chain. Many of the elements that make up these substances eventually find their way to the ocean floor where they are incorporated in ocean sediments. Phytoplankton play an important role in the carbon cycle. In fact, phytoplankton (algae and photosynthetic bacteria) conduct most of the **photosynthesis** that occurs on our planet. Carbon dioxide is one of the essential reactants for the process of photosynthesis. Carbon dioxide readily dissolves in water. This provides phytoplankton with a ready supply of carbon dioxide for photosynthesis. When carbon dioxide dissolves in water, it forms the weak acid, carbonic acid:

$$CO_2 + H_2O \rightleftharpoons H_2CO_3$$

The ocean therefore removes carbon dioxide from the air. Much of this carbon is incorporated into the bodies of marine organisms, and some remains as carbonic acid in the seawater. So what happens if the amount of carbon dioxide in the air increases because of natural or human-induced processes?

## The pH of the Ocean

pH is a measure of the hydrogen ion concentration of a solution—its acidity. For tens of millions of years, the pH of the ocean has been quite stable, held at equilibrium with the carbon dioxide in the atmosphere, at around pH 8.2—slightly basic. However, in the last two hundred years or so, its pH has begun to drop—that is, become more acidic. Currently, ocean pH is about 8.1. This may sound like a very small change, but recall that the pH scale is logarithmic, so this small change represents a 26 percent increase in acidity.

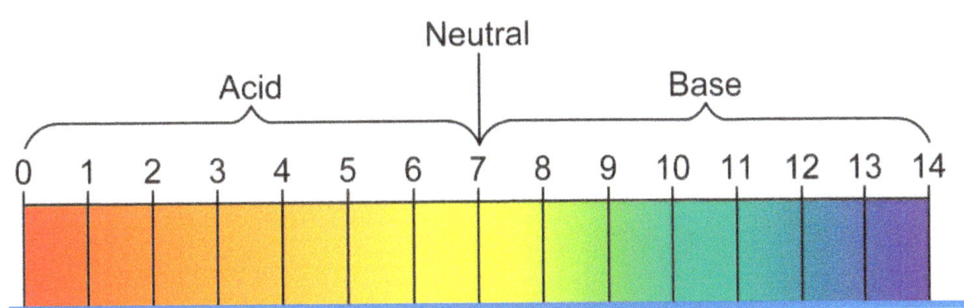


**THE pH SCALE**

Each bar on the pH scale has a numerical label indicating the pH level at that part of of the scale. Small changes on the pH scale signify large changes in acidity. Why is that?



This increase in ocean acidity is because human activities are producing carbon dioxide at unprecedented levels, and much of this ends up dissolved in ocean water. Each day, the oceans absorb over 20 million tons of carbon dioxide. At current rates of absorption, the pH of the oceans will continue to rise. By the end of the century, scientists predict that this rise could reduce ocean pH by a further 0.5 pH units. What could be the impacts of such a change in pH?

## The Impacts of Ocean Acidification

As you might expect, some photosynthetic organisms that live in the sea, such as algae and seagrasses, will benefit from higher carbon dioxide concentration. Many other organisms will be harmed, particularly those that make shells. Carbonic acid dissolves the substance—calcium **carbonate**—that many animals incorporate into their shells. Such animals include oysters, sea-urchins, crabs, and calcareous **plankton** called pteropods.

**OCEAN ACIDIFICATION**

The absorption of carbon dioxide by seawater results in the destruction of shells containing calcium carbonate. How are human activities exacerbating this problem?

Corals use calcium carbonate to build their skeletons, so coral reefs are also under threat. Ocean acidification will negatively impact most ocean food chains, including those upon which over a billion people depend upon for their main source of protein.

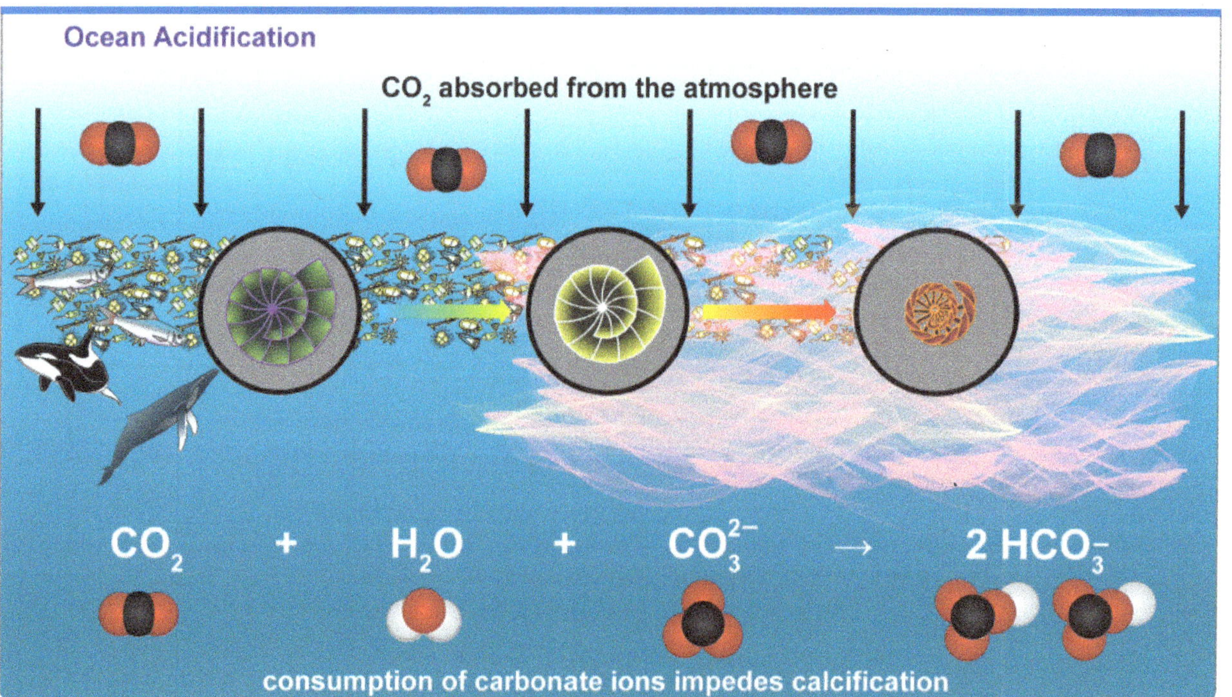
Ocean Acidification
$CO_2$ absorbed from the atmosphere
$CO_2 + H_2O + CO_3^{2-} \rightarrow 2\ HCO_3^-$
consumption of carbonate ions impedes calcification

# How Does Ocean Water Move?

## Circulation of Ocean Water

A number of factors affect the movement of water within the world's oceans. These include solar radiation, wind, **salinity** differences, currents, tides, the placement of continents, and the **Coriolis effect**.

Solar radiation is Earth's most direct surface heat source. Oceans help to distribute the sun's warmth to the whole planet. Solar radiation hits the equator region most directly. Ocean currents regulate Earth's temperature by carrying warm water away from the equator toward the poles. Deeper currents of cold water move in opposite directions to help balance the system.

Currents at or near ocean surfaces are most affected by prevailing winds. Prevailing winds, controlled by atmospheric convection currents, form bands of air that move continually in the same direction. Note that convection in the atmosphere is similar to the convection cycles that drive ocean currents and Earth's tectonic plates. In each instance, as fluids are heated, their particles gain energy and rise. As the fluids move away from the heat source, they gradually cool. Their particles lose energy and sink back down toward Earth, where they are reheated and rise again in a continuous cycle.

**WIND PATTERNS**

Prevailing winds maintain the same general direction at different latitudes and affect ocean surface currents. How is the atmosphere's circulation related to the ocean's current?



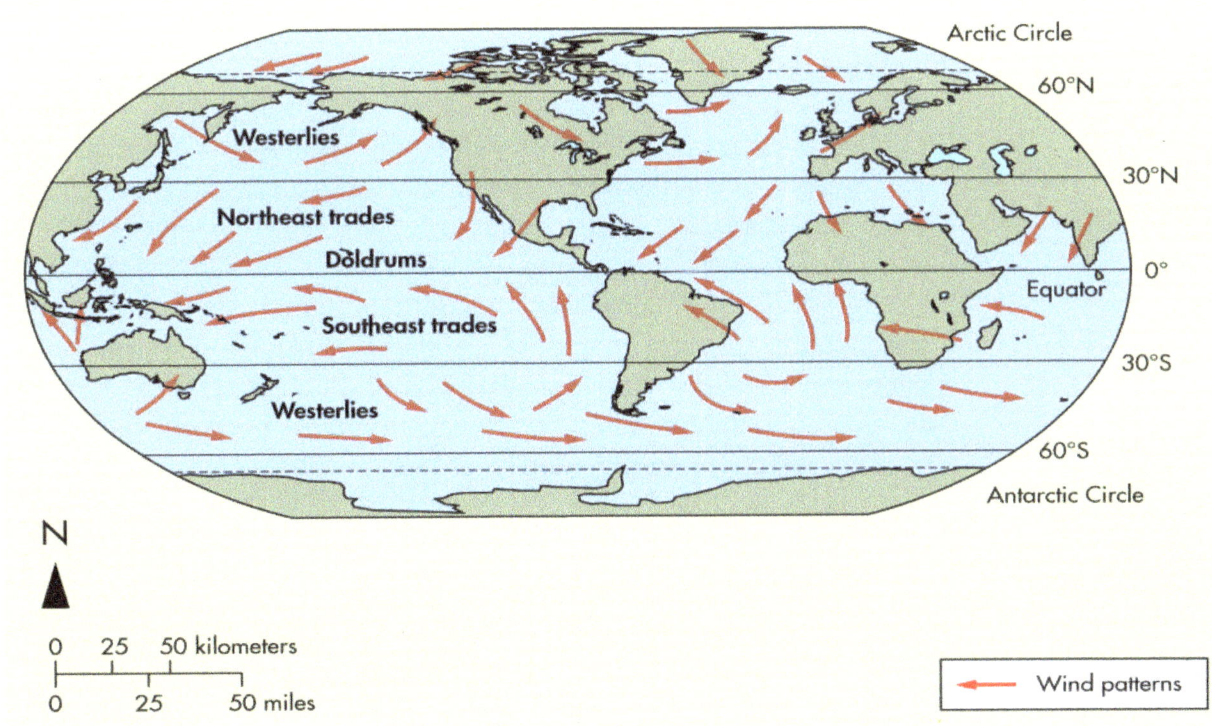

How do convection currents affect winds? As the diagram shows, each hemisphere may be divided into three distinct bands that consist of air currents blowing in alternating directions. In the tropics and polar regions, the prevailing winds blow from east to west. In the temperate regions between 30° latitude and 60° latitude, the prevailing winds blow from west to east. These winds influence the direction of surface ocean currents.

Salinity differences also help influence ocean currents. Water with a high salinity—lots of dissolved salt—is denser than water with a low salinity. Very saline water sinks to the seafloor and moves through the ocean basins as slow deep currents, perhaps not reappearing at the surface for 1,000 years. Because many currents are driven by **density** differences caused by both temperature and salt content, the circulation they create is called thermohaline circulation.

The placement of continents regulates the direction of ocean currents because water flows around large landmasses. The Atlantic, Pacific, Indian, and Arctic Oceans are considered different bodies of water or different ocean basins, because the landmasses between them cause each ocean to have its own set of currents. If there were only one ocean basin, Earth's **climate** would be much more uniform. (Indeed, scientists hypothesize this was the case during periods in the past when Earth's landmasses formed supercontinents.) The shape of ocean basins is a factor in determining global ocean circulation. Today each ocean has its own pattern of currents, which influence the climate of the ocean and the continents that surround them.

The Coriolis effect is the result of Earth's rotation. Because Earth rotates on its axis, any free-moving fluid on Earth is unable to travel in a completely straight line. In the northern hemisphere, all ocean and wind currents are deflected to the right by this effect. In the southern hemisphere, all currents are deflected to the left. This causes the curvature of ocean currents and prevailing winds. The Coriolis effect is stronger at the equator, because Earth is rotating fastest there. Although the equator and the poles complete a single rotation in the same amount of time, a point on the equator will move a much longer distance in that time than a point near the poles. Because speed equals distance traveled divided by time, the speed at which objects at the poles are rotating is much lower.

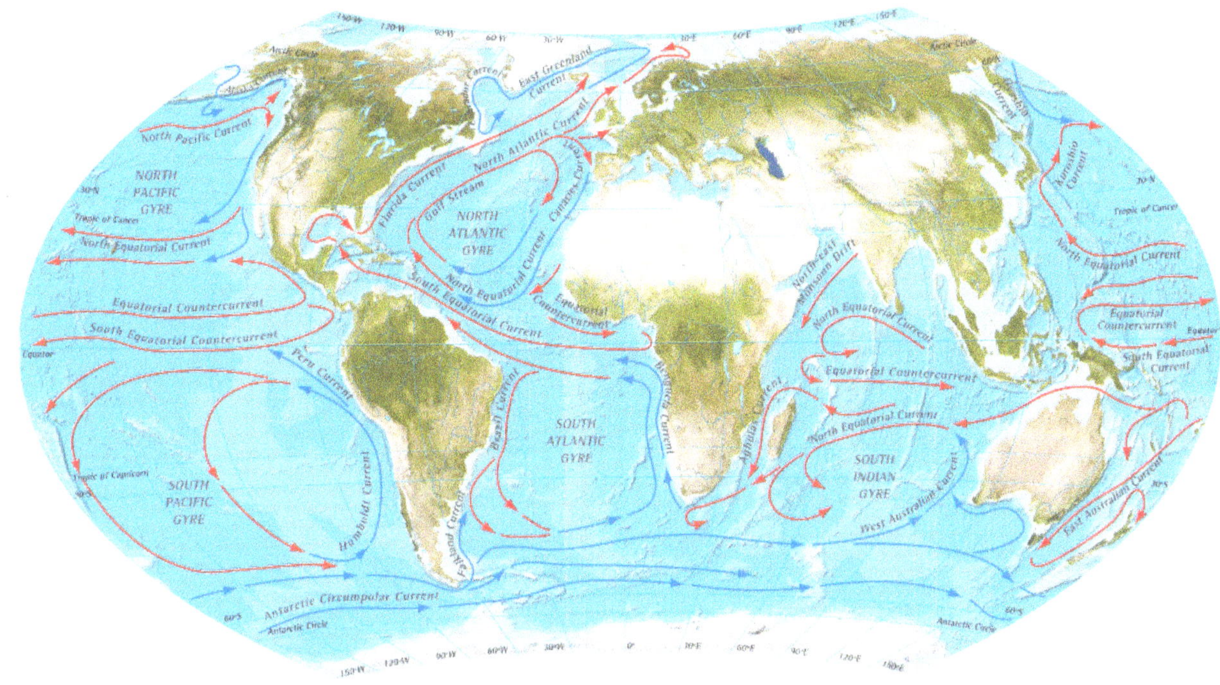


## Surface and Deep Ocean Currents

Water currents at or near ocean surfaces are usually generated by wind. These currents move quickly compared to deep ocean currents. They are mainly driven by prevailing winds, and they are deflected by the Coriolis effect.

Currents deep within Earth's oceans are driven by water density differences. Deep ocean currents are cold and move very slowly. Deep ocean currents are also connected—they do not stay in one specific ocean such as surface currents. Instead, they are part of the worldwide oceanic circulation system, often referred to as the global "conveyor belt," that helps regulate Earth's average surface temperature.

**THE CORIOLIS EFFECT**

All fluids on Earth—including prevailing winds and ocean currents—are deflected by the Coriolis effect. What direction does the water flow when you flush the toilet in the Northern Hemisphere?

As warm water moves toward Earth's poles, it becomes colder and denser. Some of this water may freeze. Because the salt does not freeze with the water, it mixes with the remaining water and increases the salinity. This cold, salty water is denser than the surrounding water, so it sinks. Warmer water moves in to take its place, and the process produces a current.

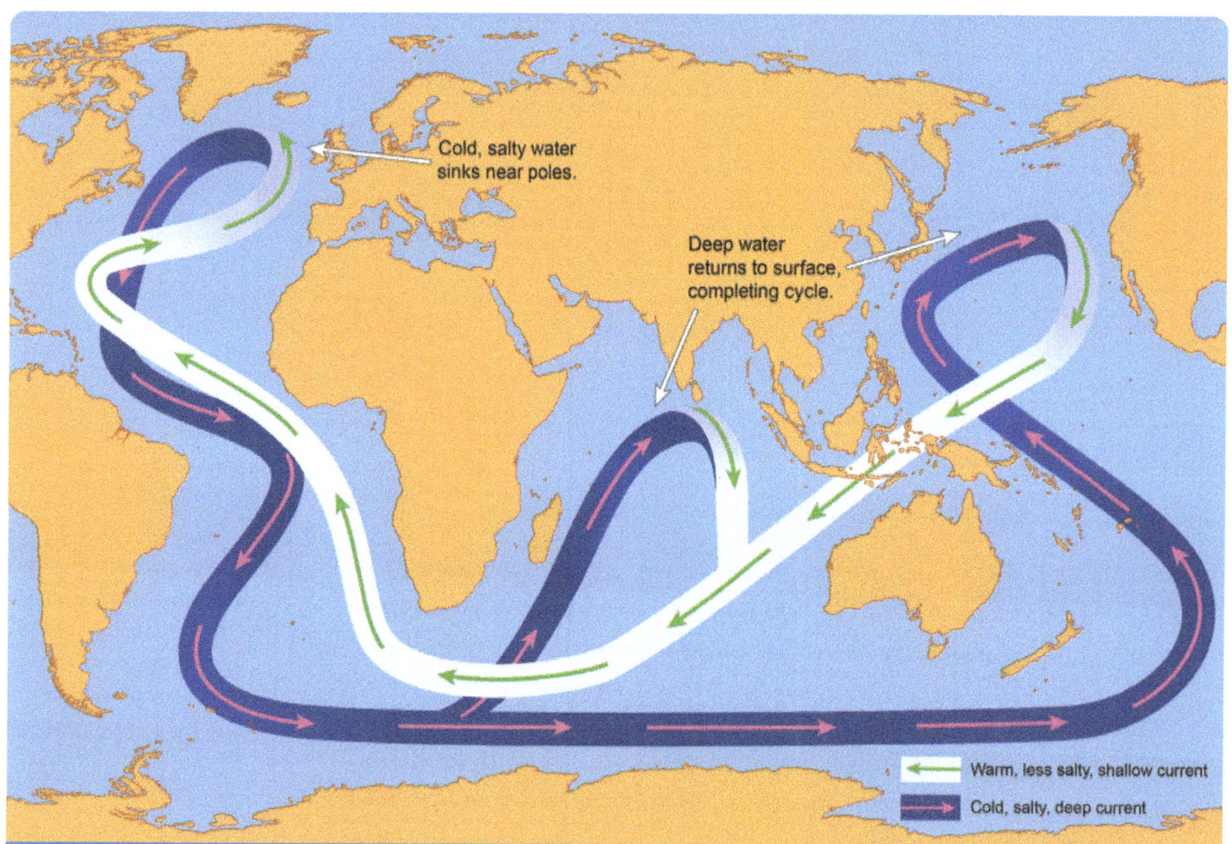
Cold, salty water sinks near poles.
Deep water returns to surface, completing cycle.
Warm, less salty, shallow current
Cold, salty, deep current

**OCEAN WATER CONVEYOR BELTC**

Ocean currents are not confined to one ocean, instead, they help form a global "conveyor belt" moving water throughout Earth's oceans. How does the chemistry of water change as it circulates around the globe?

## Upwelling

Cold water in deep ocean currents will warm up when deflected upwards by a landmass. They rise toward the surface, producing an **upwelling**. Upwellings carry nutrients as they rise to the surface. Many marine organisms are attracted to the nutrients, so the upwellings are teeming with ocean life.

Upwelling may also be driven by surface currents. When wind and surface currents push warm water away from the coast, it is replaced by cold, nutrient-rich water from below. Prevailing wind patterns contribute to this effect. These areas are often home to a great diversity of marine organisms supported by the presence of nutrient-rich waters near the surface.

## Turbidity Currents

Sometimes, along the continental shelf, underwater avalanches known as turbidity currents may occur. A disturbance, such as an earthquake, can cause sediment to rapidly travel down the continental slope. These landslides can travel at almost 100 kilometers per hour, burying anything in their paths. They finally slow down and stop as they reach the flat abyssal plain.

In 1929, an earthquake off the coast of Newfoundland generated a turbidity current. This underwater avalanche cut off 12 transatlantic telegraph cables in quick succession. The displacement of water from the turbidity current produced the only tsunami in recorded history to hit eastern North America. A tsunami is a giant wave that travels mainly under the surface of the ocean until it reaches land, where it grows quickly.

## Ocean Waves

Ocean waves are caused by winds. When wind travels across the surface of the water, it redirects the water. The water flows in the direction of the wind in the form of waves.

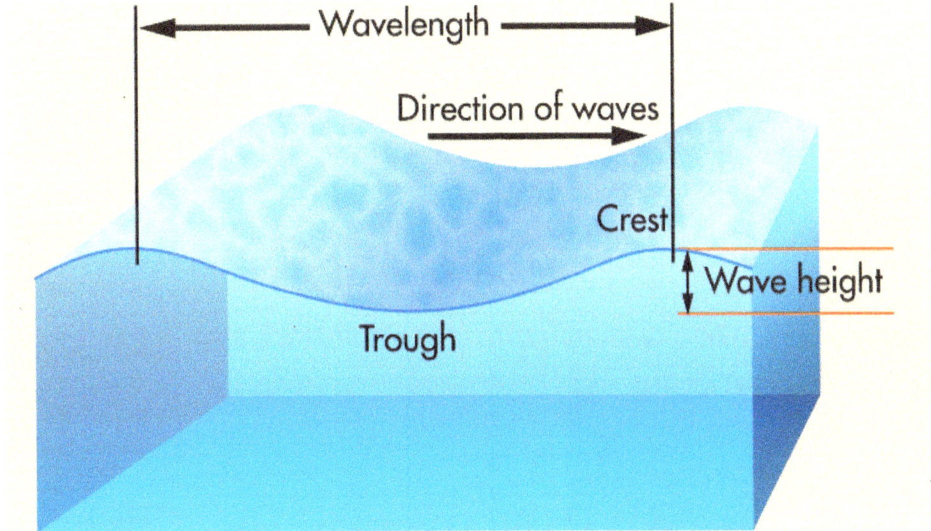


**PARTS OF A WAVE**
Wavelength is the horizontal distance between two wave crests. Wave height is the vertical distance from crest to trough of a wave. What do the trough and crest tell you about a wave?

Ocean waves, like any type of wave, are a physical form that energy takes as it travels. A number of different parameters can be measured to describe waves.

- The wave period is the amount of time between two wave crests. To be accurate, a wave's period must be measured from a fixed reference point.
- The **wave height** is a measurement of vertical distance between the wave's crest and trough. In oceans, a wave's height is twice the actual height of the wave above the ocean surface at rest.
- **Wavelength** is the horizontal distance between two wave crests, or two wave troughs.
- **Wave speed** can be calculated by dividing the wavelength by the wave period. Distance traveled divided by time elapsed gives the speed. Wave speed is typically measured in units of meters per second (m/s).

## Ocean Waves: Sample Problem

An observer is watching ocean waves from a lighthouse. He measures them carefully and finds that the distance between wave crests is 8 meters. A new wave crest passes his lighthouse every 4 seconds. What is the wave speed?

**Solution:**

Wave speed is calculated by dividing wavelength, represented by the Greek letter lambda ($\lambda$), by wave period, represented by the letter $T$. Speed is represented by $v$:

$$v = \lambda/T$$

In this example, the wavelength is 8 meters. The wave period is 4 seconds because this is the time between wave crests. Substitute these values into the equation for wave speed:

$$v = 8 \text{ m}/4 \text{ s}$$

$$v = 2 \text{ m/s}$$

The wave speed is 2 m/s.

## Waves in the Deep Ocean and near Shore

When forming waves, wind transfers energy to the water. This energy causes the movement of water molecules, which behave differently based on the depth of the water. In deep water, where the waves have little to no interaction with the ocean floor, the water molecules move down and up, in a circular pattern.

Near the shore, waves begin to interact with the ocean floor, causing water molecules to stop moving in circles. Molecular motion becomes more elliptical as the waves approach land. When the water becomes shallow enough, the circular motion becomes horizontal, toward the shore. At this point, the waves "break." The base of the wave slows down, but the crest continues at the same speed, causing the wave to tip over. This is the point where surfers begin to "hit" the waves, taking advantage of the energy at the top of the wave to propel them forward.

**WAVES HITTING A CORAL SHORE**

As the water gets shallower near shore, the bodies of waves begin to slow down, causing them to break. Where do we see the energy transfer in this picture?

Waves are capable of refraction, or bending around obstacles. This is commonly seen where points of land jut out at right angles to the wave direction. As a wave approaches the edge of the land, drag against the ocean floor causes the side closer to land to slow down while the rest of the wave continues forward. This causes a bend in the wave's overall path.

## Ocean Tides

Ocean tides are caused by gravity interactions between Earth, the moon, and the sun. The force of gravity from the sun and moon pulls on the mass of Earth's oceans. This causes sea levels to rise and fall periodically.

Gravity is a force existing between two objects, which pull on one another. Objects with more mass have stronger gravitational forces. All of the water in Earth's oceans put together has enough mass to be affected by the moon's gravity. Since it is a **liquid**, ocean water is able to change shape based on the force. Earth's water can be imagined as a sphere that changes into an elliptical shape because of the moon's gravitational pull.

Because the sun is so massive, it also has an effect on Earth's oceans, though its effect is less than the moon's because the sun is so much farther away. Distance has a much greater effect on gravitational force than mass. Gravity is affected as an inverse square of distance. If the moon were twice as far away from Earth, it would have a quarter of the gravitational force that it does now. If only the sun caused Earth's tides, they would be much less noticeable.

Tides occur on two sides of Earth at a time. There are also several different kinds of tides. During the full moon and new moon, there is a greater difference between high and low tides. At these points in the lunar cycle, Earth, the sun, and the moon are aligned and their gravitational effects are combined. This produces a stronger pull and a higher **tide** known as a **spring tide**. The low tide during a spring tide is also much lower, because more water has been pulled toward the other side of Earth.

During a first and third quarter moon, the sun and moon are at 90° angles to Earth. In these situations, the sun and moon pull on the tides from different directions. This situation produces a smaller difference between high and low tide, known as a **neap tide**.



### Consider the Explain Question

**How do the characteristics and properties of ocean water affect weather and climate?**

Go online to complete the scientific explanation.

dlc.com/ca9111s

### Check Your Understanding

**How does salinity, temperature, and dissolved gas content of ocean water affect the distribution of marine organisms in Earth's oceans?**

dlc.com/ca9112s

# in Action

## Applying Ocean Water

What happens if the ocean is polluted? How does that impact living things and systems that are affected by ocean water?

Because ocean water circulates around Earth naturally, pollution is a serious problem. Oil spills spread rapidly, for example. Because oil is generally less dense than water, most of it sits on the ocean surface following a spill. Surface currents move quickly, and wind, waves, and tides spread the oil, in some cases causing it to cover shorelines. Denser parts of the oil sink beneath the surface and are carried by deep ocean currents, making them difficult to track. Oil is not the only pollutant that is carried by the ocean currents. Plastics and other forms of debris in garbage are transported globally and threaten ecosystems at all levels.

Even in small concentrations, oil can be toxic to marine organisms. If oil damages **plankton**, the lowest parts of the marine food chain, the effects can cause food shortages for larger organisms that rely on plankton for survival. Oil coats the bodies and feathers of sea creatures and birds, making it difficult for them to swim and fly. Following an oil spill, humans do their best to clean up the pollution. An understanding of how ocean water moves is vital to cleaning up after a spill and preventing further spread of oil.

**OIL RESIDUE ON WATER**
The shimmering hues on a wet surface betray the presence of oil. What happens to the oil after it is washed away?

## STEM and Ocean Water

A mechanical **wave** transfers energy through some type of medium. The movement of ocean waves is an excellent renewable energy source. Physicists, engineers, electrical engineers, marine engineers, and other specialists have recently designed effective technologies for harnessing this energy. These electrical generation systems, located slightly offshore, use the passing energy of waves to spin turbines, which generate electricity. Because the circulation of wave energy varies with water depth, scientists have been studying which depths of ocean water generate electrical energy most efficiently.

Other scientists who study the ocean environments include oceanographers, marine biologists, and environmental engineers. For example, oceanographers, geologists, and cartographers explore and map the ocean floor. Meteorologists study ocean currents and temperatures to understand how the ocean affects our **climate**. Environmental scientists monitor ocean currents to track pollutants released in oil spills.

Marine biologists study the distribution of organisms and their feeding and breeding habits to better understand how these organisms interact with the ocean environment around them. By studying marine food chains and other biological processes, these scientists have shown that coral reefs have some of the greatest biodiversity on Earth.

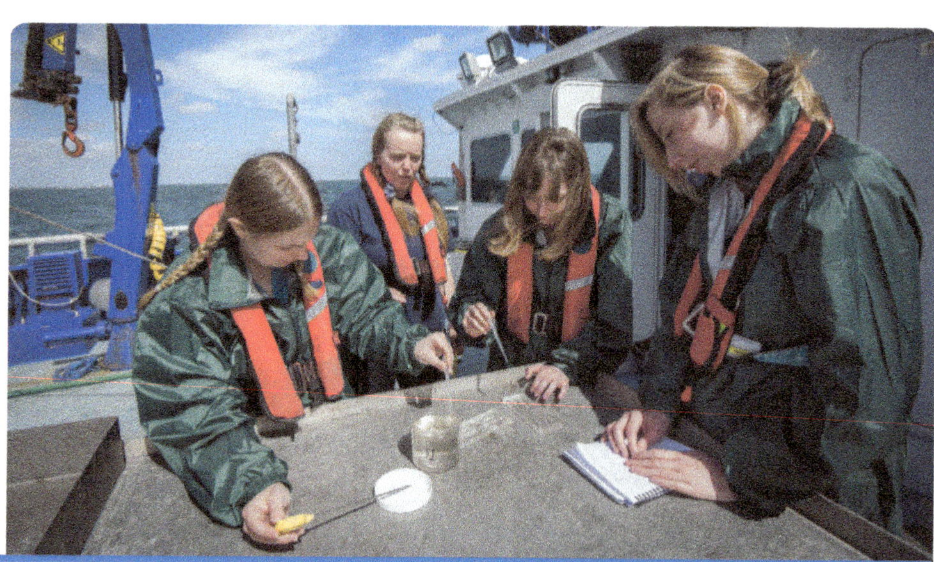

**MARINE BIOLOGISTS**
These marine biologists are sampling ocean water on a research vessel. Why do these scientists need to know about the chemical properties of ocean water?



Environmental engineers are concerned with designing tools and technologies for environmental purposes, such as harvesting wind power and hydropower for electricity. Such engineers have diverse backgrounds in electrical engineering, mechanical engineering, meteorology, and **oceanography**.

## English — A — Español

**absorption** When one substance absorbs another, it takes in the absorbed substance and distributes it throughout its bulk. For example, soil absorbs water after a rain.

**absorción** Cuando una sustancia absorbe a otra, asimila a la sustancia absorbida y la distribuye por toda su masa. Por ejemplo, el suelo absorbe agua después de la lluvia.

**acid dissociation constant (Ka)** the equilibrium constant for the ionization of an acid

**constante de disociación ácida (Ka)** constante de equilibrio de ionización de un ácido

**actinide series** the elements following actinide (Ac, atomic number 89) that fill the 5f block from period 7 on the periodic table

**serie de los actínidos** elementos que siguen al actinio (Ac, número atómico 89) y que se encuentran en el bloque 5f del periodo 7 de la tabla periódica

**activation energy** the minimum amount of energy required to initiate a chemical reaction; written as Ea and measured in kilojoules

**energía de activación** cantidad mínima de energía necesaria para iniciar una reacción química; se representa como Ea y se mide en kilojulios

**alkali metals** the group 1 (Group IA) elements: potassium (K), sodium (Na), lithium (Li), rubidium (Rb), cesium (Cs), and francium (Fr)

**metales alcalinos** elementos del grupo 1 (Grupo IA): potasio (K), sodio (Na), litio (Li), rubidio (Rb), cesio (Cs) y francio (Fr)

**alkaline earth metals** the group 2 (Group IIA) elements: beryllium (Be), magnesium (Mg), calcium (Ca), strontium (Sr), barium (Ba), and radium (Ra)

**metales alcalinotérreos** elementos del grupo 2 (Grupo IIA): berilio (Be), magnesio (Mg), calcio (Ca), estroncio (Sr), bario (Ba) y radio (Ra)

**amphoterism** the ability of a compound to act as both an acid and a base

**anfoterismo** capacidad de un compuesto para actuar como ácido o como base

**andesite** an extrusive igneous rock, light to dark gray (sometimes almost black), composed of approximately 60% silica

**andesita** roca ígnea extrusiva, de color gris claro u oscuro (a veces casi negro), compuesta aproximadamente por un 60% de sílice

**anion** an ion that has more electrons than protons and therefore has a negative charge

**anión** ión que tiene más electrones que protones, y por lo tanto, una carga negativa

**aqueous solution** a solution in which water is the solvent

**disolución acuosa** una disolución en la cual el solvente es agua

**Arrhenius acids and bases** the acid-base model in which an acid produces hydrogen ions ($H^+$) and a base produces hydroxide ions ($OH^-$) in solution

**ácidos y bases de Arrhenius** el modelo ácido-base en el cual un ácido produce iones de hidrógeno ($H^+$) y una base produce iones de hidróxidos ($OH^-$) en una disolución

**asthenosphere** The layer of soft but solid mobile rock found below the lithosphere. The asthenosphere begins about 100 km below Earth's surface and extends to a depth of about 350 km; the lower part of the upper mantle.

**astenosfera** capa de roca capaz de moverse, blanda pero sólida, que se encuentra bajo la litosfera. La astenosfera comienza aproximadamente a 100 km bajo la superficie de la Tierra y se extiende hasta una profundidad de unos 350 km; parte inferior del manto superior.

---

**atmosphere** the layers of gases that surround a planet

**atmósfera** capas de gases que rodean un planeta

---

**atom** the smallest unit of an element that has all the chemical properties of that element

**átomo** unidad más pequeña de un elemento que tiene todas las propiedades químicas de dicho elemento

---

**atomic mass unit** a unit of mass that is equal to one twelfth the mass of a Carbon-12 atom, or 1.6605 × 10-27 kg (abbreviated by using "u"

**unidad de masa atómica** unidad de masa que es igual a una doceava parte de la masa de un átomo de carbono-12, o 1.6605 × 10-27 kg (se abrevia "uma")

---

**atomic number** the number of protons in the nucleus of an atom

**número atómico** número de protones contenidos en el núcleo de un átomo

---

**atomic radius** atomic radius is a measure of the size of an atom, and is generally defined as the distance from the nucleus to the outermost electron orbit

**radio atómico** radio atómico es una medida del tamaño de un átomo y por lo general se define como la distancia desde el núcleo del átomo hasta la órbita del electrón más externo

---

**average atomic mass** the weighted average of the atomic mass of each isotope of the element, in proportion to the percentage of each isotope of the element occurring in nature

**masa atómica promedio** peso promedio de las masa atómica de cada isótopo del elemento, en proporción con el porcentaje de cada isótopo del elemento presente en la naturaleza

---

**average kinetic energy** the average amount of movement of all the particles in a system; average kinetic energy is proportional to absolute temperature

**energía cinética promedio** la cantidad promedio de movimiento de todas las partículas de un sistema; la energía cinética promedio es proporcional a la temperatura absoluta

---

**Avogadro's number** the number of particles in a mole, defined to be a quantity of 6.022 × 1023

**número de Avogadro** número de partículas en un mol, se define como una cantidad de 6.022 × 1023

---

**azimuthal quantum number (l)** the quantum number describing the orbital subshell within an atom's energy level; also referred to as the angular momentum quantum number or subsidiary quantum number

**número cuántico azimutal (l)** número cuántico que describe la subcapa orbital dentro del nivel de energía de un átomo; también se denomina número cuántico del momento angular o número cuántico secundario

# B

| | |
|---|---|
| **balanced chemical equation** an mathematical representation of a chemical process; contains chemical formulas and coefficients | **ecuación química balanceada** representación matemática de un proceso químico; contiene fórmulas químicas y coeficientes |
| **binary compound** a chemical compound that contains exactly two elements | **compuesto binario** compuesto químico que contiene exactamente dos elementos |
| **bio-fuel** a type of renewable energy source created from organisms | **biocombustible** tipo de energía renovable producida a partir de organismos. |
| **biomass** the mass of dried living matter in a given area or volume of habitat | **biomasa** masa de materia orgánica seca que se encuentra en un área dada o en un volumen de hábitat |
| **boiling point** the temperature at which the vapor pressure of a liquid equals the pressure of the surrounding environment; at this temperature, the liquid becomes a gas | **punto de ebullición** temperatura a la cual la presión del vapor de un líquido iguala la presión del entorno que lo rodea; a esta temperatura, el líquido se convierte en un gas |
| **boiling point elevation** the elevation of the boiling point of a pure liquid (solvent) caused by the addition of non-volatile solute | **elevación del punto de ebullición** elevación del punto de ebullición de un líquido puro (solvente) causado por la adición de un soluto no volátil |
| **bond polarity** the degree to which electrons are shared between atoms in a chemical bond; based on the electronegativity of atoms involved in the chemical bond | **polaridad de enlace** grado al que los electrones están compartidos entre átomos en un enlace químico; se basa en la electronegatividad de los átomos involucrados en el enlace químico |
| **Brønsted-Lowry acids and bases** the acid-base model in which an acid is a proton ($H^+$) donor and a base is a proton ($H^+$) acceptor | **ácidos y bases de Brønsted-Lowry** modelo ácido-base en el cual un ácido es un donador de protones ($H^+$) y una base es un receptor de protones ($H^+$) |
| **buffer** a substance that minimizes changes in pH when a strong acid or base is added to a solution | **amortiguador químico** sustancia que minimiza los cambios en el pH cuando se añade a una disolución un ácido fuerte o una base fuerte |
| **buffer solution** a solution that does not significantly change in pH when diluted or when relatively small quantities of acids or bases are added | **disolución amortiguadora** disolución que no cambia de manera significativa su pH cuando se diluye o cuando se le añaden cantidades relativamente pequeñas de ácidos o bases |

## C

**calorie** a unit of heat energy in the metric system

**caloría** unidad de energía calorífica en el sistema métrico

**calorimetry** the technique of measuring the energy gained or lost during a chemical reaction or physical change

**calorimetría** técnica de medición de la energía que se gana o se pierde durante una reacción química o un cambio físico

**carbon compound** compound containing chains of carbon atoms; forms the basis for organic chemistry

**compuesto de carbono** compuesto que contiene cadenas de átomos de carbono; constituye la base de la química orgánica

**carbon cycle** a natural cycle in which carbon compounds, mainly carbon dioxide, are incorporated into living tissue through photosynthesis and returned to the atmosphere by respiration, decay of dead organisms, and the burning of fossil fuels

**ciclo del carbono** ciclo natural en el cual los compuestos de carbono, principalmente el dióxido de carbono, se incorporan en los tejidos vivos a través de la fotosíntesis y regresan a la atmósfera por medio de la respiración, la desintegración de organismos muertos y la quema de combustibles fósiles

**carbon dioxide** a waste product made by cells of the body; a gas in the air made of carbon and oxygen atoms: Humans rid themselves of carbon dioxide waste by exhaling, or breathing out.

**dióxido de carbono** desecho formado por células del cuerpo; gas del aire formado por átomos de carbono y oxígeno: los humanos desechamos dióxido de carbono al exhalar o expulsar aire

**carbon reservoir** A component in the carbon cycle in which carbon is stored, such as the atmosphere, oceans, biosphere, or lithosphere. Reservoirs can serve as carbon sinks or carbon sources.

**depósito de carbono** componente en el ciclo del carbono en el cual se almacena el carbono, como la atmósfera, los océanos, la biosfera o la litosfera. Los depósitos pueden servir como sumideros de carbono o fuentes de carbono.

**carbon sink** a type of carbon reservoir, such as a forest or an ocean, known for its ability to absorb and store carbon dioxide and other greenhouse gases from the atmosphere

**sumidero de carbono** tipo de depósito de carbono como un bosque o un océano, que tiene la capacidad de absorber y almacenar dióxido de carbono y otros gases invernadero de la atmósfera

**carbonate** an ion composed of carbon and oxygen in a 1:3 ratio with a positive 2 charge ($CO_3^{2+}$); a general term for minerals and other substances that contain carbonate ions as part of their chemical structure, such as calcite, aragonite, and dolomite

**carbonato** ión compuesto por carbono y oxígeno en una proporción 1:3 con 2 cargas positivas ($CO_3^{2+}$); término general para minerales y otras sustancias que contienen iones de carbonato como parte de su estructura química, como la calcita, el aragonito y la dolomita

| | |
| --- | --- |
| **catalyst** substance that increases the rate of a chemical reaction by lowering the amount of energy needed for the reaction to occur, but is not changed by the reaction | **catalizador** sustancia que aumenta la velocidad de una reacción química al disminuir la cantidad de energía necesaria para que la reacción se produzca, pero no es modificada por la reacción |
| **cation** an ion that has more protons than electrons and therefore has a net positive charge | **catión** ión que tiene más protones que electrones, y por lo tanto, una carga positiva |
| **cell potential** a measure of the amount of electricity that a chemical battery or cell can produce (voltage) | **potencial de celda** cantidad de electricidad que puede producir una pila o celda química (voltaje) |
| **Celsius** a temperature scale in which the freezing point of water is defined as 0°C and the boiling point of water is defined as 100°C | **Celsius** escala de temperatura en la cual el punto de congelación del agua se establece como 0 °C y el punto de ebullición del agua se establece como 100 °C |
| **change of state** the process of changing matter from one form to another through heating or cooling | **cambio de estado** proceso de cambio de la materia de una forma a otra mediante calentamiento o enfriamiento |
| **chemical** a substance that has a unique composition and properties; it may exist as a solid, liquid, or gas | **sustancia química** sustancia con propiedades y composición únicas; puede existir en forma sólida, líquida o gaseosa |
| **chemical formula** a set of symbols used to describe the elements within a substance | **fórmula química** conjunto de símbolos que se usan para describir los elementos de una sustancia |
| **chemical potential energy** amount of chemical energy stored in a substance | **energía química potencial** cantidad de energía química almacenada en una sustancia |
| **chemical reaction** interaction between molecules that causes a chemical change, converting an existing substance into a new substance | **reacción química** interacción entre moléculas que produce un cambio químico y convierte una sustancia existente en una sustancia nueva |
| **chemical reactivity** the tendency of two or more chemicals to react | **reactividad química** tendencia de dos o más químicos a reaccionar |
| **chemical symbol** the one- or two-letter representation of an element; symbols are often derived from the Latin name of the element | **símbolo químico** representación de un elemento mediante una o dos letras; con frecuencia, los símbolos proceden del nombre en latín del elemento |

**climate** the current or past long-term weather conditions characteristic of a region or the entire Earth.

**clima** condiciones atmosféricas, actuales o pasadas, durante un largo periodo de tiempo, características de una región o de toda la Tierra.

---

**closed system** a system that cannot exchange matter or energy with anything outside it

**sistema cerrado** sistema que no puede intercambiar materia ni energía con nada del exterior

---

**coefficient** the number placed in front of a chemical formula to indicates the number of molecules present

**coeficiente** número que se coloca delante de una fórmula química para indicar el número de moléculas presentes

---

**colligative property** property of a solution that is dependent on the number of particles in a given volume of solvent and not on the properties of those particles

**propiedad coligativa** propiedad de una disolución que depende del número de partículas en un volumen de solvente dado y no de las propiedades de dichas partículas

---

**collision theory** faster and more frequent chemical reactions when the rate of collisions in a system is increased

**teoría de la colisión** producción de reacciones químicas más rápidas y más frecuentes cuando se aumenta la velocidad de las colisiones en un sistema

---

**combined gas law** the gas law that states that if the amount of a gas is held constant, the variables of temperature, pressure, and volume will change as described by Boyle's law and Charles's law when any one of the three variables is changed

**ley combinada de los gases** ley que establece que si una cierta cantidad de gas se mantiene constante, las variables de temperatura, presión y volumen cambiarán tal como describen las leyes de Boyle y de Charles cuando alguna de las tres variables cambia

---

**combustion reaction** a reaction in which a substance reacts with oxygen gas to produce carbon dioxide, water, and heat

**reacción de combustión** reacción en la cual una sustancia reacciona con el gas oxígeno y produce dióxido de carbono, agua y calor

---

**compound** a pure substance that is formed when two or more different elements combine; can usually be separated into component elements by chemical reaction

**compuesto** sustancia pura que se forma cuando se combinan dos o más elementos diferentes; por lo general puede separarse en los elementos que la componen mediante una reacción química

---

**compression** occurs when rocks are squeezed by external forces directed toward one another to put stress on the rock, acting to decrease its volume or shorten its dimensions

**compresión** ocurre cuando las rocas son apretadas unas junto a otras por fuerzas externas dirigidas hacia otras y que presionan la roca, dando como resultado una disminución de su volumen o un acortamiento de sus dimensiones

**concentration** the amount of a dissolved substance per unit volume of a solution

**concentración** cantidad de sustancia disuelta por unidad de volumen de una solución

**condensation** the process by which a gas changes into a liquid

**condensación** proceso mediante el cual un gas cambia a estado líquido

**condense** the physical behavior of matter, in which the temperature of a substance is lowered (energy loss) that results in; slower movement of the particles in the sample and a decrease in the space between the particles in the sample

**condensar** comportamiento típico de la materia en el cual desciende la temperatura de una sustancia (pérdida de energía) y da como resultado un movimiento más lento de las partículas de la muestra y una disminución del espacio entre dichas partículas

**conduction** the transfer of heat energy within an object, or between objects that are directly touching each other, due to collisions between the particles in the objects

**conducción** transferencia de energía térmica en el interior de un objeto, o entre objetos que están en contacto directo, debido a las colisiones entre las partículas de los objetos

**conductor** a material that allows electricity, heat or sound to pass through it easily

**conductor** material que permite el paso de la electricidad, el calor o el sonido con facilidad

**conjugate acid** part of an acid-base pair; a substance that can lose a proton to form a conjugate base

**ácido conjugado** parte de un par ácido-base; sustancia que puede perder un protón para formar una base conjugada

**conjugate base** part of an acid-base pair; a substance that can gain a proton to form a conjugate acid

**base conjugada** parte de un par ácido-base; sustancia que puede ganar un protón para formar un ácido conjugado

**conservation of energy** a law that states that energy is conserved in a closed system

**conservación de la energía** ley que establece que en un sistema cerrado la energía se conserva

**continental crust** The rocks of Earth's crust that make up the base of the continents, ranging in thickness from about 35 km to 60 km under mountain ranges. Continental crust is generally less dense than oceanic crust.

**corteza continental** rocas de la corteza terrestre que constituyen la base de los continentes, su espesor va desde alrededor de 35 km hasta 60 km bajo cadenas montañosas. Por lo general la corteza continental es menos densa que la corteza oceánica.

**continental rise** the gently sloping rise of accumulated sediments deposited by turbidity currents at the base of a continental slope

**base del talud continental** pendiente poco inclinada de sedimentos acumulados depositados por las corrientes de turbidez o de densidad en la base del talud continental

**continental shelf** the shallowest portion (approximately 130 m deep) of a continental margin, extending from the shore an average distance of 60 km into the ocean

**plataforma continental** parte menos profunda (aproximadamente 130 m de profundidad) del margen continental, se extiende desde la orilla hasta el interior del océano sobre una distancia promedio de 60 km

**continental slope** the sloping edge of the continental crust in the ocean that extends beyond the continental shelf and may be cut by sub-marine canyons adjacent to the continental shelf

**talud continental** borde de la corteza continental que formando pendiente se adentra en el océano, se extiende más allá de la plataforma continental y puede estar cortado por cañones submarinos adyacentes a la plataforma continental

**convection** the transfer of heat from one place to another caused by movement of molecules

**convección** transferencia de calor de un lugar a otro causado por el movimiento de moléculas

**convergent boundary** a tectonic plate boundary at which two tectonic plates move toward each other, causing collisions and subduction zones

**límite convergente** límite de una placa tectónica en el cual dos placas se mueven una hacia otra y producen colisiones y zonas de subducción

**coordinate covalent bond** a covalent bond where the electrons shared between atoms originate from only one of the atoms

**enlace covalente coordinado** enlace covalente en el que los electrones compartidos entre átomos se originan solo en uno de los átomos

**core** the innermost layer of Earth, comprised of the liquid outer core and solid inner core; consists mainly of iron and nickel

**núcleo** capa más interna de la Tierra, consta del núcleo externo líquido y el núcleo interno sólido; constituido principalmente por hierro y níquel

**Coriolis effect** The apparent effect of a rotating body that influences the motion of any object or fluid moving over it. In the Northern Hemisphere on Earth, air and other objects (airplanes, missiles) move toward the right. In the Southern hemisphere, air and other objects move toward the left.

**efecto Coriolis** Efecto aparente de un cuerpo que rota que influye en el movimiento de cualquier objeto o fluido que se mueve sobre él. En en hemisferio norte de la Tierra, el aire y otros objetos (aviones, misiles) se mueven hacia la derecha. En en hemisferio sur, el aire y otros objetos se mueven hacia la izquierda.

**covalent bond** a chemical link between two atoms that share electrons

**enlace covalente** unión química entre dos átomos que comparten electrones

**covalent bonding** a type of chemical bond in which electrons are shared by atoms

**enlace covalente** tipo de enlace químico en el cual los electrones son compartidos por los átomos

**critical point** a combination of temperature and pressure beyond which a liquid cannot boil

**punto crítico** combinación de temperatura y presión superior a la que un líquido no puede hervir

**crystal lattice** an ordered, three-dimensional, repeating arrangement of atoms, ions, or molecules in a crystal

**red cristalina** disposición repetitiva, tridimensional y ordenada de átomos, iones o moléculas en un cristal

**cycle** the movement of a substance through different reservoirs or stages

**ciclo** paso de una sustancia a través de diferentes lugares o etapas

## D

**decomposition** the breaking down of a compound into elements or simpler compounds as chemical bonds are broken and atoms or ions are rearranged

**descomposición** desintegración de un compuesto en elementos o compuestos más simples cuando se rompen los enlaces químicos y lo átomos o iones toman una nueva disposición

**decomposition reaction** a reaction that involves the breakdown of substances into two or more substances

**reacción de descomposición** reacción que involucra la desintegración de sustancias en dos o más sustancias

**deep ocean current** the circulation of deep water in the oceans as a result of thermal and density differences.

**corriente oceánica profunda** circulación de agua en las profundidades del océano como resultado de diferencias térmicas y de densidad

**density** an object's mass divided by its volume; a measure of the amount of matter in a given amount of space; calculated using the equation $d = m/v$

**densidad** masa de un objeto dividida entre su volumen; medida de la cantidad de materia en un espacio determinado; se calcula usando la ecuación $d = m/v$

**deposition (phase change)** phase change of matter in which a gas turns directly into a solid

**deposición (cambio de fase)** cambio de fase de la materia en la que un gas se convierte directamente a sólido

**dilute solution** a solution containing a small amount of solute relative to the amount solvent

**disolución diluida** disolución que contiene una pequeña cantidad de soluto en relación con la cantidad de solvente

**dipole** a molecule with a positive charge at one end and a negative charge at the other

**dipolo** molécula con una carga positiva en un extremo y una carga negativa en el otro

**dipole-dipole force** a strong intermolecular force that occurs between two polar molecules; the opposing charges of each dipole are attracted to one another

**fuerza dipolo-dipolo** gran fuerza intermolecular que se produce entre dos moléculas polares; las cargas opuestas de cada dipolo se atraen entre sí

**dissociation** the reversible separation of a substance into smaller components, which may be particles or ions

**disociación** separación reversible de una sustancia en componentes más pequeños, los cuales pueden ser partículas o iones

**dissolve** to add a solid material to a liquid in such a way that its particles completely disperse into the liquid, usually becoming invisible within the liquid

**disolver** añadir un material sólido a un líquido de manera que sus partículas se dispersen completamente en el líquido, por lo general estas partículas se volverán invisibles dentro del líquido

**distance** the amount of space between one point and another

**distancia** espacio entre dos puntos

**divergent boundary** a tectonic plate boundary at which two tectonic plates move away from each other

**límite divergente** límite de una placa tectónica en el cual dos placas tectónicas se mueven separándose una de otra

**double bond** a covalent bond in which two pairs of electrons are shared by two atoms

**enlace doble** enlace covalente en el cual dos pares de electrones son compartidos por dos átomos

**double displacement reaction** a chemical process in which positive ions and negative ions of two reactants switch places

**reacción de desplazamiento doble** proceso químico en el que los iones positivos y negativos de dos reactivos cambian de posición

**double-replacement reaction** a chemical reaction between two compounds where the cations in two compounds switch the anions they are bonded to, resulting in two new compounds; also called double-displacement reactions

**reacción de doble sustitución** reacción química entre dos compuestos en la que los cationes de dos compuestos intercambian los aniones a los que están unidos, dando como resultado nuevos compuestos; también se denomina reacción de doble sustitución

**ductility** the ability of a material to be stretched into a wire without breaking

**ductilidad** capacidad de un material de alargarse y formar un alambre sin romperse

## E

**electrochemical cell** a device in which an electrolyte and two electrodes are connected; includes both electrolytic cells (non-spontaneous) and voltaic cells (spontaneous)

**celda electroquímica** dispositivo en el cual están conectados un electrolito y dos electrodos; incluye celdas electrolíticas (no espontáneas) y células voltaicas (espontáneas)

**electrolysis** the process of using electricity to cause a chemical reaction that would not occur spontaneously

**electrolisis** proceso en el que se usa electricidad para producir una reacción química que no ocurriría de manera espontánea

**electrolyte** a chemical compound that forms ions in an aqueous solution and is capable of conducting an electric current

**electrolito** compuesto químico que forma iones en una disolución acuosa y es capaz de conducir una corriente eléctrica

**electrolytic cell** a cell that uses electricity to produce a chemical reaction that would not occur spontaneously

**celda electrolítica** celda que usa electricidad para producir una reacción química que no ocurriría de manera espontánea

**electromagnetic radiation** a form of energy that travels through space and exhibits wave-like behavior

**radiación electromagnética** forma de energía que viaja a través del espacio y exhibe un comportamiento similar al de las ondas

**electromagnetic spectrum** the full range of frequencies and wavelengths of electromagnetic waves

**espectro electromagnético** rango completo de frecuencias y longitudes de onda de las ondas electromagnéticas

**electromagnetic wave** a wave that can transport its energy through a vacuum

**onda electromagnética** onda que puede transportar su energía a través del vacío

**electron** a negatively charged subatomic particle that exists in various energy levels outside the nucleus of an atom

**electrón** partícula subatómica con carga negativa que existe en varios niveles de energía alrededor del núcleo de un átomo

**electron cloud** the region surrounding an atomic nuclei where electrons may be found

**nube de electrones** región que rodea el núcleo atómico y en la que se encuentran los electrones

**electron configuration** the arrangement of electrons in the energy levels and orbitals within an atom

**configuración electrónica** disposición de los electrones en niveles de energía y orbitales alrededor del núcleo de un átomo

**electron pair acceptor** a chemical species that accepts an electron pair, also known as a Lewis acid

**receptor de un par de electrones** especie química que acepta un par de electrones, también conocida como ácido de Lewis

**electron pair donor** a chemical species that donates an electron pair, also known as a Lewis base

**donador de un par de electrones** especie química que dona un par de electrones, también conocida como base de Lewis

**electronegativity** the tendency of an atom to attract electrons and acquire a negative charge

**electronegatividad** tendencia de un átomo a atraer electrones y adquirir una carga negativa

**element** a pure substance that cannot be broken down into simpler substances by chemical reactions and that is made up of only one type of atom

**elemento** sustancia pura que no puede dividirse en sustancias más simples mediante reacciones químicas y que está constituida solo por un tipo de átomos

**emission** release of energy or matter from components in a system

**emisión** liberación de energía o materia de los componentes de un sistema

**empirical formula** chemical formula showing the simplest ratio of atoms present in a compound, rather than the total number of atoms in each molecule

**fórmula empírica** fórmula química que muestra la proporción más simple de los elementos presentes en un compuesto en lugar del número total de átomos de cada molécula

**endothermic** a reaction that absorbs heat from the surrounding area

**endotérmico** reacción que absorbe calor de su entorno

**energy** the ability to do work or cause change; can be stored in chemicals found in food and released to the organism to do work

**energía** capacidad de realizar trabajo o producir un cambio; puede almacenarse en sustancias químicas que se encuentran en los alimentos y liberarse al organismo para realizar trabajo

**energy levels** the energies of electrons as they orbit the nucleus of an atom

**niveles de energía** energía de los electrones que orbitan alrededor del núcleo del átomo

**enthalpy** measure (abbreviated "H") of the total amount of energy produced by a chemical system

**entalpía** medida (simbolizada H) de la cantidad total de energía producida por un sistema químico

**entropy** measure (abbreviated "S") of the amount of energy in a physical system that is not available to do work

**entropía** medida (simbolizada "S") de la cantidad de energía en un sistema físico que no está disponible para realizar trabajo

**enzyme** proteins that catalyze specific chemical reactions in organisms by lowering the activation energy of the reaction

**enzima** proteína que cataliza reacciones químicas específicas en los organismos al disminuir la energía de activación de la reacción

**equilibrium** state of balance between all parts of a system

**equilibrio** estado de balance entre todas las partes de un sistema

**equilibrium constant (Keq)** a constant (Keq) that relates the concentration of reactants and products at equilibrium for a given chemical reaction

**constante de equilibrio (Keq)** constante (Keq) que relaciona la concentración de reactantes y productos en equilibrio para una reacción química dada

**Ernest Rutherford** New Zealand-born chemist and physicist (1871–1937) known as "the father of nuclear physics"; Rutherford discovered the proton and formulated the Rutherford, or planetary, model of the atom

**Ernest Rutherford** Químico y físico nacido en Nueva Zelanda (1871–1937), conocido como "el padre de la física nuclear"; Rutherford descubrió el protón y formuló el modelo planetario de átomo o modelo de Rutherford

| | |
|---|---|
| **evaporation** the process in which matter changes from a liquid to a gas | **evaporación** proceso en el cual la materia cambia de estado líquido a gaseoso |
| **exothermic** a reaction that releases heat to the surrounding area | **exotérmico** reacción que libera calor a su entorno |

## F

| | |
|---|---|
| **Fahrenheit** a temperature scale and a unit of temperature where the freezing point of water is defined as 32°F and the boiling point of water is defined as 212°F official temperature scale in the United States | **Fahrenheit** escala de temperatura y unidad de temperatura en la cual el punto de congelación del agua se estable como 32 °F y el punto de ebullición del agua se establece como 212 °F escala de temperatura oficial en Estados Unidos |
| **families** vertical columns of elements (groups) on the periodic table with similar valence electron configurations and similar properties | **familias** columnas verticales de elementos (grupos) en la tabla periódica con valencias, configuraciones electrónicas y propiedades similares |
| **Faraday's Law** in an electrolytic cell, the amount of a substance that is consumed or produced at an electrode is proportional to the amount of electric current that passes through the electrolyte | **ley de Faraday** en una celda electrolítica, la cantidad de una sustancia que se consume o se produce en un electrodo es proporcional a la cantidad de corriente eléctrica que pasa a través del electrolito |
| **fault-block mountain** a mountain that forms along normal faults due to the tilting, uplift, or dropping of large pieces of crust | **montaña de bloques fallados** montaña que se forma a lo largo de fallas normales debido a la inclinación, elevación o descenso de grandes trozos de corteza |
| **first law of thermodynamics** states that energy can be changed from one form to another, stored, or transferred, but cannot be created or destroyed; means that the amount of matter and energy in the Universe is fixed; also known as The Law of the Conservation of Energy | **primera ley de la termodinámica** establece que la energía puede cambiar de una forma a otra, puede almacenarse o transferirse, pero no puede crearse ni destruirse; esto significa que la cantidad de materia y energía en el universo es siempre la misma; también se conoce como "principio de conservación de la energía" |
| **folded mountain** a mountain formed by large-scale folding and uplift | **montaña de pliegue** montaña formada por un plegamiento y una elevación a gran escala |
| **force** an interaction between objects that causes a change to the objects' motion. | **fuerza** interacción entre objetos que origina un cambio en el movimiento del objeto |

**formula mass** total mass of the atoms in a compound

**masa de la fórmula** masa total de los átomos en un compuesto

**formula unit** lowest ratio of ions or atoms in a molecule represented as whole number

**fórmula unitaria** menor razón de iones o átomos de un molécula representada como número entero

**freezing** the change of state from liquid to solid

**congelación** cambio de estado líquido a sólido

**freezing point** the temperature at which a liquid turns into a solid

**punto de congelamiento** temperatura a la cual un líquido se transforma en sólido

**freezing point depression** the depression of the freezing point of a pure liquid (solvent) caused by the addition of non-volatile solute

**descenso del punto de congelación** descenso del punto de congelación de un líquido puro (solvente) producido por la adición de un soluto no volátil

**frequency** the number of waves that pass a given point during a specified period of time

**frecuencia** número de ondas que pasan por un punto dado durante un periodo de tiempo específico

**frequency** the number of times that an event is repeated within a certain period; often used to refer to the number of full cycles of a wave that pass a given point in a specific period of time

**frecuencia** número de veces que se repite un evento en un cierto periodo de tiempo; con frecuencia se usa para hacer referencia al número de ciclos completos de una onda que pasan por un punto dado en un periodo de tiempo específico

**fresh water** water that contains less than 0.2% dissolved salts; may or may not be suitable for drinking

**agua dulce** agua que contiene menos de un 0.2% de sales disueltas; puede ser o no ser potable (apta para beber)

**fresh water** water with a low salt concentration—usually 1% or less

**agua dulce** agua con una concentración baja de sal; por lo general, de menos del 1%

## G

**gamma ray** an electromagnetic wave with a very short wavelength (typically less than 10–10 m) and very high energy and frequency (typically greater than 1019 Hz)

**rayos gamma** ondas electromagnéticas con una longitud de onda muy corta (normalmente menos de 10–10 m) y con energía y frecuencia muy altas (por lo general de más de 1019 Hz)

**gas** a state of matter without any defined volume or shape in which atoms or molecules move about freely

**gas** estado de la materia que no tiene volumen ni forma definidos en el cual los átomos o moléculas se mueven casi libremente

**geochemical cycle** the cycle of stages that elements go through during geological changes

**ciclo geoquímico** ciclo de las etapas que atraviesan los elementos durante los cambios geológicos

**geometry** the mathematical study of shapes, sizes, space, and the relationships between them; in chemistry, the three-dimensional arrangement of atoms in a molecule

**geometría** estudio matemático de las formas, tamaños y del espacio, así como las relaciones entre ellos; en química, la disposición tridimensional de átomos en una molécula

**giant impact hypothesis** the generally accepted hypothesis about the origin of Earth's moon, according to which Earth was struck by a Mars-sized body, and the debris from this impact combined in orbit around Earth to form the moon

**teoría del gran impacto** hipótesis generalmente aceptada sobre el origen de la luna de la Tierra, según la cual la Tierra fue impactada por un cuerpo del tamaño de Marte y los restos de este impacto se unieron en una órbita alrededor de la Tierra y formaron la luna

**Gibbs Free Energy** the energy associated with a chemical reaction that can be used to do work

**energía libre de Gibbs** energía asociada a una reacción química que puede usarse para realizar un trabajo

**global warming** the slow increase of Earth's average global atmospheric temperature due to climatic change

**calentamiento global** lento aumento de la temperatura atmosférica promedio de la Tierra debido al cambio climático

**global wind** the atmospheric circulation around Earth that occurs in predictable patterns

**vientos planetarios** circulación atmosférica alrededor de la Tierra que ocurre siguiendo patrones predecibles

**Gondwana** one of six supercontinents that existed during the Paleozoic Era; composed mostly of modern-day South America, Africa, Antarctica, Australia, and India

**Gondwana** uno de los seis supercontinentes que existieron durante la Era Paleozoica; dio lugar a la mayor parte de lo que actualmente es América del Sur, África, Antártida, Australia e India

**gravitational force** an attractive force that occurs between all objects that have mass

**fuerza gravitacional** fuerza de atracción que se produce entre todos los objetos que tienen masa

**greenhouse gas** a gas, usually carbon-based, that contributes to global warming through the greenhouse effect, which prevents the escape of radiant heat from Earth's atmosphere

**gas invernadero** gas, por lo general a base de carbono, que contribuye al calentamiento global mediante el efecto invernadero, el cual impide que el calor radiante salga de la atmósfera terrestre

**groups** vertical columns of elements (families) on the periodic table with similar valence electron configurations and similar properties

**grupos** columnas verticales de elementos (familias) en la tabla periódica con valencias, configuraciones electrónicas y propiedades similares

## H

**half-cell** the portion of the electrochemical cell in which the oxidation or reduction reaction occurs

**media celda** porción de la celda electroquímica en la cual tiene lugar la reacción de oxidación o reducción

**halogens** the group 17 (Group VIIA) elements; consists of fluorine (F), chlorine (Cl), bromine (Br), iodine (I), and astatine (At)

**halógenos** elementos del grupo 17 (Grupo VIIA); los halógenos son: flúor (F), cloro (Cl), bromo (Br), yodo (I) y astato (At)

**heat** the transfer of thermal energy through processes such as radiation, convection, or conduction

**calor** transferencia de energía térmica por medio de procesos como radiación, convección o conducción

**heat energy** a form of energy that transfers between particles in a substance or system through kinetic energy transfer

**energía calorífica** forma de energía que se transfiere entre partículas en una sustancia o en un sistema por medio de transferencia de energía cinética

**heat of reaction** the amount of heat gained or lost in a chemical reaction; calculated by subtracting the final enthalpy (Hf) from the initial enthalpy (Hi)

**calor de reacción** cantidad de calor ganada o perdida en una reacción química; se calcula restando la entalpía final (Hf) de la entalpía inicial (Hi)

**Hess's Law** states that the change in enthalpy (?H) during a chemical process is the same, regardless of how many steps the process takes

**ley de Hess** establece que el cambio en la entalpía (?H) durante un proceso químico es el mismo, independientemente del número de pasos que tenga el proceso

**heterogeneous** made up of two or more distinct components; usually refers to a mixture in which individual substances are distinct

**heterogéneo** que está compuesto por dos o más componentes diferentes; por lo general se refiere a una mezcla en la cual se distinguen sustancias individuales

**homogeneous** made up of one uniform component; often used to describe a mixture with a consistent composition and a single phase throughout

**homogéneo** hecho de un componente uniforme; con frecuencia el término se usa para describir una mezcla con una composición uniforme y una sola fase

**hot spot** an anomalously hot, stationary point below the lithosphere where melting of the mantle and crust generates volcanism at Earth's surface

**punto caliente** lugar bajo la litosfera con más calor de lo normal y donde la fusión de materiales del manto y la corteza genera fenómenos de vulcanismo en la superficie terrestre

**hybrid bonding** bonds created from atomic orbitals that combine to form hybrid orbitals which replace simple orbitals

**enlace híbrido** enlaces creados por orbitales atómicos que se combinan para formar orbitales híbridos que remplazan los orbitales simples

**hydrogen bonding** a type of intermolecular force in which a hydrogen atom in one molecule interacts with a highly electronegative atom in another molecule

**enlace de hidrógeno** tipo de fuerza intermolecular en la cual un átomo de hidrógeno en una molécula interactúa con un átomo con una fuerte carga eléctrica negativa de otra molécula

**hydrogen ion** an ionic substance ($H^+$) formed when a hydrogen atom loses an electron

**ión de hidrógeno** sustancia iónica ($H^+$) que se forma cuando un átomo de hidrógeno pierde un electrón

**hydronium ion** an ionic substance ($H_3O^+$) formed when a water molecule combines with a hydrogen ion

**ión hidronio** sustancia iónica ($H_3O^+$) que se forma cuando una molécula de agua se combina con un ión de hidrógeno

**hydroxide ion** a negative ion ($OH^-$) formed when a water molecule loses a hydrogen ion

**ión hidróxido** ión negativo ($OH^-$) que se forma cuando una molécula de agua pierde un ión de hidrógeno

## I

**ideal solution** a solution in which the interaction of the components with one another is no different than the interaction of each component with itself

**disolución ideal** disolución en la cual la interacción de los elementos no es diferente de la interacción de cada componente con él mismo

**indicator** a substance added in small amounts to a solution in order to provide a visual reference for the pH of the solution

**indicador** sustancia que se añade en pequeña cantidad a una disolución con el fin de proporcionar una referencia visual del pH de dicha disolución

**induced dipole** a dipole in a molecule that is created by the charge of a neighboring dipole; usually produced from a molecule with an instantaneous dipole

**dipolo inducido** dipolo en una molécula creado por la carga en un dipolo cercano; por lo general se produce a partir de una molécula con un dipolo instantáneo

**inert gas** a gas that does not tend to react with other substances

**gas inerte** gas que no tiende a reaccionar con otras sustancias

**inhibitor** a substance that delays, slows, or prevents a chemical reaction; sometimes also called a negative catalyst

**inhibidor** sustancia que retrasa, hace más lenta o impide una reacción química; algunas veces también se denomina catalítico negativo

**inner core** the solid, inner portion of Earth's core, composed of an alloy of iron, nickel, and other heavy elements; rotates within the liquid outer core

**núcleo interno** parte interna y sólida de la Tierra, compuesta por una aleación de hierro, níquel y otros elementos pesados; rota dentro del núcleo externo líquido

**instantaneous dipole** a temporary dipole that occurs in an atom and causes a similar dipole in a neighboring atom; a weak intermolecular force in a nonpolar molecule

**dipolo instantáneo** dipolo temporal que ocurre en un átomo y origina un dipolo similar en un átomo cercano; fuerza débil intermolecular en una molécula no polar

**insulator** a material that does not readily conduct electricity or heat

**aislante** material que no conduce fácilmente la electricidad o el calor

**intermolecular force** interactions that occur between molecules; London forces, dipole-dipole forces and hydrogen bonding are the three main types of intermolecular forces

**fuerza intermolecular** interacciones entre moléculas; las fuerzas de London, fuerzas dipolo-dipolo y los enlaces de hidrógeno son los tres principales tipos de fuerzas intermoleculares

**inversely proportional** a relationship between two variables in which an increase in the value of one causes a decrease in the value of the other

**inversamente proporcional** relación entre dos variables en la cual un aumento en el valor de una produce una disminución en el valor de la otra

**ion** a charged atom or group of atoms formed by the addition or removal of one or more electrons

**ión** átomo o grupo de átomos con carga, se forma por la incorporación o la pérdida de uno o más electrones

**ion product constant of water** the ion product constant of water, 1 × 10-14 , is determined by finding the product of the concentrations of hydrogen ions and hydroxide ions in water at 25°C

**constante del producto iónico del agua** constante del producto iónico del agua, 1 × 10-14 , se determina hallando el producto de las concentraciones de los iones de hidrógeno y los iones hidróxidos en agua a 25 °C

**ionic bond** a chemical bond in which one atom gives up electrons (cation) to another atom that gains the electrons (anion); results from a difference in electronegativity greater than 1.7

**enlace iónico** enlace químico en el cual un átomo cede electrones (catión) a otro átomo que gana los electrones (anión); se produce por una diferencia en la electronegatividad superior a 1.7

**ionic radius** the distance between an ion's nucleus and its outer electrons (those electrons remaining after the valence electrons have been lost)

**radio iónico** distancia entre el núcleo de un ión y sus electrones más externos (los que quedan después de haberse perdido los electrones de valencia)

**ionic size** the size of an ion in a crystal lattice

**tamaño iónico** tamaño de un ión en una red cristalina

**ionization** the creation of an ion through the loss of one or more electrons from an atom

**ionización** creación de un ión mediante la pérdida de uno o más electrones de un átomo

**ionization energy** the minimum energy required to remove an electron from the ground state of an atom (also known as the binding energy)

**energía de ionización** energía mínima requerida para extraer un electrón en el estado fundamental de un átomo (también conocida como energía de enlace)

**isomer** compounds with identical molecular formulas that have different chemical structures.

**isómero** compuestos con fórmulas moleculares idénticas que tienen estructuras químicas diferentes

**isotope** an atom of the same element with the same number of protons but a different number of neutrons; has the same atomic number but different atomic mass

**isótopo** átomo del mismo elemento con el mismo número de protones pero diferente número de neutrones; tiene el mismo número atómico pero diferente masa atómica

## J

**J.J. Thomson** British physicist (1856–1940) whose cathode ray tube experiment in 1897 showed that atoms were composed of smaller particles and led him to formulate the “plum pudding” model of the atom

**J. J. Thomson** Físico británico (1856–1940) cuyo experimento con el tubo de rayos catódicos en 1897 demostró que los átomos están compuestos por partículas más pequeñas y lo condujo a formular el modelo atómico del “pudín de pasas”

**joule** a unit of measurement of energy

**julio** unidad de medida de la energía

## K

**kelvin** a temperature scale and the SI unit of temperature where the freezing point of water is 273.15°K and the boiling point of water is 373.15°K; one degree kelvin is equal to one degree Celsius

**kelvin** escala de temperatura y unidad de temperatura del SI, en la que el punto de congelación del agua es 273.15 °K y el punto de ebullición del agua es 373.15 °K; un grado kelvin es igual a un grado Celsius

**kinetic energy** the energy an object has due to its motion

**energía cinética** energía que tiene un objeto debido a su movimiento

**Kw** the ion product constant of water, 1 × 10-14, is determined by finding the product of the concentrations of hydrogen ions and hydroxide ions in water at 25°C

**Kw** constante del producto iónico del agua, 1 × 10-14 , se determina hallando el producto de las concentraciones de los iones de hidrógeno y los iones hidróxidos en agua a 25 °C

## L

**lanthanide series** the elements following lanthanum (La, atomic number 57) that fill the 6f block from period 6 on the periodic table

**serie de los lantánidos** elementos que siguen al lantano (La, número atómico 57) y que se encuentran en el bloque 6f del periodo 6 de la tabla periódica

**latent heat of fusion** the energy required to melt a unit of mass of a substance when its temperature is already at the melting point

**calor latente de fusión** energía requerida para fundir una unidad de masa de una sustancia cuando su temperatura ya está en el punto de fusión

**latent heat of vaporization** the energy required to change a liquid to vapor when its temperature is already at the melting point

**calor latente de evaporación** energía requerida para transformar un líquido en vapor cuando su temperatura ya está en el punto de derretimiento

**law of conservation of mass** a fundamental principle of science stating that mass cannot be created or destroyed in a closed system through ordinary chemical or physical means

**ley de conservación de la masa** principio fundamental de la ciencia que establece que en un sistema cerrado, la masa no puede crearse ni destruirse por medios físicos o químicos ordinarios

**law of conservation of matter** the law that states that during a chemical reaction matter cannot be created or destroyed

**ley de conservación de la materia** ley que establece que durante una reacción química no puede crearse ni destruirse materia

**Le Chatelier's Principle** concentration, temperature, volume, or pressure), then the equilibrium moves to counteract the change and a new equilibrium is established

**principio de Le Châtelier** establece que si un equilibrio se desajusta cambiando las condiciones (como concentración, temperatura, volumen o presión), el equilibrio varía para contrarrestar el cambio y establecer un nuevo equilibrio

**Lewis (electron) dot structure** a representation of an atom, ion, or molecule that uses the chemical symbol of an element to represent one atom and dots to represent one likely configuration of the valence electrons

**estructura de Lewis o diagrama de puntos** representación de un átomo, ión o molécula en la que se usa el símbolo químico de un elemento para representar un átomo y puntos para representar una configuración de los electrones de valencia

**Lewis acids and bases** according to Lewis acid-base theory, the electron pair donors and acceptors function as bases and acids in an acid-base reaction

**ácidos y bases de Lewis** según la teoría ácido-base de Lewis, los pares de electrones donadores y receptores funcionan como bases y ácidos en una reacción ácido-base

**light** a type of electromagnetic radiation, commonly referred to visible light or light which can be directly seen with the human eye

**luz** tipo de radiación electromagnética; el término se usa comúnmente referido a la luz visible o luz que el ojo humano puede ver directamente

**limiting reagent** the reactant that is completely used up in a chemical reaction, leaving some of the other reactants with nothing to react with

**reactivo limitante** reactante que se gasta completamente en una reacción química, dejando parte de los otros reactantes sin nada con que reaccionar

**liquid** a state of matter with a defined volume but no defined shape and whose molecules roll past each other

**líquido** estado de la materia con un volumen definido pero no forma definida y cuyas moléculas se deslizan unas sobre otras

**lithosphere** the part of Earth which is composed mostly of rocks; the crust and outer mantle

**litosfera** parte de la Tierra compuesta principalmente por rocas; corteza y manto exterior

**logarithmic** a scale in which the logarithm of a number with a given base value is the exponent that the base must be raised to in order to obtain the number

**logarítmica** escala en la cual el logaritmo de un número con una base dada es el exponente al cual hay que elevar la base para obtener el número

**London dispersion force** the intermolecular force between noble gas atoms or nonpolar molecules, created when an instantaneous dipole induces a dipole on neighboring atoms/molecules

**fuerzas de dispersión de London** fuerza intermolecular entre átomos de gases nobles o moléculas no polares, se crean cuando un dipolo instantáneo induce un dipolo en los átomos o moléculas cercanos

## M

**magma** melted rock located beneath Earth's surface

**magma** roca fundida que se encuentra debajo de la superficie de la Tierra

**magnetic orbital quantum number** the quantum number describing the orientation orbitals within a subshell in which electron may reside

**número cuántico magnético orbital** número cuántico que describe la orientación de orbitales dentro de una subcapa en la cual los electrones pueden residir

**magnetic spin quantum number** the quantum number describing the spin state of an electron

**número cuántico magnético de espín** número cuántico que describe el estado de espín de un electrón

**malleability** the ability of a material to be beaten into a thin sheet or to be shaped without breaking

**maleabilidad** capacidad de un material de formar una lámina sin romperse cuando se golpea

**mantle** the layer of solid rock between Earth's crust and core

**manto** capa de roca sólida entre la corteza y el centro de la Tierra

**mass** the amount of matter in an object

**masa** cantidad de materia en un objeto

**mass number** the total number of neutrons and protons in an atom (also known as nucleon number)

**número de masa** número total de neutrones y protones en un átomo (también conocido como número másico)

| | |
|---|---|
| **mass percent** the mass of a solute divided by the total mass of the solution; it can also represent the percent mass of a particular element in a molecule or compound | **porcentaje de masa** masa de un soluto dividida entre el total de la masa de la disolución; también puede representar el porcentaje de masa de un elemento determinado en una molécula o compuesto |
| **matter** material that has mass and takes up some amount of space | **materia** material que tiene masa y ocupa espacio |
| **melting** the phase change from a solid to a liquid | **fusión** cambio de fase de sólido a líquido |
| **melting point** the temperature at which matter changes from solid to liquid state | **punto de fusión** temperatura a la cual la materia cambia de estado sólido a líquido |
| **metallic character** a set of chemical properties associated with elements that are metals | **carácter metálico** conjunto de propiedades químicas asociadas a elementos que son metales |
| **metalloids** elements that lie between the metals and the nonmetals on the periodic table; these element exhibit both metallic and nonmetallic properties | **metaloides** elementos que se encuentran entre los metales y los no metales en la tabla periódica; estos elementos exhiben propiedades metálicas y no metálicas |
| **metals** elements that give up electrons easily, are malleable and ductile, and are good electrical conductors | **metales** elementos que ceden electrones con facilidad, son maleables y dúctiles, y son buenos conductores eléctricos |
| **methane** molecule comprising one carbon atom and four hydrogen atoms | **metano** molécula que tiene un átomo de carbono y cuatro átomos de hidrógeno |
| **metric** a system of measurement based on the meter and kilogram | **métrico** sistema de medida basado en el metro y el kilogramo |
| **microwave** part of the electromagnetic spectrum with wavelengths between infrared radiation and radio waves | **microonda** parte del espectro electromagnético con longitudes de onda entre la radiación infrarroja y las ondas de radio |
| **mid-ocean ridge** an oceanic rift zone that consists of long mountain chains with a central rift valley; divergent boundary | **dorsales centro-oceánicas** zona dorsal centro-oceánica que consta de largas cadenas montañosas con una fosa tectónica central; límite divergente |
| **mixture** a combination of two or more substances that do not combine chemically | **mezcla** combinación de una o más sustancias que no se combinan químicamente |

**molality** the ratio of a solute (measured in moles) to the solvent (measured in kilograms) in a solution

**molalidad** proporción de un soluto (medido en moles) en un solvente (medido en kilogramos) en una disolución

**molar mass** measure of the average molecular mass of all the molecules in an element or compound. It is the mass of the substance (typically in grams) per amount of the substance (in moles)

**masa molar** medida del promedio de la masa molecular de todas las moléculas en un elemento o compuesto; masa de una sustancia (normalmente en gramos) por cantidad de sustancia (en moles)

**molarity** the number of moles of solute divided by the volume of the solution

**molaridad** número de moles de soluto dividido entre el volumen de la disolución

**mole** an SI unit for the measure of a quantity; one mole is equivalent to Avogadro's number of a substance or 6.022 × 1023; also defined as the number of atoms in 12 g of carbon-12

**mol** unidad del SI para medir una cantidad; un mol es equivalente al número de Avogadro de una sustancia, o 6.022 × 1023; también se define como el número de átomos en 12 g de carbono-12

**mole fraction** the number of moles of solute divided by the number of moles of a solution

**fracción molar** número de moles de soluto dividido entre el número de moles de la disolución

**molecular bonding** the forces that hold atoms together in elements and compounds

**enlace molecular** fuerzas que mantienen los átomos juntos en elementos y en compuestos

**molecular formula** A chemical formula that gives the total number of atoms of each element present in a molecule

**fórmula molecular** fórmula química que da el número total de átomos de cada elemento presente en una molécula

**molecular mass** the mass in grams of one mole of molecules of a substance

**masa molecular** masa en gramos de un mol de moléculas de una sustancia

**molecular polarity** the degree to which the electrons are shared equally between atoms in a molecule

**polaridad molecular** grado al que los electrones están compartidos igualmente entre átomos en una molécula

**molecule** a group of atoms held together by chemical bonds

**molécula** grupo de átomos unidos por enlaces químicos

**monatomic ion** an ion formed from a single atom

**ión monoatómico** ión formado por un solo átomo

## N

**neap tide** shallow ocean tides that occur during first- and third-quarter moons

**marea muerta** mareas oceánicas de poca amplitud que se producen durante el cuarto creciente y el cuarto menguante de la luna

**negative charge** a charge of the same sign as that of an electron, often due to a surplus of electrons

**carga negativa** carga del mismo signo que la de un electrón, con frecuencia se debe a un exceso de electrones

**Nernst equation** an equation that relates the potential of an electrochemical cell to standard electrode potentials and effective concentrations

**ecuación de Nernst** ecuación que relaciona el potencial de una celda electroquímica con los potenciales estándar de electrodo y concentraciones efectivas

**net ionic equation** a chemical equation from which the nonparticipating (spectator) ions are excluded

**ecuación iónica neta** ecuación química en la que se excluyen los iones no participantes (espectadores)

**neutralization** a reaction in which an acid and base fully react to produce only water and salts

**neutralización** reacción en la cual un ácido y una base reaccionan completamente para producir solo agua y sales

**neutron** a subatomic particle with no electric charge that is located within the nucleus of the atom and has a mass approximately equal to 1 amu

**neutrón** partícula subatómica sin carga eléctrica que se encuentra dentro del núcleo de un átomo y tiene una masa aproximadamente igual a 1 uma

**nitrogen cycle** a process in which nitrogen in the atmosphere enters the soil and becomes part of living organisms then eventually returns to the atmosphere

**ciclo del nitrógeno** proceso en el cual el nitrógeno de la atmósfera penetra en el suelo y se convierte en parte de los organismos vivos, luego, con el tiempo, regresa a la atmósfera

**noble gas** one of six elemental gasses from group 18 of the periodic table that does not naturally react with other elements because they each contain a stable valence shell (8 valence electrons)

**gas noble** uno de seis gases elementales del grupo 18 de la tabla periódica que no reaccionan naturalmente con otros elementos porque contienen una capa de valencia estable (8 electrones de valencia)

**noble gas configuration** the electron configuration of the Group 18 element, in which the valence shell is filled with two or eight electrons

**configuración de un gas noble** la configuración electrónica de los elementos del grupo 18, en el cual la capa de valencia está llena con dos u ocho electrones

**nonmetals** elements that are poor electrical conductors and tend to accept electrons easily

**no metales** elementos que son malos conductores eléctricos y tienden a aceptar electrones con facilidad

**nonpolar covalent bond** a chemical bond between two atoms that have the same electronegativity or are identical

**enlace covalente no polar** enlace químico entre dos átomos que tienen la misma carga eléctrica negativa o que son idénticos

**non-spontaneous** A reaction that cannot occur without the input of work from an external source

**no espontáneo** reacción que no puede ocurrir sin la aportación de trabajo de una fuente externa

**nucleus (atom)** the core of an atom where most of the mass of an atom exists; contains protons and neutrons

**núcleo (del átomo)** el núcleo o centro de un átomo es donde está la mayor cantidad de masa del átomo; contiene protones y neutrones

## O

**ocean current** horizontal and vertical circulation system of ocean waters, produced by gravity, wind friction, and variation in water density

**corriente oceánica** sistema de circulación, horizontal y vertical, de aguas del océano, producida por la gravedad, la fricción del viento y las variaciones en la densidad del agua

**ocean wave** movement of water generally caused by wind blowing on the ocean surface

**ola oceánica** movimiento de agua generalmente producido por el viento que sopla sobre la superficie del océano

**oceanic crust** the portion of Earth's crust that makes up the ocean floor and is generally denser and thinner than continental crust

**corteza oceánica** parte de la corteza terrestre que conforma el fondo del océano y que por lo general es más densa y más fina que la corteza continental

**oceanography** the study of oceans

**oceanografía** estudio de los océanos

**octet rule** the rule that states that atoms tend to be the most stable when their valence shells are completely full; normally, when the valence shell contains eight valence electrons.

**regla del octeto** regla que establece que los átomos tienden a ser más estables cuando sus capas de valencia están completamente llenas; normalmente, cuando la capa de valencia contienen ocho electrones de valencia

**orbital diagram** a diagram used to illustrate the configuration of electrons within the orbitals in an atom; electrons are represented by arrows pointing up or down (denoting each electron's spin) in a boxes representing each orbital

**diagrama de orbital** diagrama que se usa para ilustrar la configuración de los electrones en los orbitales de un átomo; los electrones se representan por flechas que apuntan hacia arriba o hacia abajo (reflejando el giro de cada electrón) en recuadros que representan cada orbital

**orbital model of the atom** a model proposed in 1922 by Niels Bohr combining the planetary model of the atom with his findings about quantums of energy and electrons; in this model electrons travel in orbits around an atoms nucleus in paths that are defined by the electrons' energy

**modelo orbital del átomo** modelo propuesto en 1922 por Niels Bohr combinando el modelo planetario de átomo con sus descubrimientos sobre los cuantos de energía y los electrones; en este modelo, los electrones se desplazan en órbitas alrededor del núcleo de un átomo en trayectorias definidas por la energía de los electrones

**order of a chemical reaction** the relationship between reactants in a particular rate equation that also indicates which reactant drives the reaction

**orden de una reacción química** relaciones entre los reactantes en una ecuación cinética particular que también indica qué reactante desencadena la reacción

**osmotic pressure** the amount of external pressure required for osmosis to occur; considered a colligative property of a material

**presión osmótica** cantidad de presión externa que se requiere para que se produzca la ósmosis; se considera una propiedad coligativa de un material

**outer core** the liquid outer portion of Earth's core, composed primarily of iron and nickel

**núcleo externo** parte exterior, líquida, del núcleo de la Tierra, compuesto principalmente por hierro y níquel

**outgassing** 1. the release of previously dissolved or trapped gases from a substance or object 2. the process through which magmatic gases are released into Earth's atmosphere

**desgasificación** 1. liberación de gases previamente disueltos o atrapados por una sustancia u objeto 2. proceso mediante el cual se liberan los gases magmáticos a la atmósfera terrestre

**oxidation** a chemical reaction resulting in the loss of electrons by a metal; for example, when iron rusts

**oxidación** reacción química que resulta de la pérdida de electrones por parte de un metal; por ejemplo, cuando el hierro se oxida

**oxidation number** a whole number that represents the number of electrons that would need to be added or removed to create a neutral atom of the same element

**número de oxidación** un número entero que representa el número de electrones que se necesitaría añadir o retirar para crear un átomo neutro del mismo elemento

**oxidation-reduction reaction** a chemical reaction (redox reaction) that involves the transfer of one or more electrons from one species to another

**reacción de reducción-oxidación** reacción química (reacción redox) que implica la transferencia de uno o más electrones de una especie a otra

**oxidizing agent** the reactant that accepts electrons during an oxidation-reduction (redox) reaction

**oxidante (agente oxodante)** reactante que acepta electrones durante una reacción de reducción-oxidación (reacción redox)

**ozone layer** an atmospheric region located in Earth's stratosphere which contains ozone (O3) and absorbs harmful ultraviolet radiation emitted by the sun

**capa de ozono** región de la estratosfera terrestre que contiene ozono (O3) y absorbe radiación ultravioleta dañina emitida por el sol

## P

**Pangaea** the large supercontinent at the end of the Paleozoic Era consisting of all the land on Earth, including all seven continents and other landmasses

**Pangea** gran supercontinente al final de la Era Paleozoica que abarcaba toda la tierra de la Tierra, es decir, todos los siete continentes y las demás masas de tierra

**parts-per-million (ppm)** unit of concentration; ratio of solute to solvent in a solution

**partes por millón (ppm)** unidad de concentración; proporción de soluto respecto al solvente en una disolución

**Pauli Exclusion Principle** an important rule in quantum mechanics that states that no two electrons may have the same set of quantum numbers

**principio de exclusión de Pauli** importante regla en mecánica cuántica que establece que ninguno de dos electrones pueden tener el mismo conjunto de números cuánticos

**percent composition** the percent of the total molecular mass made up of a specific component

**composición porcentual** porcentaje del total de masa molecular que corresponde a un componente específico

**percent yield** the actual yield of a chemical reaction, based on the percentage of the theoretical yield

**rendimiento porcentual** rendimiento real de una reacción química, basado en el porcentaje de rendimiento teórico

**period** one of the horizontal rows in the periodic table

**periodo** línea horizontal de la tabla periódica

**periodic table** a chart that scientists use to organize and classify all the different elements

**tabla periódica** tabla que usan los científicos para organizar y clasificar todos los elementos

**periodic table of the elements** the chart scientists use to organize and classify all the known elements

**tabla periódica de los elementos** tabla que usan los científicos para organizar y clasificar todos los elementos conocidos

**periodic trend** the tendency of certain elemental properties to increase or decrease as one moves in a certain direction along the periodic table

**tendencia periódica** tendencia de ciertas propiedades elementales a aumentar o disminuir a medida que nos desplazamos en cierta dirección a lo largo de la tabla periódica

**pH** a scale that measures the acidity of a solution based on the concentration of hydrogen ions present

**pH** escala que mide la acidez de una disolución basándose en la concentración de iones de hidrógeno presentes en dicha disolución

**phase change** a change from one state of matter (solid, liquid, gas) to another without a change in chemical composition

**cambio de fase** cambio de un estado de la materia (sólido, líquido, gaseoso) a otro sin que se produzca un cambio en la composición química

**phase diagram** chart of pressure and temperature showing changes in states of matter

**diagrama de fase** gráfica de presión y temperatura que muestra cambios en los estados de la materia

**phase of matter** the physical forms in which matter exists, as determined by the average kinetic energy of the atoms involved; the most common phases of matter are solid, liquid, gas and plasma; also called state of matter;

**fase de la materia** forma física en la cual existe la materia, determinada por la energía cinética promedio de los átomos involucrados; las fases más comunes de la materia son sólido, líquido, gaseoso y plasma; también se denominan estados de la materia

**phosphorous cycle** the transfer of phosphorous between the biosphere, lithosphere, and hydrosphere

**ciclo del fósforo** transferencia del fósforo entre la biosfera, litosfera e hidrosfera

**photon** a bundle or quantum of electromagnetic energy that demonstrates particle-like behavior

**fotón** haz o cuanto de energía electromagnética que muestra un comportamiento similar al de una partícula

**photosynthesis** the biological process by which most plants, some algae, and some bacteria produce organic compounds for their food from water and carbon dioxide using solar energy

**fotosíntesis** proceso biológico por el cual la mayoría de las plantas, algunas algas y algunas bacterias producen compuestos orgánicos que les sirven de alimento; estos compuestos los producen a partir de agua y dióxido de carbono usando la energía solar

**Planck's wave equation** an equation that defines the relationship between the energy and frequency of a photon

**ecuación de Planck** ecuación que define la relación entre la energía y la frecuencia de un fotón

**plankton** small organisms that drift through bodies of water; include animals, plants, and bacteria

**plancton** pequeños organismos que van a la deriva a través de los cuerpos de agua; incluye animales, plantas y bacterias

**plasma** A gas-like state of matter in which most of the particles are charged ions.

**plasma** estado de la materia semejante al gas en el cual la mayoría de las partículas son iones cargados.

**plate motion** the motion of tectonic plates, which occurs at a rate of a few centimeters per year

**movimiento de placas** movimiento de las placas tectónicas que ocurre a un ritmo de pocos centímetros al año

**pOH** a scale that measures the alkalinity of a solution based on the concentration of hydroxide ions

**pOH** escala que mide la alcalinidad de una disolución basándose en la concentración de iones hidróxidos presentes en dicha disolución

**polar covalent bond** a chemical bond between two atoms that unequally share electrons between them

**enlace covalente polar** enlace químico entre dos átomos que comparten electrones de manera desigual

**polarity** the separation of opposite charge present in a molecule, which is the result of the atoms in a molecule having a difference in electronegativity

**polaridad** separación de cargas opuestas presentes en una molécula, lo cual es el resultado del hecho de que los átomos de una molécula tengan diferente carga eléctrica negativa

**polyatomic ion** an ion that is formed from tow or more atoms covalently bonded that has a charge due to the a difference in the number of protons and electrons that are available in the new molecule

**ión poliatómico** ión formado a partir de dos o más átomos unidos con enlace covalente que tiene carga debida a la diferencia entre el número de protones y electrones disponibles en la nueva molécula

**positive charge** a charge of the same sign as that of a proton, often due to a deficit of electrons

**carga positiva** carga del mismo signo que la de un protón, con frecuencia se debe a un déficit de electrones

**precipitate** solid particles that have been separated from a solution

**precipitado** partículas sólidas que se han separado de una disolución

**prefix** letters added to the front of a word to modify its meaning

**prefijo** letras que se añaden delante de una palabra para modificar su significado

**pressure** force exerted per unit area by many particles of a gas randomly striking the walls of its container

**presión** fuerza ejercida por unidad de área por muchas partículas de un gas que chocan aleatoriamente contra las paredes del recipiente que lo contiene

**principal quantum number (n)** the quantum number describing the energy level that an electron occupies

**número cuántico principal (n)** número cuántico que describe el nivel de energía que ocupa un electrón

**product** the substances that are formed during a chemical reaction; in a chemical reaction the arrows point toward the product

**producto** sustancias que se forman durante una reacción química; en una reacción química las flechas apuntan hacia el producto

**protic acid** a chemical compound that dissociates in water to form a hydrogen ion and an ionic conjugate base

**ácido prótico** compuesto químico que se disocia en agua para formar un ión de hidrógeno y una base conjugada iónica

**proton** a subatomic particle with a positive charge that is located in the nucleus of an atom and has a mass of approximately 1 amu

**protón** partícula subatómica con carga eléctrica positiva que se encuentra dentro del núcleo de un átomo y tiene una masa aproximada de 1 uma

## Q

**quantum mechanical model of the atom** the model of the atom based on the ideas of wave-particle duality and quantized matter and energy

**modelo mecánico cuántico del átomo** modelo del átomo basado en las ideas de la dualidad onda-partícula y la materia y la energía cuantizadas

**quantum mechanics** a description of the dual particle-wave behavior and interactions of energy and matter; also known as quantum physics or quantum theory

**mecánica cuántica** descripción del comportamiento dual onda-partícula y las interacciones de la energía y la materia; también se conoce como física cuántica o teoría cuántica

**quantum number** a unique set of numbers that describes the location of each electron in an atom or ion

**número cuántico** conjunto de números único que describe la ubicación de cada electrón en un átomo o en un ión

## R

**radiation** a process by which energetic electromagnetic waves move from one place to another

**radiación** proceso por el cual las ondas energéticas electromagnéticas se desplazan de un lugar a otro

**radioactive decay** a process by which an unstable atom loses energy by emitting ionized particles over a period of time

**desintegración radiactiva** proceso por el cual un átomo inestable pierde energía al emitir partículas ionizadas durante un periodo de tiempo

**reactant** the substances present before a chemical reaction occurs; in a chemical reaction the arrow usually points away from the reactant(s)

**reactante** sustancias presentes antes de que ocurra una reacción química; en una reacción química la flecha suele señalar en dirección contraria del reactante o reactantes

**reaction rate** the rate or speed at which chemical reactions occur

**velocidad de reacción** velocidad o rapidez con que ocurre una reacción química

**reactivity series** a table that orders metals based on their reactivity, usually from most to least reactive

**serie de reactividad** tabla en la que se ordenan los metales según su reactividad, por lo general van del más reactivo al menos reactivo

**redox reaction** a short-hand term for an oxidation-reduction reaction, a chemical reaction that involves the transfer of one or more electrons from one species to another

**reacción redox** término abreviado para reacción de reducción-oxidación, reacción química que implica la transferencia de uno o más electrones de una especie a otra

**reducing agent** the reactant that donates electrons during an oxidation-reduction (redox) reaction

**agente reductor** reactante que dona electrones durante una reacción de reducción-oxidación (reacción redox)

**reduction** a decrease in the oxidation state of an atom or molecule due to the gain of electrons

**reducción** disminución en el estado de oxidación de un átomo o una molécula debido a la ganancia de electrones

**residence time** the average length of time that a resource remains in a fixed location or condition

**tiempo de residencia** duración promedio del tiempo que un recurso permanece en un lugar o en una condición determinadas

**resonance structure** a Lewis structure drawn to represent all of the several possible bonding structures for the molecule

**estructura de resonancia** estructura de Lewis que se dibuja para representar todas las estructuras de enlace posibles para la molécula

**rift valley** a tectonic valley that forms by extensional stress which causes fracturing and the formation of normal faults

**fosa tectónica** valle tectónico que se forma por tensión de extensión, la cual provoca la fractura y la formación de fallas normales

**rift zone** a divergent boundary where the crust is pulled apart

**zona de fractura** límite divergente en el cual la corteza terrestre se separa

**rip current** a strong, narrow, surface current that flows out to sea from near shore

**corriente de resaca** corriente superficial fuerte y estrecha que fluye desde la orilla a alta mar

**rock cycle** a model describing the transformations of rocks from one major rock type to another

**ciclo de las rocas** modelo que describe las transformaciones de las rocas desde un tipo de roca principal a otro

**rotation** the spinning of a celestial body, such as a planet, around an axis

**rotación** giro de un cuerpo celeste, como un planeta, alrededor de un eje

## S

**salinity** the total quantity of dissolved salts in water

**salinidad** cantidad total de sales disueltas en agua

**salt** an ionic compound consisting of positive and negative ions that results from an acid reacting with a base

**sal** compuesto iónico formado por iones positivos y negativos que es el resultado de la reacción de un ácido con una base

**saltwater** a general term for water containing more than 1 part per thousand (1 g/L) of total dissolved solids

**agua salada** término general que designa agua que contiene más de 1 parte por millar (1 g/L) de sólidos disueltos totales

**saturated** having no double bonds in hydrocarbon chains

**saturado** que no tiene enlaces dobles en cadenas de hidrocarburos

**seafloor spreading** the process by which new oceanic lithosphere forms at mid-ocean ridges as tectonic plates pull away from each other

**expansión del fondo oceánico** proceso por el cual se forma nueva litosfera oceánica en las dorsales centro-oceánicas a medida que las placas tectónicas se separan una de otra

**seamount** a volcano under the ocean that rises at least 1 km from the seafloor

**monte submarino** volcán bajo el océano que se eleva al menos 1 km por encima del fondo oceánico

**second law of thermodynamics** states that energy can only transfer from a higher energy state into a lower energy state and the movement of energy stops when both states are in equilibrium

**segunda ley de la termodinámica** establece que la energía solo puede transferirse de un estado de mayor energía a un estado de menor energía y la transferencia de energía se detiene cuando ambos estados alcanzan un equilibrio

**shielding effect** decreased attraction of electrons away from the nucleus

**efecto blindaje** disminución en la atracción entre los electrones y el núcleo

**single bond** a covalent bond in which one pair of electrons is shared by two atoms

**enlace simple** enlace covalente en el cual un par de electrones es compartido por dos átomos

**single displacement reaction** a chemical process in which an element or ion moves from one reactant to another

**reacción de desplazamiento simple** proceso químico en el que un elemento o ion se mueve de un reactivo a otro

**single-replacement reactions** a reaction in which one atom or ion in a compound is replaced by another atom or ion

**reacción de sustitución simple** reacción en la cual un átomo o un ión de un compuesto es sustituido por otro átomo o ión

**solar energy** radiant energy that comes from the sun

**energía solar** energía radiante que procede del sol

**solid** the state of matter that has a defined shape and volume with molecules that are closely packed and vibrate in place

**sólido** estado de la materia que tiene forma y volumen definidos con moléculas que están muy juntas y que vibran en el sitio

**solubility** the amount of a substance (solute) that will dissolve in another substance (solvent) at a certain volume and temperature

**solubilidad** cantidad de sustancia (soluto) que se disuelve en otra sustancia (solvente) a un cierto volumen y temperatura

**solubility product constant (Ksp)** the relative solubility of a substance

**constante del producto de solubilidad** solubilidad relativa de una sustancia

**solute** the substance that is dissolved in another substance

**soluto** sustancia que se disuelve en otra sustancia

**solution** a homogeneous mixture of two substances that appear as only one state of matter

**disolución** mezcla homogénea de dos sustancias que aparece como un solo estado de la materia

**solvent** the substance that dissolves another substance

**solvente** sustancia que disuelve otra sustancia

**sp orbital** a hybrid orbital formed by the combination of an s orbital and a p orbital

**orbital sp** orbital híbrido formado por la combinación de un orbital s y un orbital p

**sp2 orbital** a hybrid orbital formed by the combination of an s orbital and two p orbitals

**orbital sp2** orbital híbrido formado por la combinación de un orbital s y dos orbitales p

**sp3 orbital** a hybrid orbital formed by the combination of an s orbital and three p orbitals

**orbital sp3** orbital híbrido formado por la combinación de un orbital s y tres orbitales p

**specific heat** amount of heat, expressed in calories, needed to raise one gram of a substance by 1°C

**calor específico** cantidad de calor, expresada en calorías, que se necesita para elevar en 1 °C la temperatura de un gramo de una sustancia

**speed of light (c)** the velocity of light in a vacuum, measured as approximately 300,000,000 meters per second

**velocidad de la luz (c)** la velocidad de propagación de la luz en el vacío es aproximadamente de 300,000,000 metros por segundo

**spontaneous** a phenomena or reaction that appears to occur without any external action

**espontáneo** fenómeno o reacción que parece ocurrir sin intervención de ninguna acción externa

**spring tide** high-amplitude ocean tides that occur during new and full moons

**marea viva** mareas oceánicas de gran amplitud que se producen durante la luna nueva y la luna llena

**standard heat of formation** the change of heat when a compound is formed; is always zero for an element in its standard state; also called enthalpy of formation

**entalpía de formación** cambio de temperatura cuando se forma un compuesto; para un elemento en estado estándar es siempre cero

**standard reduction potential** the measurement of the ability of a substance to be reduced; measured in volts

**potencial estándar de reducción** medida de la capacidad o tendencia de una sustancia a reducirse o ganar electrones; se mide en voltios

**standard temperature and pressure (STP)** the conditions of temperature at 0°C (273.15°K) and 1 atmosphere of pressure

**temperatura y presión estándar** condiciones de temperatura a 0 °C (273.15 °K) y 1 atmósfera de presión

**stoichiometry** the area of chemistry that deals with the quantities of substances that react and are produced during chemical reactions

**estequiometría** área de la química que se ocupa de las cantidades de sustancias que reaccionan y que se producen durante las reacciones químicas

**strong acid** an acid that is completely or almost completely dissociated in water

**ácido fuerte** ácido que se disocia completa o casi completamente en agua

**strong base** a base that is completely or almost completely dissociated in water

**base fuerte** base que se disocia completa o casi completamente en agua

**subatomic** of or relating to something having dimensions smaller than that of an atom

**subatómico** perteneciente a o relacionado con algo que tiene dimensiones más pequeñas que las de un átomo

**subduction** the sinking of an oceanic plate beneath a plate of lesser density at a convergent boundary

**subducción** hundimiento de una placa oceánica bajo una placa de menor densidad en un límite convergente

**subduction zone** a convergent boundary where oceanic lithosphere is forced down into the asthenosphere under the lithosphere that comprises another, less dense tectonic plate

**zona de subducción** límite convergente donde la litosfera oceánica es obligada a descender al interior de la astenosfera, bajo la litosfera, que comprende otras placas tectónicas menos densas

**sublimation** matter phase change from solid directly to gas

**sublimación** cambio de fase de la materia de estado sólido directamente a estado gaseoso

**subscript** a symbol, usually a letter or a number, placed immediately below and to the right of a written character

**subíndice** símbolo, por lo general una letra o un número, que se coloca inmediatamente abajo y a la derecha de un carácter escrito

**supercontinent** a large landmass created by the convergence of multiple continents

**supercontinente** gran masa creada por la convergencia de varios continentes

**supercontinent cycle** the cycle of the convergence and divergence of continents; the cycle of creation and break-up of supercontinents

**ciclo del supercontinente** ciclo de convergencia y divergencia de los continentes; ciclo de creación y división de los supercontinentes

**surface area** measurement of all the exposed surfaces of a substance

**área de superficie** medida de todas las superficies expuestas de una sustancia

**surface current** an ocean current at 100 meters depth or less

**corriente de superficie** corriente oceánica a 100 metros de profundidad o menos

**surface tension** a measure of the enhanced intermolecular forces at the surface of a liquid

**tensión superficial** medida de fuerzas intermoleculares aumentadas en la superficie de un líquido

**surroundings** the circumstances, objects, and conditions around a particular system that may interact with the system

**entorno** circunstancias, objetos y condiciones alrededor de un sistema particular que pueden interactuar con el sistema

**synthesis reaction** a reaction in which two or more elements or compounds combine to form a more complex compound

**reacción de síntesis** reacción en la cual dos o más elementos o compuestos se combinan para formar un compuesto complejo

**system** a related set of components that react with one another that may or may not interact with the surrounding area

**sistema** conjunto de componentes relacionados que reaccionan unos con otros y que pueden o no interactuar con el entorno

## T

**temperature** a measure of the average kinetic energy of the atoms in a system, used to express thermal energy in degrees

**temperatura** medida del porcentaje de energía cinética de los átomos de un sistema, se usa para expresar la energía térmica en grados

**tetrahedron** a structural unit found in a variety of molecules and crystals, consisting of atoms that bond to form a shape consisting of four triangular faces and four vertices; three triangular faces meet at each vertex to make six edges

**tetraedro** unidad estructural que se halla en varias moléculas y cristales, consiste en átomos que se unen para formar una figura con cuatro caras triangulares y cuatro vértices; en cada vértice se unen tres caras triangulares y forman seis aristas

**thermal energy** energy in the form of heat

**energía térmica** energía en forma de calor

**thermal equilibrium** a state that occurs when two bodies are in contact and have the same temperature and exchange no heat energy

**equilibrio térmico** estado que ocurre cuando dos cuerpos están en contacto y tienen la misma temperatura y no intercambian energía calorífica

**thermocline** layer in which temperature rapidly changes

**termoclina** capa en la que la temperatura cambia rápidamente

**tide** the regular fluctuation of sea water driven by the gravitational pull of the moon and sun on Earth's oceans and other large bodies of water

**marea** fluctuación regular del agua del mar originada por la atracción gravitacional de la luna y el sol sobre los océanos de la Tierra y otros grandes cuerpos de agua

**titration** a method for determining the concentration of an acid or base in a solution using an indicator

**titulación** método para determinar la concentración de un ácido o una base en una disolución usando un indicador

**transform boundary** a tectonic plate boundary along which plates slide horizontally past one another in opposite directions

**límite transformante** límite de placas tectónicas a lo largo de la cual las placas se deslizan horizontalmente una junto a otra en direcciones opuestas

**transition metals** elements that are made by filling either the d or f orbitals with electrons

**metales de transición** elementos que se hacen llenando los orbitales d o f con electrones

**transpiration** the loss of water vapor by plants

**transpiración** pérdida de vapor de agua en las plantas

**transport** the movement of materials

**trasporte** movimiento de materiales

**triple bond** a covalent bond in which three pairs of electrons are shared by two atoms

**enlace triple** enlace covalente en el cual tres pares de electrones son compartidos por dos átomos

**triple point** the temperature and pressure at which solid, liquid, and gas phases of a substance coexist in equilibrium

**punto triple** temperatura y presión a las cuales las fases sólido, líquido y gaseoso de una una sustancia coexisten en equilibrio

**tsunami** a giant ocean wave caused by the sudden displacement of large amounts of sea water by earthquakes, other seismic activity, or massive landslides

**tsunami** ola oceánica gigante originada por el desplazamiento súbito de grandes cantidades de agua de mar a causa de terremotos, otras actividades sísmicas o grandes deslizamientos de tierra

**turbidity current** a density current in water that carries large amounts of sediment rapidly down an underwater slope

**turbidez oceánica** corriente de densidad en el agua que transporta grandes cantidades de sedimentos rápidamente hacia abajo de una pendiente bajo el agua

## U

**ultramafic** igneous rock, lava, or magma that is composed primarily or entirely of mafic minerals

**ultramáfico** término que se aplica a rocas ígneas, lava o magma compuestos principal o completamente por minerales máficos

**ultraviolet radiation** a type of electromagnetic radiation with shorter wavelengths and higher energy than visible light waves; located on the electromagnetic spectrum between violet visible light and X rays

**radiación ultravioleta** tipo de radiación electromagnética con longitudes de onda más cortas y más energía que las ondas de luz visible; se ubica en el espectro electromagnético entre la luz violeta visible y los rayos X

**undertow** a subsurface current of water returning to the open ocean after being carried onshore by wave action along a sloping beach

**resaca** corriente de agua bajo la superficie que regresa a mar abierto después de haber llegado hasta la orilla por la acción de las olas a lo largo de una playa en pendiente

**universal solvent** substance that is able to dissolve (almost) all other substances; water is commonly referred to as the universal solvent

**solvente universal** sustancia que es capaz de disolver (casi) todas las otras sustancias; el agua se considera comúnmente como solvente universal

**upwelling** 1. the rising of cold, dense, deep seawater toward the ocean surface 2. the slow rising of magma from the mantle to Earth's surface at hot spots and along rift zones

**surgencia** 1. subida de agua de mar fría y densa a la superficie del océano 2. magma que sube lentamente desde el manto hasta la superficie de la Tierra en puntos calientes y a lo largo de zonas de fractura

## V

**valence electron** an electron that resides in the outermost shell, or principal quantum level, of an atom

**electrón de valencia** electrón ubicado en la capa más exterior, o en un nivel cuántico principal, del átomo

**valence shell** the outermost shell, or principal quantum level, of an atom

**capa de valencia** capa más exterior, o nivel cuántico principal, de un átomo

**vaporize** process of converting a substance from its liquid or solid phase to its gas phase, through the application of heat; conversion directly from solid to gas is called sublimation

**evaporar** proceso de convertir una sustancia de su fase sólida o líquida a su fase gaseosa mediante la aplicación de calor; la conversión directa de sólido a gas se denomina sublimación

**visible light** the part of the electromagnetic spectrum that can be seen by the unaided human eye

**luz visible** parte del espectro electromagnético que el ojo humano puede ver sin ayuda de ningún instrumento

**volcanic arc** a curved chain of volcanic islands created by volcanism as a result of the subduction of an oceanic plate, usually running parallel to a deep-sea trench

**arco volcánico** cadena de islas volcánicas dispuestas en curva creadas por vulcanismo como resultado de la subducción de una placa oceánica; por lo general discurre paralelo a una fosa marina profunda

**volcano** a mountain formed by extrusion and accumulation of lava and pyroclastic material

**volcán** montaña formada por la extrusión y acumulación de lava y materiales piroclásticos

**voltage** montaña formada por la extrusión y acumulación de lava y materiales piroclásticos

**voltaje** medida de la diferencia de potencial eléctrico entre dos puntos

**voltaic cell** an electrochemical cell that generates electric current using a spontaneous chemical reaction; also called a galvanic cell

**celda voltaica** celda electroquímica que genera una corriente eléctrica usando una reacción química espontánea; también se denomina celda galvánica

**volume percent** a measure of solution concentration that is the ratio of the solute's volume over the total volume of the solution multiplied by 100 to create a percentage

**porcentaje de volumen** medida de la concentración de una disolución que es la proporción del volumen de soluto sobre el volumen total de la disolución multiplicada por 100 para crear un porcentaje

**VSEPR theory** valence shell electron-pair repulsion (VSEPR) theory; states that the electron pairs in the valence shells of atoms repel each other, maximizing the angles between the atoms in the molecule

**TRePEV (teoría de repulsión de pares de electrones de valencia)** la teoría de repulsión de pares de electrones de valencia (TRePEV, o VSEPR por sus siglas en inglés) establece que el par de electrones en las capas de valencia de los átomos se repelen entre sí, maximizando los ángulos entre los átomos en la molécula

**water** a molecule that contains two hydrogen atoms and one oxygen atom; often called "the universal solvent"

**agua** molécula que contiene dos átomos de hidrógeno y un átomo de oxígeno; con frecuencia se le llama "el solvente universal"

**water cycle** the continual movement of water between the land, ocean, and the air through predictable physical processes

**ciclo del agua** movimiento continuo del agua entre la tierra, el océano y el aire mediante procesos físicos predecibles

**wave** a disturbance caused by a vibration that propagates through space and time

**onda** perturbación producida por una vibración que se propaga a través del espacio y del tiempo

**wave height** the maximum displacement that the wave experiences from its equilibrium to the crest (see also amplitude)

**altura de la onda** el desplazamiento máximo que experimenta una onda desde su equilibrio hasta la cresta (ver también amplitud)

**wave speed** the speed at which a wave travels, calculated by the frequency multiplied by the wavelength

**velocidad de onda** velocidad a la que viaja una onda, se calcula multiplicando la frecuencia por la longitud de onda

**wavelength** the distance between one peak and the next in a wave, or between one trough and the next

**longitud de onda** distancia entre una cresta de una onda y la siguiente, o entre un valle y el siguiente

**weak acid** an acid that only partially ionizes in water

**ácido débil** ácido que ioniza solo parcialmente en agua

**weak base** a base that only partially ionizes in water

**base débil** base que ioniza solo parcialmente en agua

**wind energy** electricity generated by turbines rotated by wind

**energía eólica** electricidad generada por turbinas que hace girar el viento

---

**work** the change in a system's kinetic energy due to being acted on by a force

**trabajo** cambio en la energía cinética de un sistema debido a la actuación de una fuerza sobre él

**x-ray** a form of short-wavelength, high-energy electromagnetic radiation

**rayos X** forma de longitud de onda corta, radiación electromagnética de alta energía

## A

## D

## I

## M

## P

## Q



## R

## T

## X



## Y



## Z

# ICONIC ROOMS

## KIPS BAY NEW YORK DECORATOR SHOW HOUSE

## *at* 50

ART LIFE

2020s
11
2010s
61
2000s
195
1990s
230
80s & 70s
244

# FOREWORD

Many years ago, I chose to become more involved with design industry charities because I believed that with our deep roots in the design industry, we—the New York Design Center at 200 Lex—needed to give back. The Kips Bay Boys & Girls Club was the logical choice. There was perfect synergy between the industry, the organization, and the long history of the Kips Bay Decorator Show House that this book celebrates.

The design industry and the designers themselves have always focused on creating environments that enhance people's lives, that make them better, more comfortable, and livable every day. The Kips Bay Boys & Girls Club, the most prestigious organization of its kind in America, does something similar, but even more powerful: It creates environments and programs that each year impact the lives of more than 11,000 children in the Bronx for the better. Kips Bay saves lives!

The moment I was invited to join the board, I gave Kips Bay an office in our building. This launched our efforts to build community. As board president, with Dan Quintero and Nazira Handal's help, I've continued to introduce more designers to the organization and its facilities in the Bronx. My goal is to cement forever this perfect connection between the club, the industry, and the Show Houses, which annually raise a significant portion of the Kips Bay organization's operational funding.

The Show Houses spotlight pure, superb design. Every designer in America dreams of doing a Kips Bay room because participating in the Show House can power their career trajectory. We want everyone involved to know that we're always fighting for this wonderful organization and the youngsters we support.

In my view, the design industry is one with a soul. Kips Bay is clearly a charity with a soul. I am grateful and proud to play a role in both. Yes, the rooms in the following pages are beautiful, almost beyond imagining. But all of us involved have always held the purpose of Kips Bay—improving kids' lives—closest to our hearts.

— James P. Druckman,
*President of the Board*

PAGE 2: Dining Room, Alex Papachristidis Interiors, 2016
PAGE 4: "The Andes Club," Phillip Thomas Inc., 2024
LEFT: Sitting Room, Mark Hampton, 1979

# INTRODUCTION

As a child growing up in the Bronx, I had no idea what a show house was and no clue that a design industry existed. I started attending the Kips Bay Boys & Girls Club as a nine-year-old. As a twelve-year-old, I remember helping to take tables, boxes, and chairs down to a house on the East Side of Manhattan and asking my friends, "What is this place?" Little did that twelve-year-old know that the Kips Bay Decorator Show House—the creative genius of so many designers—was going to impact my life directly and profoundly affect the organization I've now run for twenty-eight years.

Looking back, it's clear that what our organization has accomplished since it opened its doors in the Bronx in 1969 stands on the shoulders of the design industry. We started as one program. We now have ten locations serving more than 11,000 children each year. The design industry's stalwart support over these five decades and its commitment to our programs for the betterment of the young people of the Bronx has made this expansion possible.

The symbiotic relationship between Kips Bay, the Show House, the designers, and the design industry has developed naturally over time, and has been a win-win for all of us. But always, the kids of the Bronx have been the biggest beneficiaries. As the leader of Kips Bay, I take great comfort in knowing that the design industry is behind us in all our efforts. The strength of our relationship has allowed us to replicate the Show House in other cities to impact the lives of even more children and elevate the careers of even more designers.

Over the years, designers have helped us create murals throughout all our locations plus a Culinary Arts Center. Designers are now assisting us with color schemes to make our locations feel calmer. Through the Show House and our relationships with designers, textile firms, and construction companies, we've been able to tap into their expertise and utilize their skills and services to help us achieve the kind of safe and nurturing environments that improve children's lives—and that our youngsters, and all kids, deserve.

— Daniel Quintero,
*Executive Director*

**RIGHT:** "The White Orchard Room," Timothy Whealon Interiors, 2016
**PAGE 10:** "Old Guard, Avant-Garde," McMillen Inc., 2024

TIMOTHY WHEALON
Hats by Madame Paulette
HORST INTERIORS

# 20 20s

PRECEDING SPREAD & RIGHT:
"Old Guard, Avant-Garde,"
McMillen Inc., 2024

Brown SIXTY YEARS OF INTERIOR DESIGN Viking

"A Tailored Edge,"
Aman & Meeks, 2024

LALANNE

MARIO BUATTA
ADAM FUSS
INSPIRED DESIGN
JANSEN
RUHLMANN
ART
DESIGNERS AT HOME

PRECEDING: "The Club Room," Alan Tanksley Inc., 2024
RIGHT: "The Andes Club," Phillip Thomas Inc., 2024

"The Virtue of Curiosity,"
Benjamin Vandiver, 2024

PRECEDING SPREAD: "Study of Art and Beauty,"
Jeremiah Brent, 2024
RIGHT: "A Dining Room of One's Own,"
Kit Kemp Design Studio, 2024

"Steel Sanctuary,"
Shawn Henderson, 2024

"A Cotswolds Escape,"
Mikel Welch, 2024

“Sculpted Serenity,” Hollander Design Landscape Architects, 2024

PRECEDING SPREAD: "Good Spirits," Ghislaine Viñas, 2023
RIGHT & OVERLEAF: "The Parlor," David Scott Interiors, 2023

INDOCHINE
"YOU SHOULD HAVE BEEN WITH ME"

"The Library," Jay Jeffers, 2023

RIGHT: "Heart of the Vine,"
Halden Interiors, 2023
OVERLEAF: "Spring Fever,"
Timothy Corrigan Inc., 2023

LE NÔTRE
Gardens in France

“The Levant,” Georgis & Mirgorodsky, 2023

"Not Just a Kitchen,"
Wesley Moon Inc., 2023

Manu
Grains for Every Season
THE BARBUTO COOKBOOK
VEGETABLE LITERACY
VIA CAROTA
BOQUERIA
CUCINA POVERA
HARVEST
LOST KITCHEN
DEDA BLAIR
JACQUES JARRIGE
Entertaining
FRANCES ELKINS
BAKING HANDBOOK
MISSONI COOKBOOK
pasta
Claudia Roden's Mediterranean
sweet enough
SEA SALT AND HONEY
EXPRESSIONS: ART IN PASTRY
Beyond the North Wind
THE GRAMERCY TAVERN COOKBOOK
DELECTABLE
Yogurt & Whey

"Drafting Room," Neal Beckstedt Studio, 2023

GOTLANDSKA RUM
CHÂTEAU de VILLETTE

"Lady Borromeo's Grotto,"
Mary McDonald, 2023

RIGHT: "An Ethereal Primary Bath,"
Sergio Mercado Design, 2023
OVERLEAF: "The Hideaway,"
Clinton Smith Studio, 2023

THE WAY WE LIVE

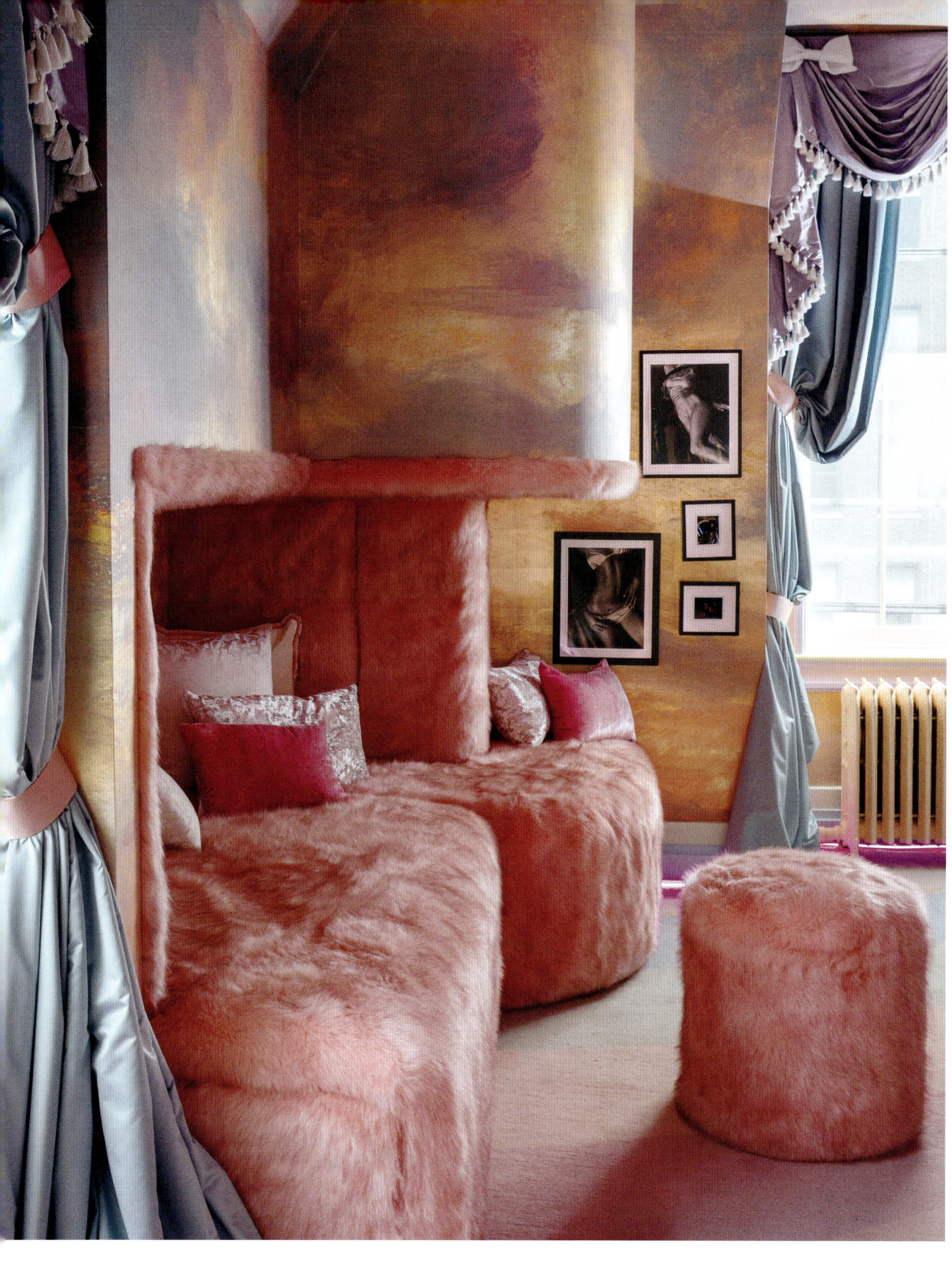

“Flight of Fantasy,” Sasha Bikoff Interior Design, 2023

“Smoke and Mirrors,” Vanessa DeLeon Associates, 2024

VOGUE LIVING HOUSES GARDENS PEOPLE
THE WORLD IN VOGUE PEOPLE PARTIES PLACES
Louis Vuitton
SUZANNE KASLER TIMELESS STYLE
INSPIRED INTERIORS SUZANNE KASLER

# 2010s

PRECEDING & RIGHT: "Sophisticated Simplicity," Suzanne Kasler, 2016
OVERLEAF: "A Room of One's Own," Eve Robinson Associates, 2019

LOUIS VUITTON
TONY DUQUETTE
Leonardo da Vinci
Matisse Cut-outs
BEATON

ORCHIDS
MODERNE
FLOWERS

A
Woman's
Drink

# ELLIE CULLMAN

Since I joined the board of the Kips Bay Boys & Girls Club almost ten years ago, I have found it to be the most meaningful board position of my adult life. It's absolutely no exaggeration to say that Kips positively influences and profoundly nourishes children's lives.

The Club is close to my heart for deeply personal reasons. I grew up in a modest home in Brooklyn. My father was born into poverty on the Lower East Side to a middle-aged mother, an immigrant from Poland, who gave birth to him on the kitchen table. My dad had to get his first job—turning the gas lights on Grand Street on and off every day—in grade school. Although he was a brilliant student and president of his junior high school, he had to drop out of school at 14 to support his aging parents. From the time my two older sisters and I were little, he emphasized the importance of education and hard work.

My dad's experience growing up makes me feel especially close to the Kips kids, who face similar challenges. What a wonderful, nurturing, character-building environment the Club provides for more than 11,000 Bronx children ages 6 to 18 every year. I wish my dad had had a Kips Bay Club to support him in his day, and I feel so fortunate now to be able to further its mission in a meaningful way.

The Coudert Learning Kitchen, a joy to design with my amazing Cullman & Kravis team, has been my favorite project so far. Our goal was to educate our kids about nutrition, but also to teach them cooking skills they could practice, take home and use every day. I knew they would really benefit. And I'm so proud that from the day we opened it, the kitchen has been filled non-stop with enthusiastic young chefs.

I am delighted to be getting a new project for the Club underway now with Lizzy Dexter, a fellow board member and Cullman & Kravis alum. We plan to design and help fund a creative arts center in the Palmaro facility, where we hope to serve our youngest members, ages 6 to 10. As our public school system continues to cut arts programs, the need to provide youngsters with an avenue for self-expression through the arts is ever more urgent.

from CLASSIC to CONTEMPORARY
Decorating Master Class
THE DETAILED INTERIOR

PRECEDING & RIGHT:
Dining Room, Cullman & Kravis, 2019

PER LA STRADA VA TUTTO FINTO

PRECEDING & RIGHT:
"Fantastic Voyage," Jeff Lincoln, 2019

DALLA PARTE DEL MEDITERRANEO PARTICOLARE

"Cherry Bitters," David Netto, 2018

**ABOVE, RIGHT & OVERLEAF:** Bedroom, Charlotte Moss, 2019

Wilder Shores Love
LESLEY BLANCH

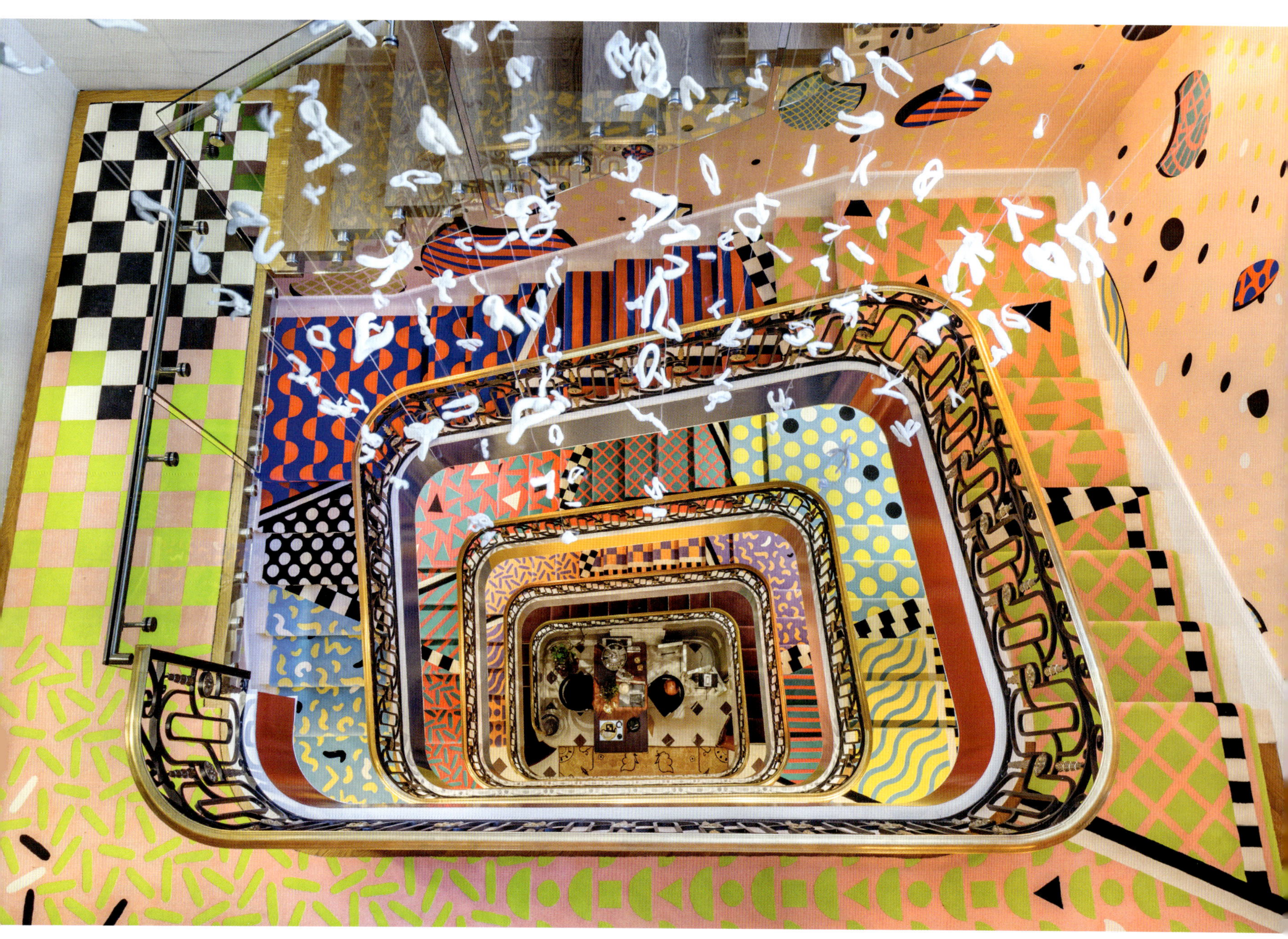

Stairwell, Sasha Bikoff Interior Design, 2018

SASHA BIKOFF
NEW YORK

# BUNNY WILLIAMS

I was first introduced to the Kips Bay Boys & Girls Club at the start of my career as a designer. I can still remember going to the very first New York Show House in 1973, and I've eagerly visited each one since. To me, the Show House annually represents the pinnacle of our industry. It is a vehicle for this country's top designers to showcase their very best work and an opportunity for all visitors to get inspired by not only the interiors, but the latest sources, artists, and vendors. It also offers a huge platform for young designers to display their talent to the public.

The Show House team works tirelessly each year to find an exceptional house that will serve as the ideal setting for this event. But as it has become more difficult to find townhouses in New York City to host the Show House each year, the organization has expanded to other cities like Palm Beach and Dallas, as it knows the importance of adapting.

I was motivated to become even more involved after visiting the Kips Bay Boys & Girls Club in the Bronx, seeing their space and being amazed at all the kids from the neighborhood who benefit from the programs that are available at the clubhouse. That visit made me want to do everything in my power for this incredible organization.

I am so humbled to be a part of the Kips Bay Boys & Girls Club and thrilled to see its impact growing year after year with the continued generosity of the design community.

PRECEDING, RIGHT & OVERLEAF:
"Gilded Knots," Bunny Williams, 2018

HORST·INTERIORS

"The Drawing Room,"
Philip Mitchell Design Inc., 2018

"The Wellness Retreat,"
Pavarini Design, 2018

"The Moonlight Room,"
Juan Montoya Design, 2018

JUAN MONTOYA

"The After-Party," Torrey, 2018

ABOVE, RIGHT & OVERLEAF: "Sleeping Beauty," Mark D. Sikes, 2018

Sotheby's

# ALEXA HAMPTON

My first memory of the Kips Bay Decorator Show House was when my third-grade class (1979?) went to visit it (no doubt, an excursion arranged by my father). He had done a dramatic dark brown library that has forever since occupied my imagination. I assume I am among the very few designers on earth who have had such an early Kips Bay revelation but, boy, am I glad for it!

That library and a beautiful bedroom my father designed for the 25th anniversary house in 1997 are such happy memories for me! And my own first foray designing a space followed in 1999. Now, 45 years later, as one of the house's co-chairs, I am still dreaming of Kips Bay and am proud to be so long a part of its community, in one way or another. It's a great event for a great cause. And the beauty of interiors comprises but a small bit of its achievements. Its real success is about people.

The mission of The Kips Bay Boy's and Girl's Club has always been "to enrich and enhance the quality of life for young people by providing educational and developmental programs, with special emphasis on youngsters between the ages 6-18 who come from disadvantaged or disenfranchised circumstances." Unlike most of us, the years have only made the club all the more robust and impressive. The courses now offered range from culinary arts, kitchen design, STEM coding, digital arts, drone operation and robotics, and many, many more on top of its already lauded programs in the theater arts. These skills are as relevant as they are exciting. And for the 11,000 plus kids in the Bronx who rely upon the club, it has never been more necessary that they be exposed to what they can do and who they can be in the future.

To this end, the design community comes together annually with our own twin desire to "enrich and enhance life" in the best way we know how: by design. Each year, the assembled class of designers, all beyond talented, put their best foot forward to entertain and delight. More importantly, all of us do this for the kids who will help New York move further into the future.

ICKS A LIFE OF DESIGN
neless Elegance
ON DECORATING

PAGES 101-103: Sitting Room, Alexa Hampton, 2018
ABOVE & RIGHT: "A Room with a View," Steilish Interior Design & Decoration, 2018

"Stairway to Savage,"
Savage Interior Design, 2017

RIGHT: Sitting Room,
Neal Beckstedt Studio, 2017
OVERLEAF: "Reimagining Mongiardino,"
Richard Mishaan Design, 2017

Die Ziege
ARTFULLY MODERN

"Living Room," Robert Stilin, 2017

"The Living Room: Soothing Surroundings Inspired by the Hudson," Susan Zises Green, 2012

PRECEDING: Dining Room,
Alex Papachristidis Interiors, 2016
RIGHT: Living Room, Victoria Hagan, 2016

RUSCHA
ROBBERT FLICK
TRAJECTORIES

"Petit Salon," Sawyer | Berson, 2016

"The Art of Modern Living,"
Eve Robinson Associates, 2016

Caravaggio

RIGHT & OVERLEAF: "Lady Lair," Phillip Thomas Inc., 2016

CRANACH
New York
FAIR 100 YEARS

# COREY DAMEN JENKINS

One Saturday morning years ago, flipping through a shelter magazine over coffee, I came across photos of the Kips Bay Decorator Show House in New York—page after page of incredible rooms curated by some of my favorite interior designers. Encouraged by their example, I began donating to the Kips Bay Boys & Girls Club and attending the annual President's Dinner. Then, I visited the organization's Bronx facilities. After witnessing this charity's dedication to its constituents, I knew I had to get more involved. When my friend Jamie Drake invited me to design a library for the 2019 New York Show House, I immediately said "yes."

Historically, libraries and dens—including this one, in a 1920s-era mansion—were designated the "Gentleman's Study," and characterized by dark stained woods, plaids, and other "masculine" motifs. But what if a woman were the library's primary user?

Our theme—"To the Lady of The House, with Love"—expanded the narrative, and the decorative vocabulary, with wall panels and moldings glossed in the palest blush color; bookcase interiors gift-wrapped in rich grass cloth; and a dramatically muraled ceiling invoking ancient Babylon's hanging gardens with a crystal-encrusted topiary of a French 1940s chandelier. A mélange of clean-lined contemporary furniture with antiques, including a showstopping mahogany desk with a Cinderella-esque acrylic chair, added layers of modernity and functionality. Stately window panels featuring layered peplums dripping with elaborate passementerie, a concept inspired by the gowns of Jean Paul Gaultier, brought the couture.

Our space pays homage to the women who run the world: from the household to the classroom to the White House—and everywhere in between. Its inspiration speaks to my personal world view of empowering minorities and voices less heard to positions of prominence and respect.

I first became aware of the Kips Bay organization when my firm was still based primarily in southeastern Michigan. Now, I am honored to serve on its board of trustees and as co-chair of its flagship fundraising event, the annual President's Dinner. This inestimable privilege enables me to get more involved with the young people of the Bronx and their families.

Our mission is to continue lifting these kids up towards greatness, encouraging them to fully utilize their God-given talents and shatter glass ceilings by changing society's long-held views through their achievements.

david hockney
Secret Knowledge
CHINA GATE
PETER BEHRENS
IRON GOLD

PRECEDING & RIGHT: "Lady's Library," Corey Damen Jenkins, 2019

LOUIS
VUITTON

RIGHT & OVERLEAF: "The White Orchard Room," Timothy Whealon Interiors, 2016

Linda McCartney Life in Photographs
TASCHEN
Oscar de la Renta
HOLBORN

RIGHT: "Le Jardin Secret," Les Ensembliers, 2016
OVERLEAF: Terrace, Hollander Design Landscape Architects, 2016

RIGHT: “Torridon Stair Gallery,” Philip Mitchell Design Inc., 2015
OVERLEAF: “Pavlos’s Retreat,” Alan Tanksley Inc., 2015

THE FALL

ALEXA HAMPTON
The LANGUAGE of INTERIOR DESIGN

**ABOVE, RIGHT & OVERLEAF:** Living Room, Branca, 2015

THE LIGHT OF NEW YORK
THE LIGHT OF PARIS
VREELAND MEMOS

LEFT: Lounge, Bennett Leifer Interiors, 2015
ABOVE: Powder Room, Scott Sanders, 2018
OVERLEAF: Kitchen, Christopher Peacock, 2019

# CHRISTOPHER PEACOCK

I first learned of the Kips Bay Decorator Show House in the mid-1990s, having launched my fledgling company in Greenwich, Connecticut, in 1992. As a transplant from the UK, I didn't truly understand or appreciate the legacy of design talent that surrounded me, but gradually I became familiar with the great decorators of that time. Names like Peter Marino, Albert Hadley, Tom Britt, Mario Buatta, Mark Hampton, and Bunny Williams graced the pages of the shelter magazines. And it was through those names that I discovered the Show House.

My first opportunity to participate came in 1999. We were in a recession, work was sparse, and I was struggling to keep my business afloat and really needed a boost. Having submitted my request to decorate the kitchen, as I had done several times before, I got the call. Only then did I understand the challenge. I had to design, make, and install a very large kitchen and be ready for showtime in less than three months. More importantly, I needed to find the money to pay for it!

The show launched on my birthday, and that very same day I was mentioned in the *Daily News*, the *New York Times*, and many other publications. The press coverage meant the world to me—this was before digital and social media—so I stood in my room every day for a month and spoke to everyone I could. My business took off almost immediately, and to this day people recall that kitchen.

The Kips Bay legacy cannot be underestimated. As designers know, if you have "done Kips," you have become a member of a very exclusive club with bragging rights to use as a springboard for the rest of your career. The Show House has provided young talent nationwide the opportunity to get exposure at the highest level. And it has created a community of designers that transcends the club itself.

Several years ago, I was asked to become a trustee of the club, and without hesitation, I joined. I have seen firsthand the difference that Kips Bay has provided to the families of the Bronx and beyond. It literally changes people's lives forever. The opportunity to repay my debt of gratitude is a gift, and I am proud and honored to do so whenever I can.

ABOVE & RIGHT: "Up in the Villa," Mark D. Sikes, 2015
OVERLEAF: "Study in Red," Georgis & Mirgorodsky, 2014

Entrance Gallery, Martyn Lawrence Bullard, 2014

Sitting Room, Carrier and Company Interiors, 2014

PRECEDING, LEFT & ABOVE: Living Room, Juan Montoya Design, 2014

RIGHT: "Young at Art,"
Young Huh Interior Design, 2019
OVERLEAF: "Salon," Darryl Carter Inc., 2014

Uncommon Vernacular
RESTORING A HOUSE IN THE CITY

PRECEDING, ABOVE & RIGHT: "Some Like It Hot," Cullman & Kravis, 2014

Kitchen, Matthew Quinn, 2014

Our Lives & Times
NEW GARDEN PARADISE
OF THE WORLD
THOMAS EAKINS
Magritte
David Sylvester

Our Lives & Times
SCOTLAND
Geographica

PRECEDING SPREAD & RIGHT:
"A Master Class in Decoration,"
Richard Mishaan Design, 2011

**ABOVE, RIGHT & OVERLEAF:** Living Room, Bunny Williams, David Kleinberg Design Associates, and Brian J. McCarthy Inc., 2012

TO THE
TERRACE

"The Lounge Suite,"
Garcia/Maldonado Inc., 2013

Pucci

"Living Room,"
Robert Stilin, 2011

LEFT: "The Stereo Lounge," Brad Ford ID, 2011
ABOVE: Primary Bath, Groves & Co., 2016

"Gentleman's Study,"
David Scott Interiors, 2012

Outside the Box
Outside the Box
Outside the Box
Outside the Box
Outside the Box

Bedroom, Amanda Nisbet Design, 2011

# 2000s

PRECEDING & RIGHT:
Primary Suite, Charlotte Moss, 2009

LEFT: "Maya Romanoff Celebrates 40 Years of Design," Amy Lau Design, 2009
ABOVE: "Midcentury Manhattan," James Rixner, 2007

RIGHT & OVERLEAF: "Nordic Light," Juan Montoya Design, 2009

# JAMIE DRAKE

The thrill of becoming a part of the Kips Bay community began for me in 2002, when I was accepted to create a room for the first time. I had been turned down twice previously, and I said to myself, "Okay, I'll try once more!" Of course, I had been attending the Show House since shortly after arriving in New York City in 1975. As a student at Parsons, seeing the work of the best of the best of the design industry in real life, in three-dimensional reality—and not just on the pages of magazines—was jaw-dropping, inspiring a few hours of wonder. My room was next to Albert Hadley's that first year, and his kind words and encouragement were extraordinarily meaningful to me.

Even better was learning about the programs the Show House funded. The kids—and we always use the alliterative Kips Kids!—whose lives were enriched by the after-school sports, music and dance courses, were embraced and supported to spend productive hours together. In the ensuing years we have helped ever more students in ever more ways: swimming, education, nutrition and cooking, basketball, baseball, skating, pottery, videography, social and emotional growth, the arts—all with the goal of preparing them for productive futures by learning teamwork and building character and integrity. These skills will not only enrich their lives, but those of their families, their communities, and the world.

Today, twenty-two years after creating my first Kips Bay room, I proudly sit on the board and co-chair both the Show House and the President's Dinner, the largest fundraisers for the club. I remain in awe of the creativity of the designers every year and am grateful for their help in funding bright futures for the 11,000 plus children that the Kips Bay Boys & Girls Club impacts every year.

EGON SCHIELE
THE RONALD S. LAUDER AND SERGE SABARSKY COLLECTIONS
SLIM AARONS · ONCE UPON A TIME
WARHOL LIVE

PRECEDING & RIGHT:
"Logical Library," Jamie Drake, 2009
OVERLEAF: Living Room,
Bunny Williams, 2009

HOW TO ESCAPE FROM
QUICKSAND
ANDY WARHOL
HOLLYWOOD STYLE

"Harem's Den," Sara Story Design, 2008

Sitting Room, Mario Buatta, 2006

MISSING!
Fisherman - Husba
and Dog.
Reward for Dog.

WIFREDO
CHÂTEAUX OF THE LOIRE
Havana in my heart
Jenkins

LEFT: "The Sitting Room," Sherrill Canet Interiors, 2010
ABOVE: "The Lounge," Sherrill Canet Interiors, 2006

ABOVE & RIGHT: "The Orange Room." Sara Bengur Interiors, 2008
OVERLEAF: "Rites of Spring," Amy Lau Design, 2007

Kitchen, Christopher Peacock, 2007

Bedroom, Jamie Drake, 2007

Bedroom, Charlotte Moss, 2006

CHANEL

Sitting Room, Celeste Cooper, 2002

"Romance in the Kitchen," St. Charles New York, 2006

# 1990s

PRECEDING: Living Room, Juan Pablo Molyneux, 1992
ABOVE: Sitting Room, Scott Salvator, 1999
RIGHT: Sitting Room, Chuck Fischer, 1995

ABOVE: Study, Sandra Nunnerley Inc., 1992
RIGHT: Staircase Landing, Richard L. Ridge & Roderick R. Denault, 1995

Salon, Feldman-Hagen Interiors, 1990

1990

ABOVE: Sitting Room, David Kleinberg Design Associates, 1998
RIGHT: Sitting Room, Clodagh, 1995

BASQUIAT DRAWINGS

Bedroom, David Barrett Inc., 1993

LEFT: Sitting Room, Susan Zises Green, 1993
ABOVE: Living Room, Michael DeSantis Inc., 1990

80s
&
70s

PRECEDING: Bedroom, Mario Buatta, 1985
ABOVE: Library, Richard L. Ridge & Roderick R. Denault, 1989
RIGHT: Dining Room, Gary Crain, 1988

Conservatory, Mark Hampton, 1988

Living Room, Noel Jeffrey, 1985

ABOVE: Meditation Room, Bromley/Jacobsen Architecture and Design, 1984
RIGHT: Media Room, Eric Bernard, 1985

Library, Juan Pablo Molyneux, 1985

Sitting Room, Barbara Ostrom Associates, 1985

Dining Room, Ruben de Saavedra, 1989

Drawing Room, Mark Hampton, 1984

Dining Room, Samuel Botero, 1984

Library, Josef Pricci, 1984

"A LADIES PAD"
BY ELLEN McCLUSKEY

# AFTERWORD

When I started with the Kips Bay Boys & Girls Club in 2011, I had no idea what a show house was and no background in design, so my learning curve to open the doors of our 2012 Show House was steep. Over the next few years, as my confidence grew, the idea of expanding the Show House to other cities started to take shape. Palm Beach came first, and, a few years later, Dallas followed.

Mounting three Show Houses a year in three different cities involves countless logistical challenges, but the energy and effort on everyone's part is more than worth it because, in the end, they help us help more kids. It's also incredibly meaningful to see how all these efforts build community.

Every designer who does a Show House room becomes a member of the Kips Bay family. They travel to support each other and make sure they're at the Show House openings and our other Kips Bay design industry events. Designers in the same Show House class, so to speak, grow very close. We've also created an alumni mentorship committee to pair designers from previous years with those currently doing a room to answer any questions, which is incredibly helpful for designers not working on their home turf.

I have been very fortunate to have a phenomenal team that works tirelessly alongside me and gives up nights, weekends, and holidays to help make the Show Houses successful. My right hand, Jeremiah Johnsen, has been a key to our success as we've expanded.

Many people think of Kips Bay Boys & Girls Club as just an after-school program. It is so much more. For some kids, the club provides the only hot meal they receive in a day. In each of our facilities, kids can get help with homework, with college readiness, and with social issues. Every year, our organization gives more than 11,000 kids a place to grow and gain the tools necessary to be able to succeed in life. This is why I cherish the design industry and their dedication to our mission—and why I love the Kips Bay Boys & Girls Club even more today than when I first saw the job posting.

— Nazira Handal,
*Director of Special Events & Corporate Partnerships*

LEFT: Room sketches clockwise from top: McMillen Inc., 1973; Susan Zises Green, 1980; Ellen L. McCluskey Associates, 1976

# DESIGNER CREDITS

# PHOTOGRAPHER CREDITS

Amy Neunsinger 96, 97, 98–99, 152, 153

Andrew Frasz 40–41

Annie Schlechter 88–89

Antoine Boots 124–25, 126, 127, 190–91

Bjorn Wallander / OTTO 154, 155

Bruce Buck 166–67

Carlos Eguiguren 254, 255

Charles Mayer 138–39

Courtesy of Kips Bay Boys & Girls Club Archive 235, 239, 240–41, 242, 243, 246, 248–49, 270

Courtesy of Kohler 58, 59

Daniel Eifert 258, 259

Dennis Krukowski 238, 247, 250–51, 256–57

Douglas Friedman / Trunk Archive 106–07

Eric Piasecki / OTTO Half title, 94–95, 160–61, 162, 163, 200–01, 202–03

Feliciano, courtesy Sandra Nunnerly, 234

Francesco Lagnese / OTTO 82, 84–85, 86–87

George Ross 110–11, 174, 175, 176–77

Gieves Anderson 44–45

Jonny Valiant 216, 217

Joshua McHugh 9, 30–31, 132–33, 134, 135, 184–85, 233

Keith Scott Morton 228, 229

Kelly Marshall 16–17, 142–43

Kris Tamburello 198, 219

Lesley Unruh 56, 57

Marco Ricca 64–65, 90–91, 114, 144, 145, 146–47, 148, 149, 158, 159, 172–73

Max Kim-Bee 192, 193

Max Kim-Bee, Courtesy of BJM 178, 179, 180, 181

Michael Mundy 4, 18–19, 76, 77, 78–79

Neil Landino 150

Ngoc Minh Ngo 10, 12–13, 164–65

Nick Johnson 168–69, 170, 171, 199, 205, 206–07, 212–13, 214, 222–23

Nick Sargent 14–15, 20–21, 22–23, 24–25, 28–29, 32–33, 34–35, 36, 37, 38, 39, 42–43, 46–47, 50–51, 52–53, 54, 55, 67, 68–69, 70–71, 72–73, 74–75, 80, 81, 101, 102–03, 129, 130–31, 273

Peter Margonelli 122–23

Peter Murdock 186–87

Peter Vitale 6, 231, 236–37

Phillip Ennis 60, 62–63, 120–21, 136–37, 189, 210, 211, 215, 220–21, 226–27, 244, 252, 253, 260–61, 262, 263, 264–65

Pieter Estersohn / Art Department 194, 196–97, 224, 225

Roger Davies / OTTO 218

Stephen Kent Johnson / OTTO 26–27, 48, 49, 104, 105, 108–09, 112–13

Steve Freihon 182–83

Thomas Loof for *House Beautiful* / Art Department 208–09

Tim Ebert 232

Tim Street-Porter 92–93

Timothy Bell 140–41, 156, 157

Trevor Tondro / OTTO 188

Tria Giovan 2, 116–17

William Waldron / OTTO 118–19